BACTERIAL BIOGEOCHEMISTRY: THE ECOPHYSIOLOGY OF MINERAL CYCLING

Second Edition

BACTERIAL BIOGEOCHEMISTRY: THE ECOPHYSIOLOGY OF MINERAL CYCLING

Second Edition

T FENCHEL[1], GM KING[2] AND TH BLACKBURN[3]

[1]Marine Biological Laboratory, University of Copenhagen, Denmark
[2]Darling Marine Center, University of Maine, USA
[3]Department of Microbial Ecology, University of Aarhus, Denmark

San Diego   London   Boston
New York   Sydney   Tokyo   Toronto

Copyright © 1998 by ACADEMIC PRESS

Academic Press
525 B Street, Suite 1900, San Diego, California 92101-4495, USA
http://www.apnet.com

Academic Press Limited
24-28 Oval Road, London NW1 7DX, UK
http://www.hbuk.co.uk/ap/

ISBN 0-12-103455-0

A catalogue record for this book is available from the British Library.

Library of Congress Cataloging-in-Publication Data

Fenchel, Tom.
    Bacterial biogeochemistry : the ecophysiology of mineral cycling/
by Tom Fenchel, Gary King, Henry Blackburn.
        p.   cm.
    Includes index.
    ISBN 0-12-103455-0 (alk. paper)
    1. Geomicrobiology. 2. Biogeochemistry. I. King, Gary (Gary M.)
II. Blackburn, T. Henry (Thomas Henry), 1933-   . III. Title.
QR103.F46   1998   97-44014
579.3 — dc21   CIP

Typeset by Laser Words, Madras, India
Printed in Great Britain by WBC Book Manufacturers Ltd, Bridgend,
Mid Glamorgan

98 99 00 01 02 03 WB 9 8 7 6 5 4 3 2 1

Cover photograph: Mass occurrence of purple sulphur bacteria (*Thiocapsa* and
Chromatium) occurring on the sediment of a protected bay during summer.

Contents

Preface   ix
Introduction   1

1 General considerations   4

1.1 Bacterial metabolism   6
    1.1.1   Dissimilatory metabolism   9
    1.1.2   Assimilatory metabolism   19

1.2 Bioenergetics of microbial metabolism   21
    1.2.1   Energetic yields of metabolic processes 22
    1.2.2   Energetics, Y_{ATP} and growth yield   24
    1.2.3   Bioenergetics and the structure of bacterial
             communities   25

**1.3 Transport mechanisms and structure of microbial
     communities**   29
    1.3.1   Uptake and excretion of solutes from bacterial cells 32
    1.3.2   Diffusion-controlled communities   36
    1.3.3   Advection and turbulence   39

References   40

2 Mineral cycles   43

2.1 Hydrolysis of organic polymers   43
    2.1.1   Substrates   43
    2.1.2   Hydrolytic enzymes   50

2.2 Comparison of element cycles   53
    2.2.1   The dominance of carbon   54
    2.2.2   Oxidation/reduction reactions   57
    2.2.3   Mobility and availability   59

References   59

3 The water column   62

3.1 Prokaryotic primary producers   63
3.2 Water-column bacteria and mineralisation   64
    3.2.1   Characteristics of the bacterial biota   64
    3.2.2   Organic matter: composition, origin and turnover 67

3.2.3   Suspended particles, their microbial biota
and coupling between plankton and sediments   73

3.2.4   Bacteria and cycling of N and P   76

References   78

4   Biogeochemical cycling in soils   85

4.1   Soil water as a master variable   88

4.1.1   Physical and chemical principles   88
4.1.2   Water stress physiology   91
4.1.3   Interactions among soil water content, water
potential and biogeochemistry   93

4.2   Responses to plant organic matter   103

4.3   Responses of soil biogeochemistry to disturbance and change   109

References   111

5   Aquatic sediments   117

5.1   Comparison of freshwater and marine sediments   117

5.2   The carbon cycle   122
5.2.1   Heterotrophy   122
5.2.2   Autotrophy/heterotrophy   127

5.3   The nitrogen cycle   128
5.3.1   Hydrolysis and ammonium production   128
5.3.2   Nitrification   130
5.3.3   Denitrification   134
5.3.4   Inorganic nitrogen incorporation   137
5.3.5   Nitrogen fixation   138

5.4   The phosphorus cycle   138

5.5   Manganese and iron   138

References   139

6   Microbial mats and stratified water columns   142

6.1   Mats based on colourless sulphur bacteria   144

6.2   Cyanobacterial mats   148
6.2.1   Structure of cyanobacterial mats   149
6.2.2   Mineral cycling in cyanobacterial mats   150

6.3   Other types of mats   156

6.4   Stratified water columns   158

References   162

7   Symbiotic systems   166

7.1   Symbiotic polymer degradation   167
      7.1.1   Symbiotic digestion in mammals   170
      7.1.2   Symbiotic fermentation in other animals   174

7.2   Symbiotic nitrogen fixation   175
      7.2.1   Symbiotic N_2 fixation in legumes   176
      7.2.2   Actinorhizal N_2-fixing symbionts   178
      7.2.3   Symbiosis with cyanobacteria   178

7.3   Autotrophic bacteria as symbionts   179
      7.3.1   Phototrophic symbionts   179
      7.3.2   Hydrogen scavengers in anaerobic protists   180
      7.3.3   Symbiotic oxidisers of reduced sulphur compounds
              and methane   181

References   183

8   Biogeochemistry and extreme environments   188

8.1   General overview and summary of hypersaline systems   189

8.2   Subsurface environments   193

8.3   Microbial diversity and high-temperature environments   195

References   198

9   Microbial biogeochemical cycling and the atmosphere   203

9.1   The atmosphere as an elemental reservoir   203

9.2   Atmospheric structure and evolution   206
      9.2.1   Atmospheric structure   206
      9.2.2   Atmospheric evolution   208

9.3   Synopsis of trace gas biogeochemistry and linkages to
      climate change   213
      9.3.1   Oxygen–organic matter–CO_2 214
      9.3.2   Oxygen–organic matter–methane   215

9.3.3 N_2-ammonia-nitrous oxide   217
9.3.4   Hydrogen sulphide (H_2S)-sulphate-dimethyl
         sulphide (DMS)   219
9.3.5   Other gases   223

9.4   Trace gas dynamics and climate change: an analysis of
       methane production and consumption   225

9.5   Summary and conclusions   233

References   235

10   Origins and evolution of biogeochemical cycles   245

10.1   Biogeochemical cycles and thermodynamics   245

10.2   Prebiotic Earth and mineral cycles   250

10.3   Theoretical perspectives on the origins of life   255
        10.3.1   The Oparin–Haldane theory   258
        10.3.2   Clays and life   261
        10.3.3   Pyrite and the origins of life   264
        10.3.4   The "thioester world"   268

10.4   Evolution of biogeochemical cycles   271

References   276

Appendix 1   284

Thermodynamics and calculation of energy yields of metabolic
processes   284

References   292

Index   293

Preface

This book treats the influence of bacterial activity on the chemical environment of the biosphere. Our approach is primarily based on physiological properties of prokaryotic organisms. In a sense the book represents a second edition of T Fenchel and TH Blackburn: *Bacteria and Mineral Cycling* (Academic Press, 1979). The structure of the old book has to some extent been retained, but the authorship has been expanded and all chapters have been completely rewritten as a consequence of many advances and discoveries during the past 20 years. We believe that the changed title now more accurately describes the scope of the book.

The topic of the book is central to many aspects of environmental science. While the book presumes some knowledge of general microbiology, biological energetics and chemistry, we still hope that it will serve as a general text for university courses at different levels. We also hope that the book, or parts of it, will prove useful for professional workers within aquatic and soil sciences, general microbiology, geochemistry and geology.

The authors acknowledge with gratitude the support of their respective wives and families, without whom endeavors such as this would be far less rewarding.

Introduction

The surface of Earth is in a state of chemical disequilibrium. If Earth had remained lifeless, its atmosphere and seas would have been closer to a state of chemical equilibrium perhaps similar to that of Mars. Of course, true equilibrium would never quite be reached due to volcanic activity and photochemical reactions in the atmosphere driven by solar illumination. In the extant biosphere the dominant process is oxygenic photosynthesis, in which light energy is harvested by plants, algae and cyanobacteria. Photic energy is used to reduce CO_2 to organic matter with H_2O as reductant; during this process photosynthetic organisms oxidise the surroundings (by producing O_2 as a metabolite). Almost all organic carbon thus produced is eventually oxidised by heterotrophic organisms, thus restoring the redox equilibrium. However, spatial heterogeneity, limitation by chemical transport and of the kinetics of reactions, and specialised susbstrate requirements of different decomposer species impose time lags resulting in a considerable complexity of mineralisation processes. The biosphere is therefore maintained in a chemical disequilibrium. Our book describes these complexities of mineralisation with special reference to the role of bacteria.

Bacteria are not the only organisms involved in mineralisation processes, but they play a quantitatively dominant role and certain key processes are carried out exclusively by specific types of bacteria. All known types of metabolism are found among the bacteria and, even in the complete absence of eukaryotes (a situation that was probably realised during a large part of the history of Earth), all major biogeochemical element cycles would probably proceed much as they do today.

The recognition that bacterial activity affects the chemical environment has developed gradually during this century. It was first related to the discovery of a wealth of novel bacterial metabolic processes and to the use of enrichment cultures ("Winogradsky columns"), work that was pioneered by SN Winogradsky and by M Beijerinck before the turn of the century. They discovered (among many other processes) chemolithotrophy and nitrogen fixation, and isolated the involved organisms. This tradition was initially carried on mainly by the "Delft school of microbiology" (including AJ Kluyver, CB van Niel, LGM Baas-Becking and their numerous students) in the early part of this century, later becoming an integral part of modern microbiology. Certain applied problems, especially soil fertility and nitrogen transformations in agricultural soils, also played a role in the early development of microbiology. Interest in microbial communities and processes of aquatic habitats developed somewhat later (in particular pioneered by CE ZoBell and his students around the middle of the century) and aquatic microbiology has developed rapidly only during the last three decades. The profound (micro)biological contribution to the chemistry of the atmosphere and of seawater has also drawn the attention of chemists as exemplified by Sillén;[35] a modern treatment of the chemistry of aquatic systems acknowledging the role of microbial catalysis is found in Stumm and Morgan[39] (see reference list for Chapter 1).

Discoveries of novel types of microbial metabolism still take place and remain an inspiration for ecologically oriented work. During the past decades, the field has been characterised by the rapid development and employment of new methods, including the application of increasingly sensitive methods of chemical analysis, the use of radioactive and stable isotopes, development of microelectrodes (which allow for a description of chemical zonation patterns and estimates of process rates on a submillimetre scale) and, most recently, methods of molecular biology from which a new picture of microbial evolution and diversity is emerging. From a conceptual point of view, an improved understanding of the spatial and temporal structure of microbial communities has also recently been achieved. The pivotal role of bacterial activity in the transformation of matter and energy in natural ecosystems has enjoyed growing recognition among ecologists and microbial ecology has become an increasingly important component of the study of aquatic ecosystems. Microbial transformations of the chemical environment play a central role in understanding early diagenesis and palaeoenvironments. They also impinge on a number of applied problems, including degradation of xenobiotic compounds, sewage treatment, eutrophication of aquatic systems, emission of greenhouse gases and ore leaching in mines.

Our book is structured into 10 chapters. The first chapter includes various general considerations serving as a background for subsequent chapters: basic relevant properties of prokaryotes, types of bacterial metabolism, bioenergetical considerations, transport mechanisms in the environment (diffusion, advection and turbulence), and functional and spatial structure of microbial communities. Together, these aspects form the basis for understanding patterns and rates of microbial processes in nature. A following chapter treats hydrolysis of polymers, the degradation of particulate organic matter and a discussion of the major element cycles (C, N, S, P, metals). The succeeding six chapters describe microbial processes in particular habitats (the water column, soils and the rhizosphere, water-saturated soils and marine and limnic sediments, microbial mats and stratified water columns, symbiotic systems and extreme environments). These chapters do not give a comprehensive treatment of all aspects of the various communities or ecosystems, but mainly emphasise general principles and controls of reaction rates. Chapter 9 considers global element cycling, the regional and global distribution of elements and the role of microbial processes in terms of controlling the abundance of gaseous phases of C, N and S in the atmosphere. The last chapter is devoted to the early evolution of life and of biogeochemical cycles. An appendix treats thermodynamic principles and redox potentials.

1

General considerations

It is not the pretence of this chapter to give a general description of prokaryote biology or physiology, but only to emphasise those aspects of bacteria which explain their role in nature and especially how they interact with the environment. For comprehensive treatments of bacterial biology we refer to two excellent modern textbooks[6,37] and for bacterial diversity, in particular, to Balows *et al*.[2]

In this book we consider "bacteria" to be synomynous with "prokaryotes" thus including the archaebacteria (Archaea) and the cyanobacteria ("bluegreen algae"). The former group, although differing from the eubacteria in important respects and although representing an independent lineage of organisms (which, in fact, may be more closely related to the eukaryotes than to the eubacteria), remain "bacteria" in terms of organisational level and basic functional properties. The cyanobacteria, while unique among the prokaryotes in their capacity for oxygenic photosynthesis, represent one among several other eubacterial groups in terms of phylogeny (see Woese).[44,45]

Bacteria are small: typical bacteria measure between 0.5 and 2 μm in diameter. A few are considerably larger: some cyanobacteria exceed 5 μm in diameter and some sulphide-oxidising bacteria may reach 20 μm or more in diameter (*Thiovulum, Beggiatoa*); *Achromatium* has been recorded to reach a size of up to 100 μm. It would seem that bacteria therefore overlap in size with eukaryotic cells (among which the smallest free-living forms measure 2–3 μm, but most are 5–100 times larger). However, in contrast to small eukaryotic cells, most of the volume of very large bacteria is constituted by vacuoles or by non-living inclusions. Most bacteria are unicellular and rod shaped (rods), spherical (cocci), comma shaped (vibrios) or helicoidal (spirilla), while the spirochaetes are long, flexible filaments. Some soil bacteria, in particular, form (fungi-like) mycelia (actinomycetes, myxomycetes), and the latter have complex life cycles including the formation of sporangia. Many bacteria form filamentous colonies (several cyanobacteria, beggiatoans and many others). Bacteria almost always have a rigid cell wall surrounding the cell membrane.

The two important characteristics of bacteria (small size, rigid cell wall) are both a necessary consequence of the absence of a cytoskeleton (which characterises eukaryotic cells). These traits explain two additionally important properties of bacteria. One is that bacteria can take up only low molecular weight compounds from their surroundings via the cell membrane and this uptake is brought about either by active (energy-requiring) transport or by facilitated diffusion. Bacteria that utilise polymers or particulate organics can do so only indirectly: the extracellular hydrolysis of the substrate catalysed by membrane-bound or extracellular enzymes is necessary before the resulting low molecular weight compounds can be transported into the cell (see Chapter 2). Bacteria cannot bring particulate material or macromolecules into their cells: the capability of phagocytosis and pinocytosis is a privilege of eukaryotes (transformation which involves uptake of single-stranded DNA being an exception). Another property is a simple correlate of small size: bacteria are extremely efficient in concentrating their substrates from very dilute solutions (see Section 1.3.1).

Finally, a consequence of small size (when comparing organisms spanning a large size spectrum) is a "high rate of living"; that is, small organisms tend to have higher volume-specific metabolic rates and shorter generation times than do larger organisms. Roughly speaking, specific growth rate constants and volume-specific metabolic rates are proportional to $(volume)^{-1/4}$ (notwithstanding that there may be variation in potential growth rates among species of similar size; see also Section 1.2). Under optimal conditions a bacterium may have a generation time of only 15–30 min; corresponding figures for a 100 µm long protozoon, a copepod and a fish would be roughly 8 h, 10 days and 1 year, respectively. Although the biomass of bacteria is not impressive in most natural habitats when compared to that of larger organisms, the impact of bacteria in terms of transformation of materials and of energy flow may be much greater. For example, seawater typically contains around 10^6 bacteria per ml of water resulting in a volume fraction somewhat less than 10^{-6}. This is comparable to the volume fraction made up by protozoa; however, the metabolic activity of the bacterial community is typically an order of magnitude higher than that of the protozoa.

Another property of importance for understanding the role of bacteria in nature is that they hold all records as "extremophiles". Some bacteria tolerate temperatures exceeding 80 °C, or even > 100 °C under hyperbaric conditions (extreme thermophiles); others thrive in concentrated brine (extreme halophiles), at a pH < 2 (acidophiles) or > 10 (alkalophiles), and some are tolerant to metal ions (e.g. As, Cu, Zn, etc.) in mM concentrations (see Chapter 8; also Brock[5] and Edwards[13]). Thus, some habitats (including hot springs, acid mine waters, hyperhaline lakes) are almost exclusively the domain of bacteria. Other habitats, while not usually considered "extreme" in the above sense, are in practice inhabited almost entirely by bacteria; such habitats include strongly sulphidic habitats (which otherwise harbour only a few specialised protists) and sediments rich in clay and silt with small pore sizes that preclude larger organisms.

Only in terrestrial habitats do the bacteria have rivals (particularly in terms of the primary decomposition of plant structural compounds). These rivals are the fungi. One reason for this is that, among all the possible types of physical and chemical environments found in nature, bacteria seem to have only one absolute requirement for activity: liquid water. Many bacteria (especially soil isolates) produce dessication-resistant spores. However, metabolic activity and growth requires water; because of this requirement, growth and metabolic activity of "terrestrial bacteria" is confined to micrometre-thick films of water which may cover mineral and detrital particles in soils, the surfaces of rocks, litter and roots, stems and leaves of living plants. Fungi can tolerate water stress to a much greater extent than bacteria, and fungal hyphae can ramify through gas-filled pores of soil as well as cellulosic walls of plants. In this respect, fungi are better adapted to life in soil and litter. The relation between fungi and bacteria is discussed further in Chapter 4.

Yet another profoundly important reason for the pivotal role of bacteria in ecosystems is their metabolic diversity. Some species of bacteria are very specialised with respect to their substrates and available metabolic pathways, but taken together the metabolic diversity of bacteria far exceeds that known for eukaryotes. Examples of metabolic processes (among several others) that are exclusively carried out by certain bacteria include methanogenesis, the oxidation of methane and of other hydrocarbons, nitrogen fixation and sulphate reduction. These are all key processes in the function of the biosphere.

Similarly, bacteria collectively have an astonishing capability to hydrolyse virtually all natural polymers as well as many "unusual compounds" such as secondary plant metabolites and compounds found in crude oil, in addition to many xenobiotics. The degradation of polymers, which is largely a question of extracellular hydrolysis, is treated in Chapter 2. Here we proceed with a discussion on bacterial metabolic diversity.

1.1 Bacterial metabolism

If bacteria can be said to have purpose, it is to increase in size and divide. Bacterial activities are directed to this end and this requires energy and a variety of substrates from the environment needed for the synthesis of cellular material. These two activities (i.e. obtaining energy and obtaining building blocks for growth) are referred to as *dissimilatory* (energy or catabolic) metabolism and *assimilatory* (anabolic) metabolism, respectively.

It is convenient to discuss these separately and we do so in the following. However, the two types of metabolism are tightly coupled in the sense that micro-organisms spend by far the largest share of the power they generate on growth (i.e. on the energetic costs of synthesis

Table 1.1   Bacterial energy budget (based on Stouthamer[38] for cells grown on glucose)

Process	% Energy (ATP) expended on each process
Synthesis:	
Polysaccharide	6.5
Protein	61.1
Lipid	0.4
Nucleic acid	13.5
Transport into cells	18.3

of macromolecules: DNA, RNA and proteins; see Table 1.1). Under most normal growth conditions there is therefore an almost linear relation between the growth rate constant (measuring the balanced increase in biomass) and the rate of power generation. Furthermore, a particular substrate may serve both as an energy source and as a carbon source. A bacterium growing aerobically on glucose will use this substrate partly as a source of energy (oxidising it to CO_2) and partly as a source for cell material (largely without changing the oxidation level of the C atoms). In other cases, the energy source and assimilated materials are different. This is trite in the case of phototrophs, but it also applies for example to sulphide-oxidising bacteria which must assimilate CO_2 or some other C source. Finally, the enzymes involved in assimilatory and dissimilatory metabolism may overlap with identical metabolic pathways serving as oxidative, catabolic pathways in some species or under some circumstances, or (running in reverse) as reductive anabolic pathways in other species or circumstances. For example, the citric acid (TCA) cycle is used in most respiratory organisms for the stepwise oxidation of acetate to CO_2. In the phototrophic green sulphur bacteria and in some archaebacteria, the citric acid cycle runs in reverse and is used as a synthetic, reductive pathway for the assimilation of CO_2. In the purple non-sulphur bacteria, the same electron transport system is used in both respiration and photophosphorylation. (These and similar examples are of considerable interest in an evolutionary context since they illuminate the origin and evolution of metabolic pathways; they also show how a relatively small number of basic pathways may lead to a relatively large metabolic repertoire; see Chapter 10.) Under all circumstances it should be kept in mind that, while the distinction between dissimilatory and assimilatory metabolism is meaningful in some contexts, the two types of metabolism are in other respects intertwined.

The terms *autotrophy* and *heterotrophy* can be applied to both the dissimilatory and the assimilatory metabolism. A heterotroph depends on organic material for energy generation and for precursors for the synthesis of cell material. Autotrophs are independent of organic material and they assimilate C-1 compounds (CO_2, CH_4 or methanol) as carbon source. *Photoautotrophs* use the energy of electromagnetic radiation for ATP generation and for obtaining reducing power for the assimilation of CO_2. *Chemoautotrophs* (lithotrophs) obtain energy by oxidising inorganic substrates (e.g. HS^-, CH_4, H_2) with an inorganic electron acceptor (e.g. O_2, NO_3^-, Fe^{3+}) and obtain cell carbon by assimilative CO_2 reduction or from reduced C-1 compounds. (Reduced C-1 compounds are sometimes considered as organic compounds; with this definition methanotrophs and methylotrophs become heterotrophs.)

The terms autotrophy and heterotrophy are rarely absolute. Many organisms may be autotrophs and heterotrophs under different circumstances. Purple non-sulphur bacteria may be strictly autotrophic in the light (using H_2 or HS^- as reductants for photosynthesis and assimilating CO_2), but they can also assimilate acetate, and in the dark they can respire oxygen using various low molecular weight organics as substrates. Some chemoautotrophs are capable of assimilating organic substrates and some sulphur bacteria can subsist in the absence of O_2 by (heterotrophic) fermentations or by sulphate reduction. Some "autotrophs" may require organic growth factors or vitamins. The concepts of autotrophy and heterotrophy may also be extended to elements other than carbon (e.g. nitrogen). A bacterium that covers its need for N by assimilation of NH_4^+ or NO_3^- is then autotrophic with respect to nitrogen whereas a requirement for organic N (e.g. amino acids) for N assimilation would be considered heterotrophic.

Finally, while the term "autotrophy" somehow implies independence of the products of other organisms, this is also a question of context. Purple sulphur bacteria, for example, are photoautotrophs (using HS^- as an electron donor in a photosynthetic process and assimilating CO_2). In most habitats, however, the reducing power of the sulphide derives from plant material originally produced by oxygenic photosynthesis and subsequently degraded under anaerobic conditions. So in an ecosystem context the purple bacteria are only a link in a detritus food chain driven by oxygenic photosynthesis. (The more exotic situation where phototrophic sulphur bacteria depend on sulphide of geothermal origin would then, to a larger extent, justify the term "autotrophy".)

In this book we are primarily concerned with microbe-mediated chemical transformations of the environment rather than with intracellular physiology. In the following narrative we therefore emphasise what is taken up and what is produced as metabolites; that is, the net results of bacterial metabolism. Our discussion of the metabolic pathways within the cells is therefore limited to what is necessary to understand the energetics of the processes and thus why certain types of metabolism are favoured over others under particular circumstances. In this context the bioenergetics of dissimilatory metabolism is of central significance and treated in some detail in Section 1.2. Accounts of the biochemistry of bacterial metabolism (a topic on which much progress has been achieved lately) can be found in Brock;[6] for phototrophic, anaerobic and chemoautotrophic types of metabolism, see also Fenchel and Finlay,[15] Schlegel and Bowien[34] and Zehnder.[46] During the past two decades a number of novel types of energy metabolism have been discovered. Also, many organisms have proven to have a wider metabolic repertoire than previously known.

1.1.1 Dissimilatory metabolism

In 1926, Kluyver and Donker[22] drew attention to the "unity of biochemistry"; more specifically they pointed out that in bacteria (and other organisms) energy-yielding metabolic processes seem always to be coupled redox reactions of the type:

$$AH_2 + B \leftrightharpoons BH_2 + A.$$

This generalisation still holds. The particular reactions must be processes that result in a decrease of free energy of the system (Gibbs free energy $\Delta G < 0$); for more details see Section 1.2 and Appendix 1. The substrates used by the organisms are taken from the environment. In some cases the reactions used by bacteria for energy conservation will also occur spontaneously outside organisms (e.g. the oxidation of sulphide by oxygen). Other reactions (e.g. the oxidation of ammonia or of many organic substrates) will take place very slowly or not at all outside the organisms because these reactions have a high activation energy. An important function of energy metabolism is, therefore, to catalyse chemical processes towards equilibrium in addition to conserving the released energy in a form that is useful to the cell.

In living cells energy is conserved first of all as adenosine triphosphate (ATP) in the reaction:

$$ADP + Pi + energy \leftrightharpoons ATP + H_2O,$$

where ADP is adenosine diphosphate and Pi stands for inorganic phosphate. The $\Delta G^{0\prime}$ (free energy change of hydrolysis under standard condition, $25\,°C$, $1\,atm$, molar concentrations, and $pH = 7$) of ATP is $-29.3\,kJ$. The cells, in turn, use ATP to power vital processes: primarily synthesis of macromolecules and active transport across the cell membrane.

There are two different methods of ATP synthesis: substrate-level phosphorylation and electron transport phosphorylation. In substrate-level phosphorylation ATP is synthesised at specific steps in the catabolism of a substrate so that $1\,mol$ of ATP is produced from the transformation of $1\,mol$ substrate (provided the free energy change of this transformation exceeds $29.3\,kJ\,mol^{-1}$). In electron transport phosphorylation there is no such strict stoichiometric coupling between the transformation of substrate molecules and ATP synthesis. In respiration (or in photosynthesis) the membrane-bound enzymes of the electron transport chain act as an H^+ (or in some cases Na^+) pump expelling these ions from the cells. The resulting electrochemical gradient creates a "proton motive force" which leads to a return flux of protons (or Na^+). This flux is coupled to ATP synthesis which is catalysed by membrane-bound ATPase molecules.

The classification of energy-yielding processes is not simple because there are many exceptions and unclear boundaries between some categories. It is now customary to distinguish between *fermentation, respiration, phototrophy* and *methanogenesis* — a scheme that is followed here.

Fermentations are energy-yielding, anaerobic processes in which the substrates are sequentially transformed by reduction–oxidation processes. No external electron acceptor is involved; that is, the redox levels of the substrate and of the metabolite(s) remain the same. Thus, fermentations represent a dismutation of the substrate molecules. In fermentations a relatively low amount of energy is conserved: fermentation of 1 mol of glucose yields 2–4 mol of ATP (depending on the type of fermentation) while respiration (with O_2 as the terminal electron acceptor) yields about 32 ATP.

A comparison of fermentation and respiration is shown in Table 1.2. None of the criteria, however, is absolute. Thus, the dismutation of $S_2O_3^{2-}$ into sulphide and sulphate (a microbial process found to take place in anaerobic sediments)[1] is technically a fermentation, but it is based on an inorganic substrate. In some succinate/propionate fermentations the step in which fumarate is reduced to succinate is coupled to electron transport phosphorylation (rather than to substrate-level phosphorylation). In some cases the excretion of fermentation products results in a proton motive force which is exploited in membrane phosphorylation. Strictly speaking, some types of fermentations in which CO_2 or H_2O is reduced do not represent a complete redox balance.

Fermenting bacteria may be facultative anaerobes (capable of oxidative phosphorylation in the presence of O_2 (e.g. *Escherichia*), aerotolerant anaerobes (e.g. *Lactobacillus*) or strict (O_2-sensitive) anaerobes (e.g. *Clostridium*).

In the fermentation of glucose to lactate or to ethanol $+ CO_2$, the yield is only 2 mol of ATP per mol substrate. This is because pyruvate (resulting

Table 1.2   Properties of fermentation and respiration

Fermentation	Respiration
Redox balance: no external electron acceptors; dismutation of substrate molecules; anaerobic processes	External electron acceptors; electron transport system with FeS proteins, ubiquinone and cytochromes; aerobic or anaerobic
Organic substrates	Organic or inorganic substrates
Substrate-level phosphorylation	Electron transport phosphorylation

from glycolysis) is used for reoxidising reduced nicotinamide adenine dinucleotide (NADH) produced during glycolysis thus restoring cellular redox balance (and producing lactate or ethanol as waste products). This is a "waste" in the sense that further fermentation of pyruvate to acetate and an additional generation of ATP would otherwise have been possible. In nature this type of fermentation is important only where easily degradable sugars accumulate at high concentrations. Lactobacilli are acid tolerant and since they acidify their environment they can maintain a competitive advantage against other types of fermenters once a large population has been established.

In mixed acid fermentation some reducing equivalents are disposed of as formate (which in most cases is broken down further to H_2 and CO_2). This allows for the oxidation of some of the pyruvate via acetyl-CoA to acetate. This last step is coupled to substrate-level phosphorylation allowing for the generation of additional ATP relative to homolactic fermentation. Other reducing equivalents (from NADH) are coupled to acetyl-CoA metabolism resulting in the production of acetate and H_2 in addition to a mixture of other compounds (butyrate, succinate, lactate, ethanol) which are less oxidised than acetate. This type of fermentation is especially known from the enterobacteria.

Another way of restoring redox balance is found in clostridial-type fermentations. Clostridia have an enzyme, pyruvate–ferredoxin oxidoreductase, which catalyses the oxidation of pyruvate by a low potential electron acceptor, ferredoxin. The ferredoxin is then reoxidised by a hydrogenase leading to the excretion of H_2. NADH is oxidised by acetyl-CoA and the final products are H_2 and butyrate. This process, which yields 3 ATP, is found in *Clostridium butyricum*. As in mixed acid fermentation, butyric acid fermentation is not very sensitive to ambient hydrogen tension (p_{H_2}).

The complete fermentation of glucose to acetate and H_2 altogether yields 4 mol of ATP and this is also realised in some *Clostridium* species. This requires the reoxidation of NADH by hydrogen evolution (proton reduction), a process that is thermodynamically possible only if ambient p_{H_2} is sufficiently low ($<\sim 10^{-4}$ atm). This normally requires the presence of other H_2-consuming bacteria such as sulphate reducers or methanogens (exemplifying syntrophic "interspecies hydrogen transfer"; see Sections 1.2 and 1.3). The so-called obligate acetate-producing bacteria ferment low molecular weight fatty acids and alcohols into acetate and H_2. They are entirely dependent on a low ambient p_{H_2} and live in obligatory syntrophic associations with hydrogen-consuming bacteria.

The so-called homoacetogens represent an entirely different type of obligate acetogens. They may ferment suitable organic materials or, in their absence, live autotrophically by reducing CO_2 with H_2 according to:

$$2CO_2 + 8H_2 \rightarrow CH_3COOH + 2H_2O.$$

When living autotrophically, the homoacetogens can create a proton gradient across their membrane which is linked to ATP synthesis; the acetyl-CoA they form is used for assimilation rather than for the generation of ATP by substrate-level phosphorylation.

Some other fermentation mechanisms should be mentioned, although they probably play a more modest role in nature. In succinate fermentation, a portion of the pyruvate derived from glycolysis is carboxylated to malate and further to fumarate which serves for reoxidation of NADH yielding succinate or propionate as end products. Altogether this process yields 3 mol of ATP per mole glucose. Fermentation of amino acids may take place by the so-called Stickland reaction in which one kind of amino acid is used to oxidise another, the end products being mainly acetate and ammonia. This process is important where high concentrations of proteins undergo anaerobic degradation.

Since organic substrates cannot be fermented further than to acetate $+ H_2$, the complete mineralisation of organic matter under anaerobic conditions depends on other physiological types of bacteria. Nevertheless, fermenting bacteria play a central role in anaerobic communities since in such systems they are the only organisms that can hydrolyse and utilise polymers (polysaccharides, proteins, etc.). All of the other important microbial players in anaerobic communities (sulphate reducers, methanogens) are capable of using only a limited number of low molecular weight substrates. Thus, terminal mineralisation depends on the activity of fermenting bacteria for the supply of substrates. A detailed account of the biochemistry of fermentations is found in Zehnder.[46]

In contrast to fermentation respiration is a form of energy metabolism that depends on external electron acceptors for substrate oxidation. The electron carrier chain (characteristically including an FeS enzyme, an ubiquinone and cytochromes, but with some variation among different groups of bacteria) catalyses the terminal oxidation steps; the energy released creates a proton motive force that generates ATP. We here classify and discuss respiration processes according to the terminal electron acceptors used.

Aerobic respiration.   This process yields the most energy of any catabolic metabolic process. A large number of bacteria oxidise organic substrates according to:

$$[CH_2O] + O_2 \rightarrow CO_2 + H_2O$$

which yields about 5 ATP per C. Species that can perform oxidative phosphorylation of organic substrates occur in many major groups of bacteria. Their great diversity is based on specialisations with respect to the substrate they can metabolise and the macromolecules they can hydrolyse (in addition, of course, to a variety of adaptations unrelated to energy metabolism).

Many aerobic respirers are facultative anaerobes which can utilise fermentative pathways or nitrate respiration in the absence of O_2.

Some more specialised aerobes depend on inorganic substrates; these are referred to as chemoautotrophs. *Hydrogen oxidisers* (knallgas bacteria) derive energy for growth and maintenance from the following reaction: $2H_2 + O_2 \rightarrow 2H_2O$. They probably play a modest role in nature since molecular hydrogen is seldomly available in aerobic habitats. One of their niches may be in cyanobacterial mats where H_2 is excreted by cyanobacteria as a byproduct of nitrogen fixation.[33] The *methanotrophs* oxidise methane (or other C-1 compounds: methanol, methylamine) according to $CH_4 + 2O_2 \rightarrow CO_2 + 2H_2O$ and include species of *Methylosinus*, *Methylocystis*, *Methylococcus* and *Methylobacter*. Methanotrophs are important in aquatic habitats as well as in soils.

Nitrifiers are extremely important in nature. Species belonging to the genera *Nitrosomonas*, *Nitrocystis* and others oxidise ammonia to nitrite according to:

$$2NH_4^+ + 3O_2 \rightarrow 2NO_2^- + 4H^+ + 2H_2O$$

and species of *Nitrobacter*, *Nitrococcus* and *Nitrospina* complete the oxidation according to:

$$2NO_2^- + O_2 \rightarrow NO_3^-.$$

Complete nitrification is thus a two-step process.

Sulphide oxidisers (or colourless sulphur bacteria) constitute an important group of bacteria, including species of *Thiobacillus* and more spectacular forms belonging to genera such as the filamentous *Beggiatoa*, *Thiotrix*, *Thioploca* and the unicellular giants *Thiovulum* and *Achromatium*; these all belong to the proteobacteria among the eubacteria. Some thermophilic archaebacteria are also sulphide or sulphur oxidisers. The complete oxidation of sulphide proceeds according to:

$$HS^- + 2O_2 \rightarrow SO_4^{2-} + H^+.$$

In addition, S^0 and $S_2O_3^{2-}$ occur as intermediate oxidation products and can also be used as substrates; the large species of sulphide-oxidising bacteria store sulphur granules as a nutrient reserve that is metabolised if the external sulphide concentration is low.

Iron and manganese oxidisers constitute a less well-known group. It is well established that *Thiobacillus ferrooxidans* oxidises ferrous iron according to:

$$4Fe^{2+} + O_2 + 4H^+ \rightarrow 4Fe^{3+} + 2H_2O$$

under acid conditions (a process of significance for ore leaching). The oxidation of reduced Mn and Fe under neutral conditions has been proposed repeatedly (e.g. for the so-called iron bacteria such as *Gallionella* and *Leptothrix* which make iron-incrusted sheaths and which are frequently found in freshwater springs). It has not, to our knowledge, been unambiguously shown that reduced Mn or Fe actually serve as substrate in a respiratory process under neutral conditions, although recent evidence suggests that this may be the case for *Gallionella* (see Section 6.3). One difficulty is that under neutral conditions the spontaneous oxidation of Fe^{2+} and Mn^{2+} is very rapid. The metals do undergo reduction–oxidation transformations in sediments, but whether these are directly mediated by bacteria or take place spontaneously due to changing redox conditions (brought about by bacterial metabolites) is still an open question; some evidence for respiratory Mn oxidation has been provided.[21] The biology and biochemistry of chemoautotrophic bacteria is treated in detail in Schlegel and Bowien.[34]

Anaerobic respiration. Anaerobic respiration is defined as the use of exogenously derived terminal electron acceptors other than oxygen. Energy yields are always lower than those obtained by oxidative phosphorylation. The important processes include *nitrate reduction* (nitrate respiration, denitrification), *sulphate reduction* (sulphate respiration), and *iron and manganese reduction*; some low molecular weight organics may also serve as terminal electron acceptors.

Nitrate or nitrite reduction is performed by many facultative anaerobes including, among others, representatives of common genera such as *Escherichia, Bacillus* and *Pseudomonas*. In these bacteria the synthesis of the key enzyme nitrate reductase is inhibited in the presence of O_2. The principal reaction is:

$$5[CH_2O] + 4NO_3^- + 4H^+ \rightarrow 5CO_2 + 2N_2 + 7H_2O.$$

However, the reduction to N_2 is often incomplete resulting in NO or N_2O as metabolites. Denitrifiers can use a wide range of organic substrates, but certain catabolic reactions that require oxygenases do not take place. The ability of sulphur bacteria to use nitrate instead of oxygen as terminal electron acceptor is probably widespread and it has been established in *Thiobacillus denitrificans* and in some filamentous sulphur bacteria. Methane oxidation through denitrification is not known. Autotrophic growth based on ammonia oxidation through denitrification has recently been discovered.[42] In addition to (respiratory) denitrification, nitrate can be reduced to ammonia in anaerobic habitats. This is brought about by fermenting bacteria which use NO_3^- as a "hydrogen dump"; the process is referred to as "nitrate fermentation" and is not a respiratory process. The quantitative importance of this process as well as of denitrification is often limited by the absence of nitrate in anaerobic habitats; in sediments and in stratified water columns denitrification is especially important immediately below the oxic zone where nitrate is generated by nitrification.

Sulphate reduction is an immensely important process in marine sediments and also contributes significantly to anaerobic metabolism in freshwater sediments. Sulphate reducers use a modest range of substrates (although the range is expanding regularly), the most important of which include low molecular weight fatty acids (acetate, butyrate, propionate, lactate), alcohols, and H_2. In nature these substrates are supplied by fermenting bacteria with which sulphate reducers often form consortia. Some of these consortia are based on interspecies hydrogen transfer. Most sulphate reducers belong to the proteobacteria; some Gram-positive eubacteria and thermophilic archaebacteria are also sulphate respirers. From a functional point of view sulphate reducers can be divided into two groups. One group does not have a citric acid cycle and the organisms use mainly H_2 or lactate as substrates (in the latter case producing acetate + sulphide as metabolites); the genus *Desulfovibrio* is an example. Members of the other group (represented by e.g. *Desulfobacter*) have a citric acid cycle and are capable of the complete oxidation of acetate according to:

$$2H^+ + CH_3COO^- + SO_4{}^{2-} \rightarrow 2CO_2 + 2H_2O + HS^-.$$

The genus *Desulfuromonas* uses elemental sulphur rather than sulphate for the complete oxidation of acetate to CO_2 and HS^-. There is some evidence to suggest that anaerobic CH_4 oxidation, which does take place in sediments, is caused by sulphate reduction, but it has not yet been possible to isolate sulphate reducers that can sustain growth on the basis of methane. Some sulphate reducers are strict anaerobes while others are tolerant to microaerobic conditions. More recently it has been shown that sulphate reduction may take place under aerobic conditions in cyanobacterial mats[16] and that some sulphate reducers may be aerobic respirers under microaerobic conditions.[11,12]

The idea that Fe^{3+} or Mn^{4+} may function as terminal electron acceptors in respiration has long been questioned due to the absence of evidence from pure cultures and since the reduction of Fe^{3+} takes place spontaneously in anaerobic environments.[14,17] There is evidence that nitrate reducers, in the absence of $NO_3{}^-$, reduce ferric iron and that this is catalysed by nitrate reductase,[19,36] but it is unclear whether this is coupled to energy conservation. However, most recently evidence has accumulated to show that respiratory iron reduction exists and that it plays a quantitative role in certain habitats.[26,27] The evidence is mainly based on growth on non-fermentable substrates (acetate) with the simultaneous reduction of the metals under anaerobic conditions. Several species (*Aquaspirillum, Geobacter, Shewanella*) have now been isolated; there is also evidence to show that oxidised forms of U and Se can serve as electron acceptors for these bacteria.

Some organic compounds may also serve as terminal electron acceptors although their quantitative role is probably limited in nature. These include fumarate (reduced to succinate), dimethyl sulphoxide (reduced to dimethyl sulphide) and trimethylamine oxide (reduced to trimethylamine). Trimethylamine oxide occurs in fish tissue and trimethylamine is an important component of the smell of fouling fish.

Methanogenesis was previously considered as a special type of fermentation. However, in some respects a very unique biochemistry is involved which distinguishes methanogenesis from fermentation as well as from respiration. Methanogenesis is performed by methanogenic bacteria, all of which belong to the archaebacteria.

There are two main types of methanogenesis. The *acetoclastic* methanogens dismutate acetate according to:

$$CH_3COOH \rightarrow CO_2 + CH_4.$$

Examples are *Methanosarcina* and *Methanosaeta*. From a formal viewpoint this is a type of fermentation. Some methanogens can also use the methyl groups of methanol or methylamines (*Methanolobus, Methanocorpusculum* and *Methanosarcina*). The *H_2/CO_2 methanogens* use H_2 as an electron donor for the reduction of CO_2 (or CO or formate) according to:

$$4H_2 + CO_2 \rightarrow CH_4 + 2H_2O$$

(e.g. *Methanobacterium* and *Methanococcus*). The process is formally a type of respiration and the organisms involved are autotrophs. However, an ordinary electron transport chain is not found. Methanogenesis is associated with special coenzymes including F_{420}, which is involved in the activation of H_2, and coenzyme M, which is involved in the terminal reduction of CH_3 groups to methane. In an oxidised state coenzyme F_{420} fluoresces in violet light, a fact that allows the microscopic identification of methanogens in mixed bacterial communities. The methanogens are strict anaerobes. The mechanism of energy conservation is still not understood in detail.

The energetics of the H_2/CO_2 methanogens would theroretically seem to be relatively favourable, but in practice cell growth rates and yields are lower than predicted. This is in part due to the fact that the organisms are autotrophs, and so must use part of their substrate (CO_2) for C assimilation. The energetics of acetoclastic methanogens are less favourable than those of the hydrogenotrophs, resulting in even slower growth rates and lower cell yields (see also Section 1.2). Relative to sulphate reducers (in terms of common substrates: H_2 and acetate) they are inferior competitors. However, in sulphate-depleted, anaerobic habitats (especially freshwater sediments, anaerobic sewage digesters, the rumen) they play a central role as H_2 scavengers and in the terminal mineralisation of acetate. They also occur as endosymbionts of some anaerobic eukaryotes.

The last type of energy metabolism to be discussed is *phototrophy*. At the biochemical (and evolutionary) level, phototrophy is closely related to (and the precursor of) respiration. The principal process is that a photoactivated chlorophyll or bacteriochlorophyll molecule delivers an electron to a primary acceptor (pheophytin: a chlorophyll or bacteriochlorophyll deprived of the Mg atom) which passes the electron on to a membrane-bound electron carrier chain (including FeS proteins, ubiquinones and cytochromes) and back to the chlorophyll. This acts as a proton pump which again is coupled to ATP synthesis by membrane phosphorylation (cyclic phosphorylation). This, of course, is quite analogous to respiration, only in phototrophs the electron source is a photoactivated chlorophyll rather than NADH. In phototrophs electrons can also be used for the production of reduced nicotinamide adenine dinucleotide phosphate (NADPH), which is in turn used for the reductive assimilation of CO_2 (see Section 1.1.2). In this case an external electron donor is necessary according to the general scheme:

$$H_2A + CO_2 \rightarrow [CH_2O] + A$$

where H_2A (the reductant) represents, for example, H_2O, H_2S, S^0 or H_2.

The "reaction centre" consists of a chlorophyll/bacteriochlorophyll molecule surrounded by light harvesting molecules including carotenoids which absorb light in the 400–550 nm range), phaeophytin and the electron transport chain. The wavelengths exploited are within the range of 400–1000 nm, but different groups of organisms have different chlorophylls and accessory photosynthetic pigments, so there is a differential utilisation of the light spectrum among different types of phototrophs. Reaction centres reside in the membranes of intracellular vesicles or tubular structures which either constitute invaginations of the cell membrane (in purple bacteria and in cyanobacteria) or are made up of special (non-unit) membranes (in green bacteria).

It is customary to distinguish between oxygenic and anoxygenic photosynthesis. Oxygenic photosynthesis is the most important and all-dominant process in the biosphere, since it fuels practically all biological activity on Earth. In oxygenic photosynthesis the reaction centres include two coupled photosystems (I and II) of which photosystem II is involved in the oxidation of H_2O to O_2.

Among the prokaryotes only the cyanobacteria (including the prochlorophytes) perform oxygenic photosynthesis. They have chlorophyll *a*; phycobilins (phycocyanin and phycoerythin) serve as accessory pigments which allow them to utilise the orange and green regions of the spectrum. Cyanobacteria are important in shallow-water sediments and dominate "cyanobacterial mats" (especially the filamentous forms *Oscillatoria, Lyngbya, Microcoleus* and *Spirulina,* but also unicellular forms (see Chapter 6)). They are also important as primary producers in the plankton, but here it is mainly the unicellular types (*Synechococcus*) that dominate (however, mass occurrence of colonial forms is common in eutrophic waters); cyanobacteria also occur as symbionts of eukaryotes (see Chapter 7). Sediment-dwelling cyanobacteria are frequently exposed to sulphide which inhibits photosystem II. However, photosystem I remains functional and many forms then use anoxygenic photosynthesis with HS^- as electron donor.[8] In the dark and under anaerobic conditions some cyanobacteria seem to be capable of fermentation or sulphate reduction.[31]

The prochlorophytes are small unicellular forms; they also contain chlorophyll *b* (like green algae and higher plants). Originally discovered as symbionts in the tunic of some ascidians, they have more recently proven to be important in marine plankton (see Section 3.1).

Bacteria performing anoxygenic photosynthesis are found within four unrelated groups of eubacteria; these are characterised by (among other features) their photosynthetic pigments and by their substrates. The purple bacteria have bacteriochlorophyll *a* (with an infrared absorption maximum at 825–890 nm) or bacteriochlorophyll *b* (1020–1040 nm); different groups also have different combinations of carotenoids. The purple sulphur bacteria (e.g. *Thiocapsa*, *Thiopedia*, *Chromatium*) use reduced sulphur as an electron donor according to:

$$2H^+ + 2HS^- + CO_2 \rightarrow [CH_2O] + 2S^0 + H_2O$$

and

$$5H_2O + 2S^0 + 3CO_2 \rightarrow 3[CH_2O] + 2SO_4^- + 4H^+.$$

The cells often store elemental sulphur. They occur and sometimes dominate in sulphidic habitats exposed to light. They are autotrophs and to a varying degree O_2 tolerant (some are strict anaerobes) and this, as well as their variable tolerance to sulphide concentration, explains species diversity in nature. The term "purple non-sulphur bacteria" (e.g. *Rhodopseudomonas*, *Rhodospirillum*) has proven to be imprecise in that these organisms are also capable of using HS^- as electron donor, although they are more sensitive to exposure to high sulphide concentrations than are the purple sulphur bacteria. The term non-sulphur bacteria is perhaps also justified in that they do not — in contrast to the purple sulphur bacteria — store elemental S in their cells. They can also use H_2 and some organic compounds as electron donors, and most recently it has been discovered that some can use Fe^{2+}.[43] These metabolically versatile organisms can appear as microaerobic respirers (in the dark) and they can assimilate low molecular weight compounds such as acetate.

The green sulphur bacteria (e.g. *Chlorobium*, *Pelochromatium*) have bacteriochlorophylls *c* or *d* (infrared absorption maxima at 745–755 and 705–740 nm, respectively). They are autotrophic, strictly anaerobic organisms which use sulphide as an electron donor, but they only oxidise it to S^0 which is excreted. In nature, they typically occur beneath the purple sulphur bacteria in sediments or in anoxic water columns.

In freshwater they are usually found in physically close, syntrophic associations with heterotrophic sulphur reducers. The green non-sulphur bacteria (*Chloroflexus*) are filamentous gliding forms found in microbial mats. They can live as photoautotrophs using HS^- and CO_2 (again, they were named before it was discovered that they could use sulphide) or as aerobic heterotrophs. The heliobacteria (*Heliobacter*) are strict phototrophic anaerobes which have been isolated from anoxic soils; not much is known about their ecological importance.

A unique and quite different form of phototrophy is found in halobacteria (a group of archaebacteria that live in hyperhaline habitats). These are normally aerobic heterotrophs. However, under suitable conditions (anoxia, light) they develop a purple membrane consisting of a photosensitive protein, bacteriorhodopsin. When exposed to light it facilitates proton export from the cell, creating a proton motive force used by ATPase in the cell membrane for ATP synthesis.

1.1.2 Assimilatory metabolism

In environments with available sugars and amino acids most of the constituents of bacterial cells (Table 1.3) are readily available and the reductive assimilation of C and N and the synthesis of the basic units of biological molecules is then unnecessary. More commonly, however, the basic building blocks and organic nitrogen, in particular, are unavailable. Therefore a few organics must be used to synthesise a diversity of

Table 1.3   Constituents of bacteria

(A) Elemental composition

Element	% of dry weight
C	55
O	20
N	10
H	8
P	3
S	1

(B) Chemical composition of bacteria (data from Brock[6])

Constituents	%	% of dry weight
Water	70	–
Protein		55
Polysaccharide		5
Lipids		9
DNA		3
RNA		20
Monomers (sugars, amino acids, etc.)		3.5

Table 1.4   Principal mineral forms of C, N, S and P

Elements	In oxic environments	In anoxic environments
C	CO_2	CO_2, CH_4
N	NO_3^-, NO_2^-, N_2	NH_4^+, N_2
S	SO_4^{2-}	S^0, HS^-
P	PO_4^{3-}	PO_4^{3-}

cell metabolites, and the organisms must assimilate N in an inorganic form. Autotrophs assimilate all necessary elements in an inorganic form (Table 1.4). The chemical reduction of elements (e.g. C^{4+} in CO_2 to C^0 in $[CH_2O]$) and the subsequent synthesis of glucose, amino acids and nucleotides require energy (in addition to the energy consumption for synthesis of macromolecules and for active transport). Organisms that depend on autotrophy therefore have a lower growth rate and yield per unit of ATP synthesised than if monomeric units are assimilated from the environment (see also Section 1.2.3).

The assimilation of C-1 compounds occurs by several mechanisms. The most important pathway for CO_2 reduction and assimilation is the Calvin cycle (ribulose biphosphate cycle) in which CO_2 reduction is catalysed by the key enzyme, ribulose biphosphate carboxylase (RuBiSco), as a reaction between CO_2 and ribulose biphosphate to form two C-3 molecules. The Calvin cycle is found in organisms with oxygenic photosynthesis (cyanobacteria), in purple bacteria and in most chemoautotrophs. A different pathway (a reverse citric acid cycle) is used by the green sulphur bacteria. Some anaerobic autotrophs (H_2/CO_2 acetogens, sulphate reducers growing on H_2 and SO_4^{2-}, methanogens) synthesise acetyl-CoA as a precursor for cell materials. Methanotrophs (aerobic oxidisers of C-1 compounds such as CH_4, methanol, CO) oxidise part of their substrate to HCHO (formaldehyde) which is used as a precursor for organic compounds via either the "serine pathway" or the "ribulose monophosphate pathway".

Most bacteria can synthesise amino acids by the assimilation of NH^+ which does not require further reduction; ammonia is the dominant N compound in anaerobic environments and it may also occur under some circumstances in large quantities in aerobic habitats. Urea may also serve as a N source for many bacteria. In aerobic habitats NO_3^- is frequently the dominating form of inorganic N and most aerobes are capable of assimilatory nitrate reduction. Nitrate is reduced stepwise (with NADPH as electron donor), first (catalysed by nitrate reductase) to nitrite and then further via hydroxylamine to ammonia.

Nitrogen fixation (the assimilatory reduction of N_2) is exclusively found among certain prokaryotes. The basis for this is the presence of nitrogenase which is a complex consisting of two enzymes — dinitrogenase and dinitrogenase reductase; both enzymes contain Fe and the latter also contains Mo. Dinitrogenase reductase is O_2 sensitive and nitrogen fixation can take place only under anaerobic conditions. The activation energy of N_2 is very high and the process requires a low redox potential making it energetically costly. For reasons not understood, the process involves the reduction of protons so that H_2 is produced simultaneously with the reduction of N_2 (to $2NH_4^+$) and the entire process requires 18–24 ATP per N_2 reduced. Nitrogenase is not very specific and catalyses the reduction of other compounds (e.g. acetylene to ethylene, a process that is exploited for quantifying nitrogenase activity).

Nitrogen fixation is found among many anaerobic bacteria (the phototrophic purple and green bacteria, *Clostridium*, and many sulphate reducers and methanogens). It has also been found in many facultative anaerobes which can fix N_2 under anaerobic conditions (*Bacillus*, *Klebsiella*, etc.). Species of *Azotobacter* can maintain an intracellular anoxic environment under aerobic conditions due to copious mucus secretion combined with high rates of O_2 consumption in excess of energetic needs. Many cyanobacteria are nitrogen fixers under anaerobic or microaerobic conditions. In some colonial forms N_2 fixation takes place in specialised cells (heterocysts) in which only photosystem I is functional; thus cyclic phosphorylation provides ATP for N_2 fixation, but O_2 is not produced. Many plants have symbiotic N_2-fixing bacteria: *Rhizobium* in legume root nodules is the most well-known example, but other plants (e.g. *Myrica* and *Alnus*) have associations with N_2-fixing actinomycetes and the cycads, the aquatic fern *Azolla* and some lichens have symbiotic N_2-fixing cyanobacteria. Symbiotic N_2-fixing bacteria in the hindgut of termites are also important for the N supply of these insects (see also Section 7.2).

Inorganic sulphur can be assimilated in the form of HS^- (in anaerobic habitats) and this does not require reduction. Under aerobic conditions sulphate is the predominant precursor for S in biomass. After transport into cells sulphate is activated (at the cost of ATP) to form APS (adenosine 5'-phosphosulphonate) which is then reduced to S^{2-} via sulphite. Phosphorus does not change valence in biological processes and is generally assimilated in the form of PO_4^{3-} (possibly after extracellular hydrolysis of organic P).

1.2 Bioenergetics of microbial metabolism

Predicting which bacterial processes predominate under given circumstances requires an understanding of the energetics of dissimilatory metabolism. There are two aspects involved: (i) considerations based on chemical thermodynamics, and (ii) kinetic constraints of chemical reactions. Kinetic constraints imply that certain processes, which are possibly based on thermodynamics, do not occur spontaneously since a high activation energy is required. Thus, thermodynamic considerations alone would suggest that oxidation of N_2 with O_2 could provide a possible way of making a living for bacteria. However, the $N \equiv N$ bond is strong so the process requires a considerable activation energy and is therefore not realised. We first of all discuss the bioenergetic processes on the basis of equilibrium considerations. These are for several reasons (which are discussed below) only approximative, but they do provide an heuristic insight with respect to the distribution of different types of bacterial metabolism. The thermodynamic background for the following section is presented in Appendix 1.

1.2.1 Energetic yields of metabolic processes

The standard free energy of a given process can be calculated from the free energy of formation[40] (Table A.1.1) according to:

$$\Delta G^{0'} = \Sigma \Delta G_f^{0'} \text{(products)} - \Sigma \Delta G_f^{0'} \text{(reactants) (see Appendix 1).}$$

Calculations of standard free energy changes approximate the energetics of particular metabolic reactions. Thus, based on values in Table A.1, the free energy change of hydrogen oxidation with four different electron acceptors can be calculated as (expressed as kJ per mol H_2 oxidised):

$$2H_2 + O_2 \rightarrow 2H_2O; \Delta G^{0'} = -238 \text{ kJ}$$

$$5H_2 + 2NO_3^- + 2H^+ \rightarrow N_2 + 6H_2O; \Delta G^{0'} = -224 \text{ kJ}$$

$$4H_2 + SO_4^{2-} + H^+ \rightarrow 4H_2O + HS^-; \Delta G^{0'} = -38 \text{ kJ}$$

$$4H_2 + CO_2 \rightarrow CH_4 + 2H_2O; \Delta G^{0'} = -33 \text{ kJ}.$$

Clearly, aerobic oxidation and methanogenesis are the energetically most favourable and the least favourable processes, respectively. Quantitatively, however, the above picture is only approximate. For example, the ATP yield of nitrate respiration is actually only about 50% that of O_2 respiration because the mechanism by which hydrogen oxidation is coupled to nitrate reduction is energetically less efficient than for oxygen respiration.

In general, the efficiency of energy conservation is not high. For the aerobic degradation of glucose: ($C_6H_{12}O_6 + 6O_2 \rightarrow 6CO_2 + 6H_2O$), $\Delta G^{0'} = -2877 \text{ kJ mol}^{-1}$. The process is known to yield 32 mol of ATP. The hydrolysis of ATP has a free energy change of about -29 kJ mol^{-1} and so the efficiency of energy conservation is only $29 \times 32/2877$ or about 32%. The remaining 68% is lost as metabolic heat.

Another problem is that the calculation of standard free energy changes assumes molar concentrations. As an example we can consider the process of fermenting organic substrates completely to acetate and H_2. As discussed in Section 1.1.1, this requires the reoxidation of NADH (produced during glycolysis) by H_2 production. From Table A.2 we have $E_0' = -0.32 \text{ V}$ for NAD/NADH and $E_0' = -0.41 \text{ V}$ for H_2O/H_2. Assuming a p_{H_2} of 1 atm we have from equations (A.1.5 and A.1.6) that $\Delta G^{0'} = +17.4 \text{ kJ}$

which shows that the reaction is impossible. If we assume instead that p_{H_2} is 10^{-4} atm ($Q = 10^{-4}$) we find that $\Delta G^{0\prime}$ is ~ -5. Thus at an ambient $p_{H_2} < 10^{-4}$ the process is feasible. In natural systems, maintenance of such a low p_{H_2} requires the presence of H_2-consuming bacteria.

This requirement also applies to the fermentation of ethanol and fatty acids to acetate $+ H_2$ by obligatory acetogens (see Section 1.1). In Fig. 1.1 the dependency on p_{H_2} of two such fermentations is shown. Also shown is the dependency of two types of H_2 consumers (sulphate reducers and methanogens) on p_{H_2}; obviously they are favoured by a high hydrogen pressure. The graph suggests that such anaerobic systems (with syntrophic interspecies hydrogen transfer) will equilibrate at a hydrogen pressure between 10^{-6} and 10^{-5} atm (sulphate reducers being able to maintain a somewhat lower H_2 tension than methanogens). This result accords with actual measurements of p_{H_2} in anaerobic habitats.

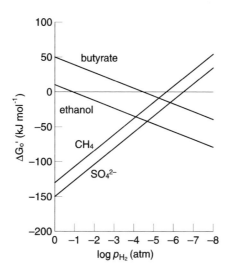

Figure 1.1   The standard free energy changes (pH 7) for the fermentation of ethanol and butyrate to acetate and H_2 and for the oxidation of H_2 by sulphate reduction or by H_2/CO_2 methanogenesis as a function of log p_{H_2}. Coexistence of H_2 producers and H_2 consumers will make both processes energetically feasible and maintain an ambient p_{H_2} within the range of 10^{-6}–10^{-5} atmospheric pressure

While the ambient concentrations of substrates or electron acceptors in principle always affect the energetic yield of metabolic processes, this is especially important in the case of H_2 metabolism. In most other cases standard free energy calculations as presented above will provide reasonable indications of reaction energetics and which reactions are possible.

1.2.2   Energetics, Y_{ATP} and growth yield

The outcome of competition between different bacteria is determined by several factors including the kinetics of substrate uptake and the efficiency with which substrate utilisation is coupled to growth. Various other adaptive traits may also determine the actual outcome under different sets of circumstances, but an important component of competitive ability is the maximisation of ATP yield from energy metabolism.

In accordance with the fact that most energy is spent on growth processes (Table 1.1), it has been found that there is usually proportionality between power generation and growth. More specifically, it has been found that about 10 g dry weight of organic matter are produced for each mole of ATP synthesised through dissimilatory metabolism ($Y_{ATP} = 10\,\text{g mol}^{-1}$). This value has been obtained from different bacteria.[3] Although not universally applicable (see below) it serves as a useful generalisation.

In order to visualise this, growth yields can be calculated for two organotrophic heterotrophs using the same substrate, but with different types of dissimilatory metabolism. By "growth yield", we mean the ratio between organic material incorporated into cell material and the total amount of organic substrate consumed (for dissimilatory + assimilatory metabolism). Consider aerobic and fermentative bacteria growing on glucose. The former can gain 32 ATP per mol glucose (= 180 g). Assuming that $Y_{ATP} = 10\,\text{g mol}^{-1}$ then the aerobic bacterium can synthesise 320 g cell material by dissimilating 180 g glucose. The cell must therefore altogether consume a total of $180 + 320$ g glucose in order to produce 320 g cell material and the yield becomes $320/(180 + 320) = 0.64$; that is, 64% of the substrate consumed is recovered as cell material. A similar calculation for the anaerobic fermenter (and assuming 3 mol ATP per mol dissimilated glucose) results in a growth yield of only 14%. Such figures are close to what has actually been measured. Thus, under similar circumstances (similar rate of glucose uptake), the aerobe should be able to multiply about four times faster than the anaerobe. This principle applies to other situations: for example, in anaerobic environments sulphate reducers outcompete methanogenic bacteria as long as sulphate is available (cf. Fig. 1.1).

In reality Y_{ATP} is not always $10\,\text{g mol}^{-1}$. A constant value of Y_{ATP} usually holds over a wide range of growth rates (growth rate constant is linearly proportional to power generation) but deviations occur at very low growth rates because maintenance energy (i.e. energy used for the maintenance of cell integrity not directly related to growth) then becomes a relatively larger fraction of the total power generation. Various (in part artificial) growth conditions or limiting availability of specific nutrients can also decrease Y_{ATP}. A more important factor is the cost of biosynthesis as a function of the available forms of C for assimilatory metabolism. Thus, if only 2- or 3-C compounds are used or are available, Y_{ATP} is decreased, and if the C source is CO_2 an even larger amount of energy is spent on reductive assimilation at the cost of Y_{ATP}. Finally, some adaptive traits such as energy-requiring N_2 fixation will also lower cell yields and growth rate constants.

1.2.3 Bioenergetics and the structure of bacterial communities

At this point it is important to discuss some general properties of bacterial communities. Classical bacteriology has long held that only pure laboratory cultures were worthy of study. This explains why the interdependency of bacteria has been ignored and microbial functional biology was incompletely understood. The following discussion of bacterial communities first emphasises flows of energy and material; spatial and temporal patterns of bacterial communities are treated in more detail in Section 1.3.

In one respect aerobic heterotrophic micro-organisms and the communities they form differ fundamentally from their anaerobic counterparts. This difference is apparent from Section 1.1. Individual aerobic bacteria can, in almost all cases, mineralise their substrates completely. Different species are specialised with respect to which (if any) polymers they can hydrolyse, and to which particular low molecular weight organics they transport and dissimilate/assimilate. However, almost all have the complete enzymatic machinery (glycolytic pathway, citric acid cycle, electron transfer chain) to effect a total mineralisation of carbohydrates and amino acids with metabolic end products including bacterial cells + CO_2 + H_2O (+ mineral N, etc.). Furthermore, due to a relatively efficient energy metabolism, a large fraction of the organic material metabolised ends up as cell material. In contrast, the relatively lower energetic efficiencies of anaerobic metabolism mean that a much larger fraction of the substrates is dissimilated (turns up as metabolites) rather than being incorporated into cell material. Furthermore, no single type of anaerobic bacterium (excluding denitrifiers) seems capable of complete mineralisation of, for example, glucose. Anaerobic mineralisation takes place stepwise, involving a sort of "food chain" or "food web" composed of several different functional types of bacteria. It could logically be asked why (and in analogy with some aerobes) there are no methanogens that can hydrolyse cellulose into glucose, then degrade it via the glycolytic pathway, ferment it on to hydrogen and acetate and finally produce CH_4 + CO_2. In fact methanogens use only a very limited number of substrates (acetate, a few C-1 compounds, or CO_2 + H_2) and they are entirely dependent on the metabolites of fermenting bacteria (and conversely the fermenters depend on methanogens or sulphate reducers for the removal of H_2). So far, it seems, no convincing explanation why this should be so has been suggested, but it does have a profound effect on the organisation of anaerobic microbial communities.

We first look at communities of fermenting bacteria (Fig. 1.2). In anaerobic communities a variety of fermenting bacteria are solely responsible

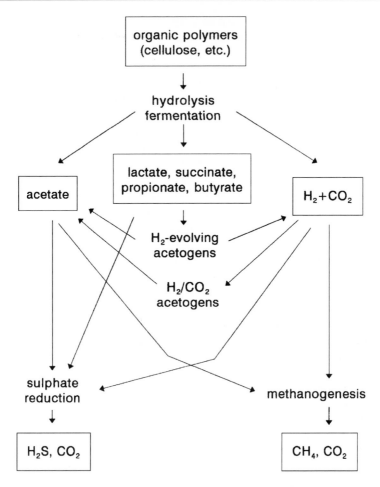

Figure 1.2   The organisation of a community of fermenters. The final products (principally H_2 and acetate) are then oxidised and mineralised by sulphate reducers or methanogens under anaerobic conditions. The H_2 consumption of the latter types of bacteria drives the fermentation to completion[15]

for the hydrolysis of organic polymers yielding different fatty acids, alcohols and H_2 as metabolites. In sulphate-containing habitats these are further degraded by sulphate reducers which are responsible for the terminal anaerobic mineralisation. In the absence of sulphate, other types of fermenters are necessary for the conversion of all substrates (butyrate, propionate, etc.) into acetate $+ H_2$ (which can be used by methanogens for terminal mineralisation). These fermenters are known as obligate acetogens.

The relative roles of H_2/CO_2 methanogens and homoacetogens (H_2/CO_2 acetogens) have drawn some attention. It seems that under acid conditions the latter are competitively superior and play the role of H_2 scavengers. The resulting acetate they produce is then degraded by acetoclastic methanogens. In more neutral environments the H_2/CO_2 methanogens dominate H_2 consumption.[20,30]

As previously discussed (see Section 1.2.1) interspecies H_2 transfer is a crucial feature of anaerobic mineralisation: only the efficient removal of H_2 through methanogenesis or anaerobic respiration allows for the completion of the fermentation processes. Otherwise the environment becomes acid and further mineralisation is inhibited. This hydrogen transfer is often dependent on the physical proximity of pairs of H_2-producing and H_2-consuming species, and is defined as "syntrophy" (see also Section 1.3).

Fermentation holds a special place in that it does not, in principle, change the redox potential of the environment because this type of energy generation is only based on a dismutation of organic molecules without changing the overall reduction–oxidation state. Fermentation represents the basis of anaerobic degradation, but since external electron acceptors are not involved, fermentation does not feature in Fig. 1.3 or can be considered as remaining at the level of CH_2O. (Direct measurements of electrode potentials will, however, indicate that a culture of fermenting bacteria, for example, reduces the environment. This is because the substrates of fermenters do not usually show electrochemical activity (e.g. cellulose) whereas some of the metabolites (H_2) will lower measured electrode potentials; see also below.)

The information provided in Fig. 1.3 is largely equivalent to that presented in Table A.2, but clearly indicates which processes are thermodynamically possible. The figure also presents a simplified biosphere model. The driving force is oxygenic photosynthesis which creates the potential energy constituted by free O_2 together with reduced organic material ($[CH_2O]$). Part of the energy of the organic material will be released through fermentation, but most will be released via oxidation–reduction processes (respiration) involving external electron acceptors. As long as O_2 is available, oxidative phosphorylation will be responsible for mineralisation. When oxygen is depleted, the energetically less favourable NO_3^- reduction will take over followed by the reduction of oxidised iron and manganese; thereafter sulphate reduction will predominate as an electron acceptor and eventually (as sulphate becomes depleted) H_2/CO_2 methanogenesis takes over. This *redox sequence* describes and explains the temporal succession of the degradation of organic matter and the spatial distribution of processes in general terms. For example, when going downwards from the surface of aquatic sediments, different electron acceptors are sequentially depleted: first O_2, then NO_3^- and eventually SO_4^{2-} (see also Chapters 5 and 6). While the energetics explain this pattern, the quantitative importance of the individual processes is largely determined by the availability of the different electron acceptors: in most

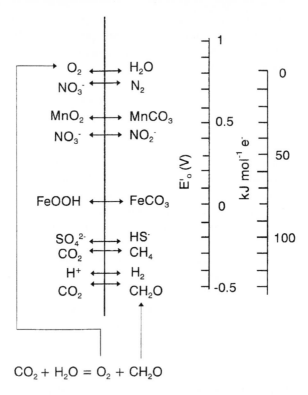

CO$_2$ + H$_2$O = O$_2$ + CH$_2$O

Figure 1.3   The standard redox potentials (pH 7) of some important redox couples and the free energy changes of processes involving two redox couples (i.e. respiratory processes or H$_2$/CO$_2$ methanogenesis). The graph can also be considered as a biosphere model: oxygenic photosynthesis creates the chemical potential and chemical equilibrium (mineralisation) is restored through a number of redox proceses carried out in a variety of organisms[15]

habitats aerobic respiration and sulphate reduction predominate in this respect.

Figure 1.3 also shows that the reduced products of anaerobic mineralisation processes (H$_2$, CH$_4$, HS$^-$, NH$_4^+$) are ultimately oxidised by other electron acceptors; eventually everything is in principle oxidised by O$_2$ so that chemical equilibrium is restored through the concerted action of many different types of bacteria. A particular complication is denitrification leading to N$_2$ which represents an incomplete reoxidation: N$_2$ is ultimately oxidised either via electrical discharges in the atmosphere (leading to nitrogen oxides), or in a rather more convoluted way via biological N$_2$ fixation → ammonia → microbial nitrification.

It is useful to conclude this section with a brief discussion of the relation between redox potentials as used in a (theoretical) bioenergetic context (see Appendix 1) and in terms of potentials that can actually be measured in aquatic habitats with a platinum electrode. Data like those presented in Table A.2 are in most cases not measured directly, but have been calculated indirectly from values of standard free energies. Many such couples can, however, be measured directly in the laboratory with an arrangement as shown in Fig. A.1. It would seem natural to use redox potentials in natural waters to achieve a direct picture of the microbial processes, that is to provide a picture as presented in Fig. 1.3 using empirical data from nature.

This was attempted early[18,28] and since then such measurements have frequently been presented in the literature to characterise aquatic sediments, in particular. A vertical profile in a sediment will characteristically show values around +0.4 V in the supernatant oxic water and the superficial oxic sediment layer; at some depth the potentials fall more or less steeply to values around -0.15 V in the anaerobic and sulphidic zone. This partially illustrates the ideas presented in Fig. 1.3 and measurements of electrode potentials do yield a somewhat crude picture of the ongoing microbial processes and the chemical environment.

It has, however, proven extremely difficult or impossible to interpret measured electrode potentials in terms of the exact chemical environment. There are several reasons for this. Some redox couples equilibrate very slowly with the electrode, so readings tend to drift and impurities on the Pt electrode may affect the measured potential to a larger extent than the system in the environment. Thus, the expected potential of around 0.8 V in oxygenated water is never obtained. Some redox pairs are electrochemically inactive, or they do not equilibrate so that several systems may simultaneously affect the potential. Complexity of the chemistry of some elements is also a problem. The Fe^{3+}/Fe^{2+} couple has a standard potential of +0.77 V. However, iron chemistry is complicated: the Fe ions combine with a variety of different ligands in aquatic environments and the redox potential of the Fe(III)/Fe(II) couple can assume a wide range of (mainly lower) values. The theory and practical measurements of redox potentials are discussed in detail in Stumm and Morgan.[39]

1.3 Transport mechanisms and structure of microbial communities

Bacteria exchange materials with the environment at a rate that depends on physical transport mechanisms. There are three different types of physical transport mechanism: *advection, turbulence* and *diffusion*. Advection refers to the orderly transport of molecules in the form of currents. Turbulence (like molecular diffusion) is a random motion dominated by eddies of different sizes. At the scale of the individual bacterial cell, only molecular diffusion plays a role. At this scale objects are surrounded by a viscous "diffusive boundary layer" in which turbulent motion is absent and advective flow approaches zero close to the bacterial surface. In fact, to a bacterium water appears syrup-like due to internal friction. Advection and turbulence are important for shaping certain microbial communities and are discussed briefly later. However, in our context molecular diffusion is the most important mechanism and we discuss it first and in most detail.

Diffusion is the statistical outcome of random molecular motion. In one-dimensional diffusion (Fig. 1.4, above) a concentration gradient results in a net flux of material from higher to lower concentrations. Within a sufficiently short time interval the probabilities for a molecule to move either to the right or to the left are identical and so the net flux in the x-direction (J amount of molecules passing a unit area per unit time) is proportional to the concentration gradient according to:

$$J = -D \, dC/dx, \tag{1.1}$$

where the negative sign indicates that net flux is in a direction from higher to lower concentrations. The equation is referred to as Fick's first law. The constant D is the diffusion coefficient with the dimension $L^2 T^{-1}$; it is a characteristic property of the solute (largely determined by molecular size), the solvent (especially viscosity) and temperature. In water, dissolved low molecular weight compounds have a diffusion coefficient of about 10^{-5} cm^2 s^{-1}. The unit of D shows that the time it takes to transport materials by diffusion is proportional to the square of distance; thus, rapid transport of solutes requires small distances and steep chemical gradients. This is of paramount importance for understanding the spatial structure of microbial communities (cf. Table 1.5).

In order to express the change in concentration at a point in space, imagine a volume with thickness $x_2 - x_1$ and an area of unit size perpendicular to the gradient (Fig. 1.4, below). The change in concentration within the box must be proportional to the difference in net flux into and out of the box ($J_{x_2} - J_{x_1}$) so that $dC/dt = (J_2 - J_1)/(x_2 - x_1)$. In the limit as $x_1 - x_2 \to 0$ we have $dC/dt = -dJ/dx$, or after substituting eqn 1.1:

$$dC/dt = D \, d^2C/dx^2 \tag{1.2}$$

which is referred to as Fick's second law. It is apparent from eqn 1.2 that for steady state $dC/dt = 0$ and if the diffusing substance is conservative

Table 1.5   The time (T) for transport of solutes (exemplified by O_2; $D = 2 \times 10^{-5}$ cm^2 s^{-1}) for different distances and (L) calculated as $L^2/(2D)$. From Fenchel and Finlay.[15]

L	1 μm	10 μm	100 μm	1 mm	1 cm	10 cm	1 m
T	0.25 ms	25 ms	2.5 s	4.2 min	6.9 h	29 days	7.9 years

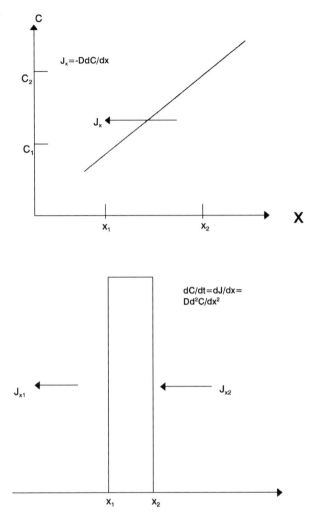

Figure 1.4   Above: illustration of Fick's first law. Since the concentration at x_2 is larger than at x_1 correspondingly more molecules will move to the left from x_2 than molecules moving to the right from x_1. Consequently, flux is proportional to the steepness of the gradient. Below: illustration of Fick's second law. The change in concentration with time inside the box (thickness: $x_2 - x_1$, unit area perpendicular to the gradient) will be the difference between the net influx from the right and the net outflux to the left

(is neither produced or consumed) then the gradient must be linear. With this background we can discuss some problems of microbial ecology. For a thorough and formal treatment of the physics of diffusion and for the application to biological problems see Berg[4] and Cussler;[10] a discussion of diffusion and bacterial physiology can also be found in Koch.[24]

1.3.1 Uptake and excretion of solutes from bacterial cells

Bacterial activity depends on the uptake of dissolved substrates; the substrates are transported to the cell by a diffusive flux driven by the concentration gradient which is caused by uptake at the cell surface. We first consider uptake by a spherical cell with radius R (Fig. 1.5). We assume it is "diffusion limited" in the sense that the cell maintains the substrate concentration (C) at zero at its surface. The bulk concentration of the substrate (far from the cell) is C'. Distance from the cell centre is denoted r. If we imagine a concentric spherical shell (with $r > R$) then the flux through a unit area will be, according to eqn 1.1, $J = -D\,dC/dr$ and the entire flux through the sphere (which is equal to the uptake V by the cell at steady state) is given by $V = 4\pi r^2 J = -4\pi r^2\,dC/dr$. Boundary conditions are that $C(R) = 0$ and $C(r \to \infty) = C'$. Furthermore (since steady state is assumed so that the fluxes through the imaginary sphere are independent of r), $dC/dr = \text{const.} \times r^{-2}$. We can now guess a solution: the expression for the concentration as a function of r:

$$C(r) = C'(1 - R/r) \tag{1.3}$$

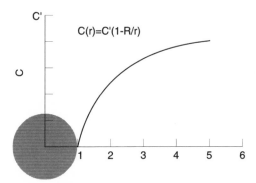

$$C(r) = C'(1 - R/r)$$

r (distance from centre of cell in units of R)

Figure 1.5   The concentration gradient of a substrate around a spherical cell in the diffusion-limited case

satisfies the boundary conditions and so $dC/dr = -C'R/r^2$ and by substitution we have:

$$V = 4\pi RDC', \tag{1.4}$$

suggesting that in the diffusion-limited case the uptake by the cell is only a function of cell size. Dividing eqn 1.4 by C' we obtain an expression for the "clearance" of the cell ($= 4\pi RD$); that is, the volume of water the cell can clear for substrate per unit time. If we further divide the expression with cell volume ($4/3\pi R^3$), asssuming that the needs of the cell are proportional to volume, we then obtain:

$$E = 3R^{-2}D. \tag{1.5}$$

This is a meaningful measure of the competitive efficiency or affinity in a nutrient-limited situation; the bacterium with the highest value of E will be able to extract most substrate per unit cell volume. Clearly, smaller size improves efficiency. Thus, the only strategy for bacteria to improve their circumstances under substrate limitation is to decrease in size. We return to the question of to what extent eqns 1.4 and 1.5 are realistic descriptions, but first we extend eqn 1.4 to cover situations that are not completely diffusion limited.

Complete diffusion limitation is an extreme case. An alternative extreme is that uptake is limited only by the uptake mechanism so that uptake is independent of C' and $C(R) = C'$. Assume a "transport coefficient" k where k can be considered a measure of the density of uptake sites in the cell membrane and the time constant during which they are occupied by transporting a molecule across the membrane. The maximum uptake rate at a very high C' is given by $V_{max} = k4\pi R^2$. V_{max} has dimensions of substrate per unit time, and V_{max}^{-1} measures the time for transport of one unit of substrate during which interval no additional substrate can be transported through the cell membrane. Taking this into account, eqn 1.4 now becomes: $V = 4\pi RC'D[1 - V/V_{max}]$, where the bracket measures the fraction of time the uptake sites are not occupied. Solving for V we have $V = [4\pi R^2 kC']/[(kR/D) + C']$. This is identical to the Monod (Michaelis–Menten) kinetics which is usually written as:

$$V = V_{max}C'/(K_m + C'), \tag{1.6}$$

where the "half saturation constant" $K_m = kR/D$; it represents an *ad hoc* constant measuring the ratio between uptake limitation and diffusion limitation. The relation is shown in Fig. 1.6; the slope at the origin is $E \times$ cell volume (eqn 1.4) and the uptake approaches V_{max} at very high values of C'. To the extent that the growth rate constant is proportional to substrate uptake (which is generally true) then the function also describes the growth rate as a function of substrate concentration when the y-axis is multiplied by the growth yield coefficient.

We may now consider the consequences of eqns 1.4–1.6. We consider an aerobic spherical bacterium ($R = 1\,\mu m$) with a volume of $1.33 \times 10^{-12}\,ml$ and, assuming 30% dry weight, that it contains $4 \times 10^{-13}\,g$ organic material. If $D = 10^{-5}\,cm^2\,s^{-1}$ then, according to eqn 1.5, $E = 1.1 \times 10^7\,h^{-1}$. Suppose the bacterium takes up its own weight (d.w. organic material) in $1\,h$ (thus allowing an aerobe to divide every $1.5\,h$ or so). According to eqn 1.4 (and taking care to use correct unit conversions) only $9\,\mu g$ organic substrate per litre water of are required.

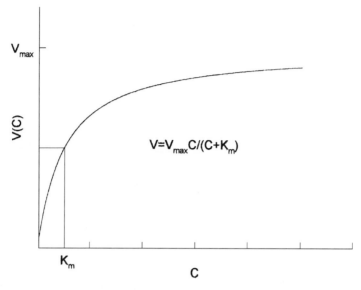

Figure 1.6   The Monod equation for the uptake (or growth) of a bacterium as a function of ambient substrate concentration. The slope at the origin represents the diffusion-limited case (uptake proportional to *C*) and the uptake-limited case is when *C* is very large and uptake is invariant with substrate concentration

The extent to which this is actually realised has been studied experimentally with *Escherichia coli.*[23,25] It has been found that the value of E (eqn 1.5) (and thus of the slope at the origin (eqn 1.4)) is, in fact, 50–100 times lower than predicted. It has also been found that, after prolonged growth under nutrient limitation, cells improve somewhat in this respect at the expense of a decrease in V_{max}. The explanation offered for this is that the cell wall of bacteria constitutes a diffusion barrier and that changes in cell wall structure may somewhat modify the efficiency of coping with very dilute substrate concentrations. Bacteria that occur in very oligotrophic waters have not been studied in detail in this respect, but limited data do suggest similar values for E.[7] However, even if we accept the values for E within this range (1–$2 \times 10^{-6}\,h^{-1}$), bacteria are still amazingly effective in exploiting very dilute resources: something like 0.5–1 mg organic substrate per litre can support growth with generation times of about 1 h (also our calculations are based on a very large bacterium; most bacteria found in seawater are considerably smaller than 2 µm across and should perform accordingly better).

It is an empirical fact that when different bacteria are compared there is a correlation between E and V_{max} such that bacteria with a high V_{max} (i.e. a high capacity for rapid growth at high nutrient concentrations) tend to have a low value of E (poor competitor under nutrient limitation) and vice versa.

This dichotomy undoubtedly contributes to natural diversity: some species are specialised for rapid exploitation of temporally and spatially patchy occurrences of high substrate concentrations whereas others can effectively compete for very dilute substrate concentrations, but are incapable of exploiting higher substrate levels in terms of a very rapid growth. These two extremes have been referred to as "zymogeneous" and "autochtonous", respectively.

We may now turn to the opposite problem: the excretion of metabolites from cells. This is of interest because the accumulation of some metabolites (notably H_2 in fermenters) affects metabolic pathways and because some bacteria depend on the metabolites of other species. The solution to this problem is quite analogous to the considerations leading to eqns 1.3–1.4. If metabolite production is denoted P then for $r > R$ we have $P = 4\pi r^2 D\, dC/dr$. The gradient is $C(r) = [C(R) - C']R/r + C'$, so that $dC/dr = -[C(R) - C']Rr^{-2}$ and the concentration at the surface $C(R) = P/(4\pi RD) + C'$.

Figure 1.7 is based on a hypothetical spherical fermenter ($R = 1\,\mu m$) which is assumed to produce $10\,\text{fmol}\ H_2\,h^{-1}$. The ambient p_{H_2} is assumed to be 10^{-5} atm ($\sim 1\,Pa$) which corresponds to $C' \approx 8\,\text{pmol cm}^{-3}$. It is seen that the H_2 concentration immediately around the cell increases by a factor of about 30 and decreases rapidly within 2–3 cell diameters from the surface. This has implications for obligate syntrophic pairs: to function efficiently they must be juxtaposed within aggregates as shown schematically in Fig. 1.8 in which, for example, the cocci can be considered H_2-evolving and the rods H_2-consuming bacteria. This phenomenon is well documented. The efficiency of anaerobic degradation in sewage sludge or in lake sediments is strongly enhanced by aggregation. Most of the methanogenesis takes

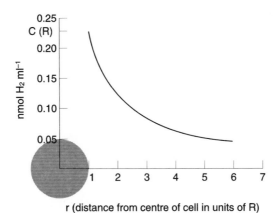

Figure 1.7 The H_2 concentration gradient around a spherical cell ($R = 1\,\mu m$) which is supposed to produce $10\,\text{fmol}\ H_2\,h^{-1}$; ambient H_2 concentration is assumed to be $8\,\text{pmol ml}^{-1}$

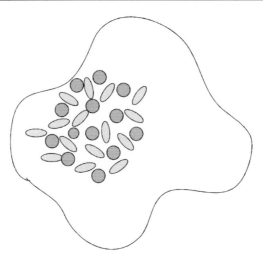

Figure 1.8   The juxtaposition of cells of a pair of syntrophic bacteria on a detrital particle

place on flocculated material in a way in which almost no H_2 is lost; the bulk fluid shows much lower methanogenic activity, and measurements of H_2 concentration and turnover in the bulk fluid underestimate the actual H_2 transfer.[9,32,41] Similar physical juxtaposition is also known from pairs of sulphate/sulphur-reducing and sulphide-oxidising bacteria. *Chlorochromatium aggregatum* is a compound organism consisting of a large motile rod (assumed to be a sulphur-respiring bacterium) which is covered by cells of the green sulphur bacterium *Chlorobium*. *Pelochromatium* represents an aggregate of green sulphur bacteria and sulphur-reducing bacteria.[29] The physical closeness of different bacteria with complementary metabolic demands (sulphide oxidisers/sulphate reducers, photo- and chemoautotrophs/various heterotrophs) probably plays an essential role, particularly in habitats dominated by particulate material (such as sediments, soils, and microbial films and mats; see also Chapters 5–7).

1.3.2 Diffusion-controlled communities
In certain microbial communities there is no or almost no transport due to turbulence or advection. Microbial mats provide the best example (see Chapter 6); in general, aquatic sediments (see Chapter 5) can also be considered diffusion dominated, but "bioturbation", that is the mechanical activities of animals (mixing of the superficial sediment layers or advective flow of oxygenated water generated by burrow-dwelling worms, bivalves and crustceans) may complicate matters. In both cases, we apply diffusion considerations as developed below. Such considerations often make it possible to estimate fluxes of materials and reaction rates solely on the basis of chemical gradients. Soils (see Chapter 4) are complicated in that

they contain both a liquid and a gas phase. Diffusion coefficients of gases are about 10^4 times higher than those of solutes in water; consequently gases (like O_2) easily reach all depths in unsaturated soils. Diffusion is still important within individual water-saturated soil aggregates, but patterns are more complex than the simple structures of (undisturbed) aquatic sediments.

Idealised sediment systems (realised in microbial mats) can be described as communities which are entirely controlled by one-dimensional diffusion. Oxygen or other electron acceptors are supplied from above while reduced metabolites (resulting from the anaerobic degradation of buried organics) are supplied from below. Anaerobic conditions beneath the surface arise because the diffusive transport of O_2 cannot meet the demand of reaction rates. As a result, a vertical zonation pattern of processes and of available electron acceptors develops according to the redox sequence discussed in Section 1.2.3 (see also Fig. 1.3). We will consider here only the O_2 gradient in the surface layer of a sediment; the general principles can then be applied to other gradients as exemplified (in the context of particular habitats) in Chapters 5 and 6.

Figure 1.9 shows a schematic presentation of an O_2 concentration gradient immediately above and below the surface of an aquatic sediment; for a shallow productive sediment the entire vertical axis of the figure will represent 2–4 mm. The bulk of the water column is mixed and has a constant O_2 concentration (typically corresponding to atmospheric saturation). At 0.5–1 mm above the surface (depending on turbulence of the overlying water column) there is a *diffusive boundary layer* in which there is no turbulence and all vertical transport is diffusive. Since the sediment consumes O_2 a concentration gradient forms and, assuming a neglible O_2 consumption or production in the water, the gradient will be linear. The O_2 flux to the sediment can therefore be calculated directly from Fick's first law (eqn 1.1) (diffusion coefficients for dissolved O_2 at different temperatures can be looked up in tables).

At and beneath the surface the situation becomes somewhat more complicated. Immediately beneath the surface the gradient becomes steeper. This is because the diffusion coefficient in sediments is somewhat lower than in the water and, since the fluxes immediately above and immediately beneath the surface must be identical, the gradient must change according to eqn 1.1. Oxygen is also consumed beneath the surface. We assume a homogeneous rate of O_2 uptake R (amount of oxygen consumed per unit volume of sediment per unit time). Depth in the sediment is denoted z ($z = 0$ at the surface). According to Fick's second law (eqn 1.2) we have: $dC/dt = D\,d^2/dz^2 - R$ (D now represents the diffusion coefficient in the sediment). Assuming steady state the derivate vanishes so that $d^2C/dz^2 = R/D$. Integrating twice we then have $C(z) = z^2 R/(2D) + az + b$, where a and b are constants. The boundary conditions are $C(0) = C'$ and $C(L) = 0$,

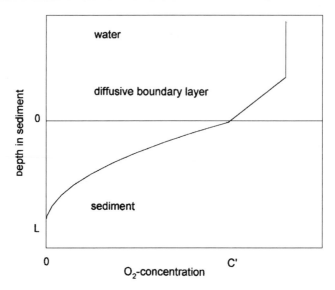

Figure 1.9   The oxygen gradient above and in the superficial layer of a sediment. In real shallow-water sediments L typically varies between 1 and 5 mm; in sediments from deeper, oligotrophic waters L may measure several centimetres

$dC(L)/dz = 0$ (L is the depth at which O_2 disappears and the last condition secures that C cannot become negative). Substituting these relations and solving for a and b we have $a = (2C'R/D)^{1/2}$ and $b = C'$ so that:

$$C(z) = z^2 R/(2D) + z(2C'R/D)^{1/2} + C'. \tag{1.7}$$

This shows that the O_2 gradient in the sediment is parabolic. The gradient is downwards concave because there is a net consumption; if it were downwards convex this would indicate that there is a net oxygen production (as happens if there is a high rate of photosynthesis, cf. Figs 6.1 and 6.6).

Setting $C(z) = 0$ in eqn 1.7 we have:

$$z = L = (2C'D/R)^{1/2} \tag{1.8}$$

so that R can be calculated directly from the depth of the anoxic zone. The flux is given by $J = RL\phi$ (where ϕ is the porosity of the sediment) and substitution then shows that $J = 2D\phi C'/L$; this allows for an estimate of the O_2 flux which is independent of the O_2 gradient in the diffusive boundary layer above the sediment.

Some of the assumptions in the above description can be questioned. It is not obvious that R remains absolutely constant with depth. In particular, it is assumed that R remains constant until $C = 0$; this is certainly not the case when p_{O_2} is $< 0.5\%$ atm.sat., but since this is at the detection limit for O_2 electrodes the effect is not observed. In practice the model gives a relatively close description of actually measured O_2 gradients.

Similar calculations not implying one-dimensional diffusion can be made. The oxic zone surrounding (cylindrical) worm burrows will (assuming identical values of C' and R) be thinner than that beneath the (plane) sediment surface due to the curvature of the burrows. A version of eqn 1.2 with spherical coordinates (and using considerations similar to those presented above for one-dimensional diffusion) can be used to calculate the minimum size of an oxygen-consuming sphere (e.g. a detritus or soil particle) immersed in an oxygenated environment which can maintain an anoxic interior (see Fig. 4.6). If the thickness of the oxic layer is L, the solution is (in analogy to eqn 1.8) that $L = [6C'D/R]^{1/2}$ which is the size that just allows for an anaerobic centre. Assuming C' is atmospheric O_2 saturated ($\sim 200 \, \text{nmol cm}^{-3}$) and that O_2 consumption $R = 1000 \, \text{nmol} \, O_2 \, \text{ml}^{-1} \, \text{h}^{-1}$ (a reasonable value for organic detritus), and $D = 1.5 \times 10^{-5} \, \text{cm}^2 \, \text{s}^{-1}$, then this works out to $L = 2.5 \, \text{mm}$; that is, detrital particles in a fully oxic environment can represent anaerobic microniches (with associated anaerobic microbial processes) if they exceed a few millimetres in diameter.

1.3.3 Advection and turbulence

In Section 1.3.1 it was implied that advection and turbulence play no role in the nutrient uptake of a bacterium (in contrast with, for example, a fish which enhances uptake of O_2 over the gill surface by ventilation, or nutrient uptake by seaweeds that is improved by turbulence; a mechanism based on a reduction of the diffusive boundary layer). Turbulence vanishes totally at a scale of about $0.5 \, \text{cm}$ when the mechanical energy is converted into heat. As a result bacteria cannot perceive turbulence. But since some bacteria are motile it could be asked whether advective flow may enhance nutrient uptake. A crude answer can be obtained by the following reasoning. The time scale for advective transport is L/V where V is velocity and L is length. The time used for diffusive transport is L^2/D, which follows from dimensional considerations. The dimensionless ratio between these expressions VL/D is then a measure of the relative importance of the two types of transport: if it is $\gg 1$ advective transport dominates and, conversely, if it is $\ll 1$ diffusion dominates. Taking a motile bacterium, swimming velocity (V) may be $5 \times 10^{-3} \, \text{cm s}^{-1}$, a relevant distance is $10^{-4} \, \text{cm}$ and $D = 10^{-5} \, \text{cm}^2 \, \text{s}^{-1}$. The ratio then works out to 5×10^{-2}, suggesting that motility has no measurable effect on nutrient uptake.

Water currents and turbulence however, may, be of paramount importance for the nutrient supply of microbial communities attached to surfaces. Examples are microbial films in streams and springs. In such cases the relevant dimensions and current velocities for advection are sufficient to be important.

Turbulence is a complicated topic. It is responsible for mixing and transport processes in the water column of lakes and seas. It is often quantified in terms of a "turbulent diffusion coefficient" (in analogy with molecular diffusion) but this is only an approximate description. Turbulence is not isotropic in natural water bodies; that is, horizontal turbulent mixing is much greater than vertical mixing. While it does not affect the individual bacterium directly it does affect microbial communities in terms of mixing and homogenisation of chemical gradients. An example of the effect can be seen in Fig. 6.9. Vertically stratified lakes or marine basins with anoxic deep water show microbial and chemical stratification patterns that are quite similar to those of sediments. However, even in vertically stabilised water columns turbulent mixing is about three orders of magnitude higher than molecular diffusion. Consequently, zonation patterns, which in sediments span only a few millimetres, will span > 1 m in the water column.

References

1. Bak F, Cypionka H (1987) A novel type of energy metabolism involving fermentation of inorganic sulphur compounds. *Nature* **326**: 891–892.
2. Balows A, Trüper HG, Harder W, Sheifer K-H (eds) (1991) *The Prokaryotes*. Vols I–IV. Springer, New York.
3. Bauchop T, Elsden SR (1960) The growth of micro-organisms in relation to their energy supply. *J Gen Microbiol* **23**: 457–469.
4. Berg HC (1983) *Random Walks in Biology*. Princeton University Press, Princeton, New Jersey.
5. Brock TD (1978) *Thermophilic Micro-organisms and Life at High Temperature*. Springer, New York.
6. Brock TD (1996) *Biology of Micro-organisms* (7th edn). Prentice-Hall, Englewood Cliffs, New Jersey.
7. Button DK (1986) Affinity of organisms for substrate. *Limnol Oceanogr* **31**: 453–456.
8. Cohen Y, Jørgensen BB, Revsbech NP, Poplawski R (1986) Adaptation to hydrogen sulfide of oxygenic and anoxygenic photosynthesis among cyanobacteria. *Appl Environ Microbiol* **51**: 398–407.
9. Conrad R, Phelps TJ, Zeikus JG (1985) Gas metabolism evidence in support of the juxtaposition of hydrogen-producing and methanogenic bacteria in sewage sludge and lake sediments. *Appl Environ Microbiol* **50**: 595–601.
10. Cussler EL (1989) *Diffusion. Mass Transfer in Fluid Systems*. Cambridge University Press, Cambridge.
11. Dannenberg S, Kroder M, Dilling W, Cypionka H (1992) Oxidation of H_2, organic compounds and inorganic sulfur compounds coupled to the reduction of O_2 or nitrate by sulfate-reducing bacteria. *Arch Microbiol* **158**: 93–99.
12. Dilling W, Cypionka H (1990) Aerobic respiration in sulfate reducing bacteria. *FEMS Microbiol Lett* **71**: 123–128.

13. Edwards C (ed.) (1990) *Microbiology of Extreme Environments.* Open University Press, Milton Keynes.

14. Fenchel T, Blackburn TH (1979) *Bacteria and Mineral Cycling.* Academic Press, London.

15. Fenchel T, Finlay BJ (1995) *Ecology and Evolution in Anoxic Worlds.* Oxford University Press, Oxford.

16. Fründ C, Cohen Y (1992) Diurnal cycles of sulfate reduction under oxic conditions in cyanobacterial mats. *Appl Environ Microbiol* **58**: 70–77.

17. Ghiorse WC (1988) Microbial reduction of manganese and iron. In [46], pp. 305–321.

18. Hutchinson GE, Deevey ES, Wollack A (1939) The oxidation–reduction potential of lake waters and their ecological significance. *Proc Nat Acad Sci* **25**: 87–90.

19. Jones JG, Gardener S, Simon BM (1984) Reduction of ferric iron by heterotrophic bacteria in lake sediments. *J Gen Microbiol* **130**: 45–51.

20. Jones JG, Simon BM (1985) Interaction of acetogens and methanogens in anaerobic freshwater sediments. *Appl Environ Microbiol* **49**: 944–948.

21. Kepkay PE, Nealson KH (1987) Growth of a manganese oxidizing *Pseudomonas* sp. in continuous culture. *Arch Microbiol* **148**: 63–67.

22. Kluyver AJ, Donker HJ (1926) Die Einheit in der Biochemie. *Chem Zelle Gewebe* **13**: 134–190.

23. Koch AL (1971) The adaptive responses of *Escherichia coli* to a feast and famine existence. *Adv Microbiol Physiol* **6**: 147–217.

24. Koch AL (1990) Diffusion. The crucial process in many aspects of the biology of bacteria. *Adv Microbiol Ecol* **11**: 37–70.

25. Koch AL, Wang CH (1982) How close to the theoretical diffusion limit do bacterial uptake systems function? *Arch Microbiol* **131**: 36–42.

26. Lovley DR (1993) Dissimilatory metal reduction. *Ann Rev Microbiol* **47**: 263–291.

27. Nealson KH, Saffarini D (1994) Iron and manganese in anaerobic respiration: environmental significance, physiology, and regulation. *Ann Rev Microbiol* **48**: 311–343.

28. Pearsall WH, Mortimer CH (1939) Oxidation–reduction potentials in waterlogged soils, natural waters and muds. *J Ecol* **27**: 483–501.

29. Pfennig N (1980) Syntrophic mixed cultures and symbiotic consortia with phototrophic bacteria: a review. In: Gottschalk G, Pfennig N, Werner H (eds) *Anaerobes and Anaerobic Infections.* Gustav Fischer, Stuttgart, pp. 127–131.

30. Phelps TJ, Zeikus JG (1984) Influence of pH on terminal carbon metabolism in anoxic sediments from a mildly acidic lake. *Appl Environ Microbiol* **48**: 1088–1095.

31. Richardson LL, Castenholz RW (1987) Enhanced survival of the cyanobacterium *Oscillatoria terebriformis* in darkness under anaerobic conditions. *Appl Environ Microbiol* **53**: 2151–2158.

32. Schink B (1992) Syntrophism among prokaryotes. In [2], Vol I, pp. 276–299.

33. Schlegel HG (1989) Aerobic hydrogen-oxidizing (knallgas) bacteria. In [34], pp. 305–329.

34. Schlegel HG, Bowien B (eds) (1989) *Autotrophic Bacteria*. Science Technical Publishers, Madison/Springer; Berlin.
35. Sillén LG (1966) Regulation of O_2, N_2 and CO_2 in the atmosphere: thoughts of a laboratory chemist. *Tellus* **18**: 198–206.
36. Sørensen J (1982) Reduction of ferric iron in anaerobic, marine sediments and interaction with reduction of nitrate and sulfate. *Appl Environ Microbiol* **43**: 319–324.
37. Stanier RY, Ingraham JL, Wheelis ML, Painter DR (1987) *General Microbiology* (5th edn). MacMillan, London.
38. Stouthamer AH (1973) A theoretical study of the amount of ATP required for microbial cell material. *Antonie van Leeuwenhoek* **39**: 545–565.
39. Stumm W, Morgan JJ (1996) *Aquatic Chemistry* (3rd edn). John Wiley, New York.
40. Thauer RK, Jungerman K, Decker K (1977) Energy conservation in chemotrophic anaerobic bacteria. *Bact Rev* **41**: 100–180.
41. Thiele JH, Chartrain M, Zeikus JG (1998) Control of interspecies electron flow during anaerobic digestion: role of flock formation in syntrophic methanogenesis. *Appl Environ Microbiol* **54**: 10–19.
42. van de Graaf AA, de Bruin P, Robertson LA, Jetten MSM, Kuenen JG (1996) Autotrophic growth of anaerobic ammonium-oxidizing micro-organisms in a fluidized bed reactor. *Microbiology* **142**: 2187–2196.
43. Widdel F, Schnell S, Heisig S, Ehrenreich A, Assmus B, Schink B (1993) Ferrous iron oxidation by anoxygenic phototrophic bacteria. *Nature* **362**: 834–836.
44. Woese CR (1987) Bacterial evolution. *Microbiol Rev* **51**: 221–271.
45. Woese CR (1991) Prokaryote systematics: the evolution of a science. In [2], Vol I, pp. 3–18.
46. Zehnder AJB (ed.) (1988) *Biology of Anaerobic Micro-organisms*. John Wiley, New York.

2

Mineral cycles

The mineralisation of organic matter to non-organic molecules is essentially the degradation of particulate, insoluble polymers. The first step is the hydrolysis of the complex polymers to soluble molecules, often to the component monomeric units. It is after this rate-limiting step that the soluble molecules can be taken up by bacterial cells and processed further. We first describe some aspects of the hydrolysis of organic molecules and later discuss the individual mineral cycles (C, N etc.).

The role of bacteria in mineral cycles is clearly heterotrophic; first the hydrolytic step, followed by uptake and metabolism. Although some bacteria are photoautotrophs, only the cyanobacteria contribute significantly to CO_2 fixation, especially in the oligotrophic oceans. In a perfect heterotrophic system there would be no accumulation of organic matter, but this situation is seldom found; there is often a net accumulation.

2.1 Hydrolysis of organic polymers

2.1.1 Substrates

The substrates are principally the products of primary production; that is, the fixed carbon molecules made mainly by the eukaryotic photoautotrophs, plants and algae. Fixed carbon is found in proteins and polynucleotides, but more generally in structural polysaccharides. The proteins and polynucleotides are easily hydrolysed and the constituent monomers are mineralised, as is described in the individual element cycles (see Section 2.2). The structural polysaccharides are more resistant to hydrolysis. They have a variety of bonds between different monomers, as seen in Table 2.1.

Cellulose is the most common polysaccharide in land plants and is one of the most difficult to degrade. It is composed of glucose molecules linked through β-1,4 bonds to form flat, rigid, ribbon-like chains with crystalline regions. Oligopolymers from cellulose, as short as six glucose units, are insoluble in water. Chitin is also very insoluble, rigid and resistant to enzyme attack. Unlike cellulose, it is found in arthropods and fungi rather than in plants. It is composed of N-acetylglucosamine monomers linked through β-1,4 bonds. Chitosan is more flexible and less highly organised, lacking the N-acetyl substitution on many glucosamine molecules. Mannan

Table 2.1   Polysaccharide substrates

Substrate	Bond	Monomer	Structure
Cellulose	β-1,4	Glucose	Ribbon-like, rigid
Chitin	β-1,4	N-acetyl glucosamine	Ribbon-like, rigid
Chitosan	β-1,4	Glucosamine 30%	Acetylated, ribbon-like, rigid
Mannan	β-1,4	Mannose	Flexible ribbons
Xylan	β-1,4	Xylopyranose	Twisted ribbons
Peptidoglycan	β-1,4	N-acetyl glucosamine + N-acetyl muramic acid	Ribbon-like, rigid
Amylose	α-1,4	Glucopyranose	Non-rigid, linear
Amylopectin	α-1,4, α-1,6	Glucopyranose	Non-rigid, branched
Laminarins	α-1,3	Glucopyranose	Non-rigid
Pectin		Galacturonic acid	Non-rigid
Agar		Galacturonic acid + galactose	Non-rigid

and xylan are also less rigid and organised, even though the monomers are linked β-1,4. The xylopyranose units are often substituted with glucuronyl, acetyl or arabinofuranoside residues which reduce the compaction of their polymeric structure. Peptidoglycan, which is found in eubacterial cell walls, is the β-1,4 polymer of alternating molecules of N-acetylglucosamine and muramic acid. The polysaccharide chains are cross-linked through short peptide chains of unusual amino acids. The final structure is very rigid and often highly resistant to enzyme hydrolysis.

Starch, like cellulose, is composed of glucose monomers. However, it is relatively susceptible to enzyme attack due to its loosely organised polymeric chains. It has two components: amylose and amylopectin. The former is composed of a long linear chain of α-1,4 linkages; the latter has shorter chains of α-1,4 linkages with branches of α-1,6 linkages. Starch is often a storage compound in the seeds of land plants. Laminarins, which are polymers of α-1,3 linked glucose units, are found in seaweeds. Agar is also found in marine algae and contains galacturonic acid and galactose. It is relatively resistant to degradation by non-marine bacteria, but does not accumulate in the sea, indicating a rapid breakdown. Pectin also contains galacturonic acid residues in a loose polymeric structure. It is found in land plants and is easily degraded.

Efficiency of degradation.   The presence of accumulations of plant residues is usually indicative of the slow turnover times of large original inputs. This slow turnover can be the result of inherent resistance to enzyme attack, but it can also be due to other factors. The substrate can bind to inorganic ions, silts or to other organic residues, which protect it from degradation. The plant residue can be naturally linked to other molecules, for example cellulose which is often intimately associated with lignin. The lignin prevents the access of cellulases. As lignin itself is principally attacked by reacting with oxygen catalysed by a mono-oxygenase, lignified cellulose is not readily decomposed in anaerobic environments. It is not known to what extent oxidants other than oxygen (nitrate, metal oxides and sulphate) can alleviate the inhibition. Even short exposure to air appears to increase the degradation of marine detritus, an effect that may not depend on the continuing presence of oxygen.

The efficiency of hydrolysis is increased by physical maceration of plant fibres. Frequently, chewing by animals achieves this effect, but physical abrasion by waves or wind can have the same result. The creation of a larger surface area generates more sites for enzyme attack. Containment of fibres within the animal digestive system is also important in promoting a rapid hydrolysis (see Chapter 7). In warm-blooded animals, there is the increased advantage of an elevated temperature to stimulate degradation of plant fibres.

There may be another consequence from animal/microbial symbiotic relationships: the generation of animal, rather than microbial, biomass. This would have the effect of producing soft, easily decomposed flesh as opposed to relatively more resistant microbial cells. This scenario should not be exaggerated, but much of the energy in the plant residues is diverted to the animal via the short-chain fatty acid end products of microbial anaerobic digestion. If the digestion were not anaerobic, a much larger quantity of microbial biomass would be produced, due to the greater efficiency of carbon incorporation during aerobic respiration: 50% compared to approximately 20% for fermentation.

Microbial biomass degradation. This incorporation of substrate carbon into microbial biomass may have very significant effects on the efficiency of the heterotrophic process (i.e. on the complete mineralisation of detritus). Half the detrital carbon may flow through the microbial biomass, creating substrates that were not originally present in the plant detritus. It is a curious fact that in both soils and sediments, irrespective of the nature of the detrital input, the organic matter that accumulates or get buried tends to have a molar C:N ratio of approximately 10.[3] It is suggested that this is due to the accumulation of partially degraded bacterial cell wall material (peptidoglycan), relatively rich in nitrogen. This is not a new concept in soil microbiology, where for many years it has been recognised that microbial biomass is an important component of the soil. The incorporation of soluble inorganic nitrogen and phosphorus nutrients into microbial biomass (immobilisation) is a consequence of the differential ease of hydrolysis of different plant residues. Because proteins and polynucleotides are readily hydrolysed and subsequently mineralised (to NH_4^+, PO_4^{3-} and CO_2), there is a surplus of mineral N and P together with residual structural polysaccharide material of low mineral content. Slow hydrolysis of this polysaccharide and metabolism of the component monomers demand an incorporation of N and P into microbial cell biomass. To some extent, the microbial biomass acts rather like a catalyst, the quantity increasing or decreasing in response to substrate availability. Immobilisation would occur during a phase of increase; during steady state conditions, growth and death of cells might be expected to be equal, thus leading to no demand for a new N and P supply. However, this situation probably does not exist, due to the non-mineralisation of some of the structural N-containing material in the microbial cells. It should be remembered that the mineralisation of bacterial biomass always contributes to the pools of inorganic N and P: these do not originate solely from the degradation of dead detritus.[2] Presumably, non-degradeable microbial cell structures enter the humic reservoir.

The C:N ratio of the particulate organic detritus being mineralised is often inferred from the ratio of the mineral products, for example $\Sigma CO_2 / \Sigma(NO_3^- + N_2 + NH_4^+)$. This inference is valid if there is neither net accumulation nor degradation of microbial biomass. However, ecosystems are not usually in a steady state, and net synthesis or breakdown of microbial biomass can be taking place. This complicates the picture, as the ratio of mineral products is no longer an exact reflection of the substrate's N:C ratio, defined as N_s.[2] If x is substrate carbon metabolised, N_c is N:C ratio of microbial biomass synthesised and E is efficiency of carbon incorporation into microbial biomass, then:

$$x(1 - E) = \text{carbon mineralised}$$

$$x(N_s - EN_c) = \text{nitrogen mineralised}$$

$$(N_s - EN_c)/(1 - E) = \text{N:C ratio of mineral products.}$$

There will be no net mineralisation of nitrogen if EN_c is greater than N_s. N_c is ~ 0.2 for bacteria (C:N = 5) and therefore there will be no net nitrogen mineralisation from substrates with C:N ratios > 10 when E = 0.5, nor from substrates with C:N ratios greater than 16.7 when E = 0.3. This relationship is illustrated in Fig. 2.1 which shows substrate C:N plotted against the C:N ratio in the mineral products, when E equals 0.5 and 0.3. These two efficiency values roughly represent aerobic and anaerobic metabolisms. At the higher E value of 0.5, there is a marked difference between C:N in substrate and mineral product, reaching a maximum difference approaching a C:N in substrate of 10; for example at a substrate C:N of 9, the C:N ratio of the mineral products would be 45.0 and 13.7 when biomass accumulated at efficiencies of 0.5 and 0.3, respectively. The microbial biomass may be remineralised later at the site of its production, or it can be mineralised at another site, to which it has been transported by its own motility or by a predator. If this occurs, the C:N ratio of the mineral products will never be equal to the C:N ratio of the particulate substrate, even when integrated over a long time. There are instances (see Chapter 5) where

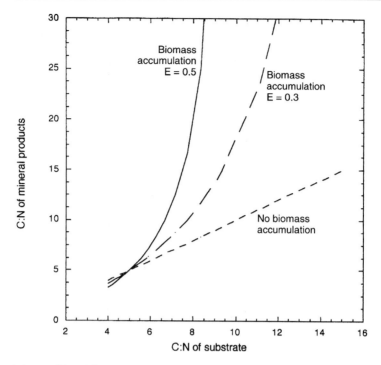

Figure 2.1   Plot of C:N in substrate versus C:N in mineral products at different efficiencies of carbon incorporation, with and without microbial biomass accumulation

microbial biomass may be synthesised at one sediment site and mineralised in another zone. It is suspected that this may be quite a common occurrence in soil and aqueous ecosystems.

The transitory accumulation of microbial biomass will also affect "biological oxygen demand" (BOD) determinations. It is assumed that O_2 uptake is equivalent to the quantity of degradable organic matter present in a test, but it is clear that this is an underestimate, as half the organic matter may be incorporated into microbial cells. The same applies to the chloroform fumigation procedure for the assay of soil biomass.[9] Most of the soil biomass is killed by chloroform and the cell detritus is metabolised by other micro-organisms, consuming O_2 and producing CO_2. However, microbial biomass is also produced, accounting for up to half the killed cells, thus underestimating the original biomass.

The recycling of the nutrients bound in the microbial biomass may be very important in promoting further degradation of nutrient-low polymeric substrate. There is some evidence that browsing of bacteria can stimulate this degradation,[6] but presumably excessive predation of the hydrolytic population would result in a reduced degradation rate. It is likely that the availability of nutrients limits the rate of mineralisation. The addition of inorganic nitrogen can increase the mineralisation rate of cellulose or leaf litter. It is possible that, just as the availability of N and P limits primary production, their availability may also control mineralisation. Thus, immobilisation and release of nutrients may control the relation between decomposition and production. The large steady-state pool of dead organic matter in many ecosystems may reflect the fact that plant tissues have lower nutrient contents than the decomposer organisms and that the primary producers and the decomposers compete for essential nutrients. The large pool of dead organic matter may, however, reflect the low rate at which certain structural plant polymers can be hydrolysed even under optimal conditions, thus explaining the fact that decomposition lags behind production.

Humic material and hydrocarbons.   Humics and hydrocarbons are the products of bacterial activity and in this way are different from the structural polysaccharides, proteins and polynucleotides in the primary producers. Humic substances are normally recovered from soils and sediments by alkali extraction. The acid-insoluble fraction of this extract is termed humic acid and the acid-soluble fraction is termed fulvic acid. The alkali-insoluble fraction of soil is termed humin. There are no sharp distinctions between these fractions, the differences mainly being the molecular weight and the attached side groups; humin is believed to consist of humic and fulvic acids bound to mineral matter.

The core of humic substances is made up of aromatic rings; these originate from lignin residues, phenols and quinones synthesised by microorganisms and later polymerised with nitrogenous compounds to form the humic substances. Experiments with ^{14}C-labelled substrates (microbial carbohydrates, cellulose, glucose and wheat straw) added to soils showed that a part of the labelled carbon rapidly appeared in the humic fraction.[13] This probably represented microbially produced amino-N which was incorporated into the humic fraction. Humic material has a very long turnover time in natural soils (~ 1000 years); cultivation of a soil may result in a more rapid degradation. In general, there is a tendency for humic substances to accumulate in cold climates. Under anaerobic conditions, in water-logged soils, swamps and sediments, the mineralisation of resistant plant and microbial residues including waxes, resins, cork substances and lignin is very inefficient resulting in low pH values and peats, which often preserve the original structure of plant tissue. Here again is seen the importance of O_2 in degradation of some compounds. Low pH values and other factors in peat can lead to the inhibition of bacterial degradation so that even animal tissues may be preserved. The nature of these factors is poorly known. In addition to humic substances, peat contains lignin, some cellulose and bitumen, which consists of waxes, paraffins and resins. Over geological time, through abiological processes, peat may turn into lignite (brown coal) and eventually into hard coal. The changes leading from peat to anthracite coal imply an increase in carbon content from about 55 to 94% and this carbon can probably not return to the biosphere unless it is combusted directly to CO_2. Fossil fuels in the form of hydrocarbons (petroleum and natural gas), however, may be mineralised through the activity of micro-organisms. This could be expected since many hydrocarbons (e.g. methane, terpenes, camphor, carotenoids, paraffins) are produced by living organisms.

Crude oils contain normal paraffins (ranging in length from 1 to 30 carbon atoms), isoparaffins, branched paraffins (e.g. phytane, pristane), cycloalkanes, aromatic hydrocarbons and steranes; these occur in various proportions together with some non-hydrocarbons, for example metalloporphyrins. Many micro-organisms (species of the genera *Pseudomonas, Flavobacterium, Alcaligenes, Acromobacter, Nocardia, Mycobacterium, Arthrobacter, Micrococcus* and *Brevibacterium* in addition to some yeasts) can utilise hydrocarbons aerobically. Nitrifying bacteria can also oxidise aromatics; *Nitrosomonas europaea* can oxidise benzene.[11] There is also evidence that sulphate-reducing bacteria are involved in hydrocarbon oxidation[4,16] and nitrate-reducing bacteria can degrade *p*-xylene.[8] Different bacteria show some degree of substrate specificity with respect to different types of hydrocarbon. In general, unbranched carbons are more easily degraded than branched ones; these are again more easily degraded than cyclic hydrocarbons. Aromatic compounds are very slowly degraded.[7] Attempts have been made to increase the rate of oxidation of oil spills, by seeding the spills with cultures of microbes known to degrade components of the oil. The results of such efforts have been disappointing, but there is reason to hope that genetically manipulated strains with enhanced degrading powers may be useful in this context. Furthermore, the addition of N- and P-containing nutrients, which are partially oil soluble, should promote the growth of oil-degrading micro-organisms.

The decomposition rates of some important litter components in soil are shown in Fig. 2.2. Sugars are degraded very quickly, whereas phenols are degraded very slowly; waxes, lignin, cellulose and hemi-celluloses have intermediate rates. In fact, the decomposition rates vary considerably according to the type of terrestrial ecosystem. Thus, in tropical forests, mineralisation rates of up to 40% per month of the detritus pool have been recorded, in contrast to less than 20% per year in tundra.[20] Similar ecosystem and geographical differences occur in the decomposition of detritus in marine sediments, which are also composed of various components that are degraded at different rates by first-order kinetics.[12] The mineralisation of detritus in marine sediments and soils is probably similar, except for the more resistant components of vascular plants which are less common in marine detritus. The high concentration of SO_4^{2-} in seawater affects marine sediment mineralisation (see Chapter 5).

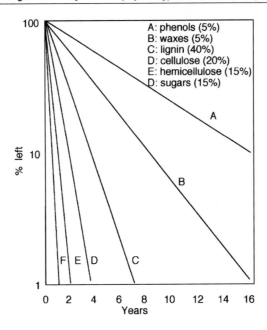

Figure 2.2   An idealised presentation of the breakdown rates of some important litter components in soil. The numbers in parentheses indicate the approximate quantitative importance of the components in litter[17]

Measurement of the rates of mineralisation and organic accumulation is usually possible for both terrestrial and marine soils and sediment; the sum of the two proccesses gives an estimation of the input of organic matter. Ecosystem heterogeneity, however, makes it difficult to extrapolate these calculations to larger areas and there is often considerable uncertainty as to the input of organic matter to particular systems, for example marine basins.

2.1.2 Hydrolytic enzymes

All hydrolysis of polymers must take place outside the cell membrane, because no large molecules can pass through the membrane into the cell. However, the reverse process is possible: hydrolytic enzymes synthesised within the cell can pass to the outside. These are defined as exoenzymes, because of their extracellular location (not to be confused with exo-acting enzymes which attack the ends of polymers, for example exogluconases which attack the ends of cellulose chains). Exoenzymes are directed through the membrane by amino-(N)-terminal extensions (signal peptides of 15–40 amino acid residues) to the proteins.[14] Exoenzymes, however, are not usually released immediately from the parent cell. In Gram-negative bacteria, the enzymes are generally retained in the periplasmic space between the cytoplasmic membrane and the outer cell membrane.

In Gram-positive bacteria, the exoenzymes are held by the cytoplasmic membrane, often in specialised structures (e.g. cellulosomes). Some members of the Archaea also produce exoenzymes, but little is known of their location with respect to the cytoplasmic membrane. At least some exoenzymes are released immediately from their cell-bound location and are free to diffuse away. Most enzymes become free after some time, which can be due to cell lysis. The complete liberation of exoenzymes can be seen by the zones of hydrolysis surrounding hydrolytic colonies on protein, starch or cellulose agar plates. In some situations there must be advantages in the complete liberation of hydrolytic enzymes, for example the tissue-degrading enzymes of a skin pathogen. In most situations, it is an advantage that close contact be maintained between cell, enzyme and substrate resulting in a short diffusional path from the products of hydrolysis to the cell surface, thereby improving the chance for the cell to reap the rewards of the degradation process.

Control of synthesis. The synthesis of hydrolytic enzymes appears to be under the same type of operon control that regulates intracellular enzyme production. Positive control, negative control and more general catabolite repression have been observed. The main difference is that the large molecular weight substrate cannot act as inducer. It seems that most exoenzymes are inducible: a small constitutive synthesis always occurs, but the products of hydrolysis act as inducers.

There are probably differences in regulation in different types of micro-organisms, but the distinctions are not obvious. All types of micro-organisms produce a variety of hydrolytic enzymes. Archaea seem to produce a restricted selection: cellulase production has not been reported. In general, many micro-organisms produce proteinases and nucleases. Proteins and polynucleotides are the first polymers to be degraded in detritus. The less rigid polysaccharides, especially those not having a β-1,4 bond, are relatively easily hydrolysed. These include starch, laminarins and pectin (Table 2.1). Cellulose, chitin and peptidoglycan are the most resistant polymers, with chitosan, mannan and xylan more susceptible. Both aerobic and anaerobic micro-organisms hydrolyse all types of polysaccharide, but some of the best investigated systems are in the anaerobic bacteria. The prevalence of anaerobic digestion may be a reflection of the extent to which detrital degradation occurs in anaerobic environments, particularly where water saturation restricts the access of O_2. In soil, which tends to be well aerated, detrital breakdown is predominantly by fungi. Unfortunately, little is known about the types of enzymes that slowly degrade humic material, but it may be surmised that a variety of enzymes must be involved and that they may be relatively non-specific. Different degrees of specificity occur; many cellulases are quite specific, but a xylanase can also have weak cellulase activity. This topic, and other aspects of polysaccharide hydrolysis, has recently been reviewed in depth.[18,19] There is an amazing complexity in many of these hydrolytic enzymes. Sometimes they contain one active catalytic domain, but often there are more than six domains. One at least is the main catalytic domain, another usually is a substrate-binding domain, but the other domains can have a variety of other hydrolytic activities. Glycoside hydrolases belong to more than 50 families of related domains, based on amino-acid sequences. The complexity of the situation is illustrated in the variety of polysaccharide-hydrolysing enzymes, produced by the rumen anaerobe *Butyrivibrio fibrisolvens* H17c, as summarised by Warren.[19] There is an endoglucanase, a cellodextrinase, a xylanase, a β-glucidase and a xylosidase (with arabinofuranosidase activity). *Fibrobacter succinogines, Ruminococcus flavefaciens, Ruminococcus albus* and *Prevotella ruminicola* produce a similar multiplicity of enzymes. These bacteria probably have their enzymes arranged in an organised manner at the cell surface, similar to the cellulosome of *Clostridium thermocellum.*[18]

The cellulosome.   Typically, the hydrolysis of cellulose to monomers requires the action of at least three different types of enzyme: endo-1,4-β-glucanase, exo-1,4-β-glucanase or cellobiohydrase, and β-glucosidase or cellobiase. The endocellulase attacks the amorphous regions of the cellulose, producing water-soluble oligosaccharides, which are then hydrolysed to cellobiose and glucose. The hydrolysis of intact cellulose is best accomplished when these three activities operate in unison, which occurs when the enzymes are organised in a cellulosome, as in *Clostridium thermocellum*. There may be at least 10 proteins with endoglucanase activity in the cellulosome.[5] The isolation of multiple proteins with endoglucanase activity can sometimes be the result of protease fragmentation of a smaller number of enzymes. In addition, there are two β-glucosidases, three xylanases, a lichenase, a laminarase and a cellobiohydrase belonging to a large number of enzyme families. There seems to be no doubt that a large number of different polypeptides are present in the cellulosome, that these have a wide variety of enzyme domains, and that multiple domains can be present in a single polypeptide. This is not altogether surprising, as this complexity of enzymatic domains has already been discussed, but what is astounding is the complex arrangement of these polypeptides within the cellulosome. The component enzymatic polypeptides and cellulose-binding polypeptides have unique sites (dockerins, with approximately 22 amino-acid segments repeated twice) which bind to complementary sites (cohesins) on a scaffolding protein. There are nine of these catalytic domain-binding sites and a similar site for the attachment of the cellulose-binding domain. This structure is illustrated in Fig. 2.3. It is likely that the binding sites are arranged so that the enzyme domains are positioned precisely in relation to the cellulose bonds allowing simultaneous cutting to occur. The cellulosome does not traverse along the cellulose chain and the exact

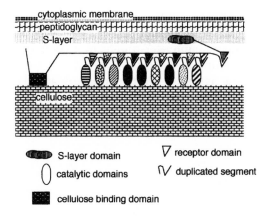

Figure 2.3 Probable organisation of the *Clostridium thermocellum* cellulosome. The main scaffolding has nine binding sites (cohesins) which bind duplicated segments (dockerins) of the catalytic domain polypeptides. The scaffolding polypeptide has a cellulose-binding domain which attaches very firmly to the cellulose substrate. In addition, it has a dockerin, which binds to another polypeptide protruding from the S layer of the cell wall. The cellulosome is thus locked to the cell and to the substrate, which it can then hydrolyse[18]

mechanism of progressive cellulose hydrolysis is unknown. Binding to the cellulose is initially very tight. Cellulosomes are themselves arranged in complex polycellulosomes. The binding of dockerins to cohesins seems to be random and the individual polypeptides with catalytic domains are not coded for polycistronically: the polypeptides are synthesised individually, but possibly under some overall control.

It is impressive that 'primitive' organisms such as clostridia have evolved such a complex and intricate organelle, but presumably this is the price they have to pay in order to live on cellulose, which has a structure designed to repel enzymatic attack.

2.2 Comparison of element cycles

Adhering to the concept that bacteria are mainly of interest and importance because of their heterotrophic life style, it follows that element cycling is mainly driven by heterotrophic carbon metabolism. Some chemoautotrophic processes are of interest, but not in the context of assimilative CO_2 reduction.

Our foremost interest in element cycling is the liberation from detritus of those elements that are essential for the primary producers, so that new biomass may be synthesised. In this context, it is clearly C, N, P and Fe that are of primary concern. S is rarely a limiting nutrient and Mn is seldom found in cellular biomass. S and Mn are, therefore, of less interest except in the context of oxidation/reduction interactions involving the other elements. Most of the major components in the element cycles have been presented in Section 1.1 and the ecological significance of the processes is discussed in subsequent chapters. We present a comparative summary of carbon, nitrogen and sulphur cycling in Figs 2.4, 2.5 and 2.6, respectively. With reference to these figures, a comparison is made between the element cycles, with emphasis on the essential elements and the factors that control their availability, after the initial mineralisation has occurred.

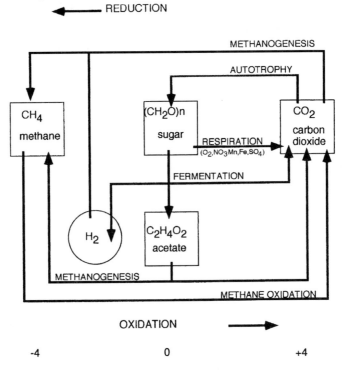

Figure 2.4   The carbon cycle. The starting point of the cycle is the product of polymer hydrolysis, for example a sugar with oxidation/reduction potential of 0

2.2.1 The dominance of carbon

The carbon cycle is unique in its dominance over the other element cycles. To some extent P and N can be mineralised independently of C, often resulting in preferential liberation of these elements from a detritus, which becomes progressively nutrient (P and N) impoverished. The mineralisation of C, however, results in major changes at the site of detrital degradation, particularly in the consumption of oxidants. Oxygen being the chief oxidant, due to its prevalence in the atmosphere and the high energy yield associated with its use, disappears first at sites where its access is limited, usually by restricted diffusion through water.

Where the other oxidants are available, they are consumed in the order: NO_3^-, Mn^{4+}, Fe^{3+} and SO_4^{2-}. When all oxidants have been consumed or are not available (depending on the ecosystem) detrital carbon, having predominantly an oxidation/reduction potential of zero, is converted to a mixture of CO_2 and CH_4. Thus, once particulate organic detritus has been hydrolysed, the processing of the soluble hydrolytic products is almost inevitable and usually rapid. The products of carbon catabolism are, therefore, CO_2 plus CH_4 in reducing environments, otherwise only CO_2. Methane can later be oxidised by O_2, probably indirectly by SO_4^{2-}, but apparently not by NO_3^-, Mn^{4+} or Fe^{3+} (evidence is lacking but the reactions are thermodynamically favourable). The products of carbon mineralisation are very limited, and it is unlikely that inorganic C is a major limiting factor in primary production. However, the consumption of oxidants in the process of carbon oxidation creates conditions that have a profound effect on the other cycles.

2.2.2 Oxidation/reduction reactions

Unlike carbon, whose mineralisation results in the production of CO_2, there are possibilities for more varied products in the mineralisation of organic N. The first product of organic N mineralisation is NH_4^+, which is the most reduced form of N. Organic N is also in this reduced form, unlike organic C which has an intermediate state of oxidation/reduction. The N cycle resembles the C cycle in having an eight electron difference between the most reduced and most oxidised inorganic molecules: NH_4^+ to NO_3^- and CH_4 to CO_2. Ammonium can be used by most photoautotrophs and its production thus fulfils the objectives of heterotrophic mineralisation (i.e. to regenerate a nutrient in a mineral form that can be the start of new biomass production). However, in an aerobic environment NH_4^+ is usually oxidised to NO_3^- via the intermediates illustrated in Fig. 2.5 on p. 56. It is unusual for any of these intermediates to build up to high concentrations. There is no extracellular pool of hydroxylamine, but frequently there is a production of nitrous oxide as a byproduct. Typically, the oxidation of NH_4^+ is performed by chemoautotrophic bacteria via a mono-oxygenase. The oxidation process, from NH_4^+ with an oxidation state of -3 to NO_3^- (+5), supplies little energy to the lithotrophic nitrifiers, and consequently cell yield is low. Heterotrophic nitrification occurs, mostly due to fungi at low pH, and with no net energy yield. Nitrification is a dissimilatory reaction (i.e. no product is incorporated into cell biomass). Denitrification is similarly a dissimilative reaction, the reverse of nitrification, in which NO_3^- is first reduced by nitrate reductase (reduced C or S as electron donor) to NO_2^-, which is reduced further to N_2 or N_2O. Nitrate is used instead of O_2 as an electron acceptor. Recently, it has been discovered that denitrification of NH_4^+ is possible in the presence of low O_2 concentrations.[15] There are also indications that Mn^{4+} can oxidise NH_4^+ to N_2 , and that NO_3^-

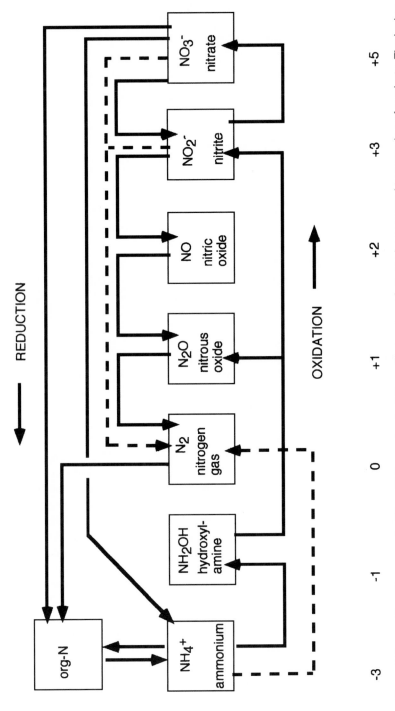

Figure 2.5   The nitrogen cycle. Via oxidative and reductive reactions, ammonium can be converted to a number of products. The broken line indicates that N_2 can arise by reactions other than classical denitrification

can be reduced by Mn^{2+} to N_2 (B Sundby, personal communication). It has long been suspected that NO_3^- could oxidise NH_4^+ to N_2, in the absence of O_2.[1] Fermentative bacteria can also reductively dissimilate NO_3^- to NH_4^+. As a result of the various oxidation and reduction reactions outlined (Fig. 2.5), the NH_4^+ originating from organic N degradation can be converted to a variety of products. Depending on the relative availability of O_2 and e-donors, the ratio of products (NH_4^+, N_2, NO_3^-) varies, even if the total amount of mineral N is constant. These complex inter-relationships in aquatic sediment biogeochemistry are discussed in Sections 5.3.2 and 5.3.3.

Phosphate does not undergo oxidation–reduction reactions and does not have a cycle comparable to N and C. Iron, another essential nutrient for the primary producers, does have a cycle moving between Fe^{2+} and Fe^{3+}. Reduced iron is very easily oxidised by O_2 at neutral pH values. Reduced carbon can be oxidised by Fe^{3+} under anoxic conditions (Fig. 2.4). This can sometimes be of importance in sediments, where Fe^{3+} is present in quantity. The degree to which Fe^{3+} is involved in other oxidations (e.g. reduced S and N) is unknown. Manganese also has two main oxidation states (Mn^{2+} and Mn^{4+}). Organic C can be oxidised by Mn^{4+}, where this is found in quantity under anoxic conditions. It has been suggested that Mn^{4+} can also oxidise NH_4^+ (Fig. 2.5). Reduced Mn can be oxidised spontaneously by O_2 and possibly by NO_3^-. The involvement of Mn in other oxidation–reduction reactions (e.g. in the S cycle) is unknown.

In common with the C and N cycles, S is found in multiple oxida- tion states spanning eight electrons (Fig. 2.6 on p. 58). Organic S is in the most reduced form (-2), and is in this respect similar to organic N. The most important organic S compounds are the amino acids cysteine and methionine. Degradation of organic S yields HS^-. This can be oxidised by O_2 chemically or biologically. It can also be oxidised by NO_3^- biologically, both oxidations yielding energy for cell synthesis. It is probable that, Fe^{3+} and Mn^{4+} also can oxidise HS^-, but the ecological significance is doubtful. The end product of oxidation is SO_4^{2-}, but there can be an accumulation of intermediates, particularly S^0 and thiosulphate. However, it is SO_4^{2-} that is of major significance, because of the enormous reservoirs in the ocean and its importance in anoxic respiration of organic C, yielding HS^- and CO_2. The anoxic oxidation of CH_4 has been linked to SO_4^{2-} reduction. Sulphate can be produced by the dissimilation of thiosulphate, a reaction that also produces HS^-.[10]

There are, therefore, many oxidation and reduction reactions involving the essential nutrients C, N and Fe which determine the ratios of the forms in which the elements are finally made available to the primary producers. Because the latter are predominantly aerobes (plants, algae, etc.) it is inevitable that they must be able to use the oxidised form of the elements: CO_2, NO_3^- (or N_2), Fe^{3+} and SO_4^{2-}. The photoautotrophs have

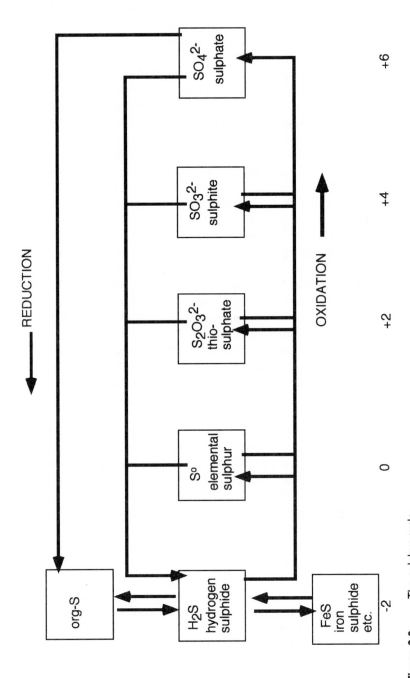

Figure 2.6   The sulphur cycle

little difficulty in reducing these oxidised elements to the oxidation state found in biomass, but other factors affect the availability of the elements.

2.2.3 Mobility and availability

Diffusive mobility of molecules in the N cycle determines their availability to some extent. Ammonium is bound to organic and inorganic particles by ion exchange. This decreases the mobility of NH_4^+, prevents it being washed out of soils, and slows its diffusion to the oxic interface, in saturated soils and sediment. Nitrate is not retained in soils by ion exchange and can easily be washed through the soil and lost to the roots of the primary producers. Dinitrogen, which is a gas, can rapidly leave the site of production. This is not of significance as N_2 is not available to most primary producers.

Sulphide reacts with Fe to produce insoluble FeS. This compound is very insoluble and thus prevents the diffusion of both Fe^{3+} and HS^-. The availability of Fe may be decreased by this binding. The non-diffusibility has very significant consequences for the oxidative status of soils and sediments, which can affect nitrification and denitrification reactions (see Sections 5.3.2 and 5.3.3.). When reduced Fe (either FeS or pyrite) is oxidised, usually by bioturbation or advective transport of oxic water, it forms insoluble ferrohydroxides, which prevent the recycling of Fe from sediments to the overlying water and the primary producers. Of equal, or perhaps more, significance is the trapping of inorganic P in these Fe complexes, preventing its recycling, until the sediments become more reduced. This is not a problem for the primary producers with roots in soil. Most aerobic species with a requirement for Fe have developed mechanisms for the solublisation of Fe^{3+} complexes.

Restriction of mobility may slow the rate at which elements are made available for new biomass synthesis, but in general anything that is not permanently buried is recycled in one of a number of oxidation states. The burial of C has implications for global warming, but the burial of N, P and Fe has more immediate influence on primary production. However, new N, P and Fe are continually being made available by the dissolution of rocks, keeping the cycles in balance.

References

1. Bender Ml, Fanning KA, Froelich PN, Heath GR, Maynard V (1977) Interstitial nitrate profiles and oxidation of sedimentary organic matter in the eastern equatorial Atlantic. *Science* **198**: 605–609.
2. Blackburn TH (1988) Benthic mineralization and bacterial production. In: Blackburn TH, Sørensen J (eds) *Nitrogen Cycling in Marine, Coastal Environments.* Wiley, Chichester, pp. 175–190.

3. Blackburn TH (1991) Accumulation and regeneration: processes at the benthic boundary layer. In: Mantoura RCF, Martin J-M, Wollast R (eds) *Ocean Margin Processes and Global Change*. Wiley, Chichester, pp. 181–195.

4. Elsgaard L, Prieur D, Mukwaya GM, Jørgensen BB (1994) Thermophilic sulfate reduction in hydrothermal sediment of Lake Tanganyika, East Africa. *Appl Environ Microbiol* **60**: 1473–1480.

5. Felix CR, Ljungdahl LG (1993) The cellulosome: the extracellular organelle of *Clostridium. Ann Rev Microbiol* **47**: 791–819.

6. Fenchel T, Harrison P (1976) The significance of bacterial grazing and mineral cycling for the decomposition of particulate detritus. In: Anderson JM, Macfadyen A (eds) *The Role of Terrestrial and Aquatic Organisms in the Decomposition Process*. Blackwell Scientific Publications, Oxford, pp. 285–299.

7. Goetz FE, Jannasch HW (1993) Aromatic hydrocarbon-degrading bacteria in the petroleum-rich sediments of the Guaymas Basin hydrothermal vent site — preference for carboxylic acids. *Geomicrobiol J* **11**: 1–18.

8. Häner A, Höhener P, Zeyer J (1995) Degradation of *p*-xylene by a denitrifying enrichment culture. *Appl Environ Microbiol* **61**: 3185–3188.

9. Jenkinson DS, Powlson DS (1976) The effects of biocidal treatments on metabolism in soil. 1. Fumigation with chloroform. *Soil Biol Biochem* **8**: 167–177.

10. Jørgensen BB, Bak F (1991) Pathways and microbiology of thiosulfate transformation and sulfate reduction in a marine sediment (Kattegat, Denmark). *Appl Environ Microbiol* **57**: 847–856.

11. Keener WK, Arp DJ (1994) Transformations of aromatic compounds by *Nitrosomonas europaea. Appl Environ Microbiol* **60**: 1914–1920.

12. Kristensen E (1994) Decomposition of macroalgae, vascular plants and sediment detritus in seawater — use of stepwise thermogravimetry. *Biogeochemistry* **26**: 1–24.

13. Martin JP, Haider K, Farmer WJ, Fustec-Mathon E (1974) Decomposition and distribution of residual activity of some ^{14}C-microbial polysaccharides and cells, glucose and wheat straw in soil. *Soil Biol Biochem* **6**: 221–230.

14. Priest FG (1992) Synthesis and secretion of extracellular enzymes in bacteria. In: Winkelmann G (ed.) *Microbial Degradation of Natural Products*. VCH, Weinheim, pp. 1–26.

15. Robertson LA, Dalsgaard T, Revsbech N-P, Kuenen JG (1995) Confirmation of 'aerobic denitrification' in batch cultures, using gas chromatography and N-15 mass spectrometry. *FEMS Microbiol Ecol* **18**: 113–119.

16. Rueter P, Rabus R, Wilkes H, Aeckersberg F, Rainey FA, Jannasch HW, Widdel F (1994) Anaerobic oxidation of hydrocarbons in crude oil by new types of sulphate-reducing bacteria. *Nature* **372**: 455–458.

17. Stout JD, Tate KR, Molloy LF (1976) Decomposition processes in New Zealand soils with particular respect to rates and pathways of plant degradation. In: Anderson JM, MacFadyen A (eds) *The Role of Terrestrial and Aquatic Organisms in Decomposition Processes*. Blackwell Scientific Publications, Oxford, pp. 97–114.

18. Tomme P, Warren RAJ, Gilkes NR (1995) Cellulose hydrolysis by bacteria and fungi. *Adv Microbiol Physiol* **37**: 1–80.
19. Warren RAJ (1996) Microbial hydrolysis of polysaccharides. *Ann Rev Microbiol* **50**: 183–212.
20. Witkamp M, Ausmus BS (1976) Processes in decomposition and nutrient transfer in forest systems. In: Anderson JM, Macfadyen A (eds) *The Role of Terrestrial and Aquatic Organisms in Decomposition Processes*. Blackwell Scientific Publications, Oxford, pp. 375–396.

3

The water column

The quantitative role of water-column bacteria has not been fully appreciated until recently. The presence of suspended bacteria in natural waters has, of course, been long known. Until the 1960s, however, the method of choice for quantifying bacteria was based on colony counts on nutrient agar plates. Today, we know that this method yields estimates that are probably only 1% or less of the bacteria that are actually present in natural waters. It was also generally believed that available organic substrates are too dilute to sustain significant bacterial growth. Models of carbon flow in the water column (including phytoplankton and zooplankton) therefore relegated bacteria to a minor role or ignored them altogether. However, during the past 30 years it has become apparent that bacteria constitute 20–30% of plankton biomass and that they process a variable, but substantial, part of primary production. The observations that led to a contemporary picture of plankton bacteria include measurements of O_2 uptake in water samples from which larger plankton organisms had been removed,[86] the rapid turnover of ^{14}C-labelled low molecular weight compounds[41,42] and direct (microscopic) counts of bacteria.[28,43] Early observations documented the role of bacteria in the mineralisation processes following algal blooms.[97] It was also discovered that prokaryote oxygenic photosynthesis may be quantitatively important.[84] Methods were developed for estimating *in situ* growth rates of heterotrophic bacteria.[19,30,39] Finally, it was shown that protozoan phagotrophy results in a rapid turnover of bacterial populations that is compatible with the new estimates of bacterial population growth.[4,26] Thus, a new picture of water-column carbon flow emerged that incorporated the role of microbes.[103] During the past 10 years a vast amount of literature on plankton microbial ecology has appeared.

The "microbial loop" describes the current view of the role of bacteria in planktonic food chains;[6] see also Fig. 3.1. The essential observation is that a large fraction of primary production is not directly consumed by phagotrophs, but is "lost" in the form of detritus or dissolved organic matter, which is then utilised by bacteria. The resulting bacterial production is in turn consumed by phagotrophs (mainly small heterotrophic protists) and thus channelled into the "classical" plankton food chains. The relative importance of the microbial loop varies among aquatic systems and over time; as a generalisation its role is highest under oligotrophic conditions

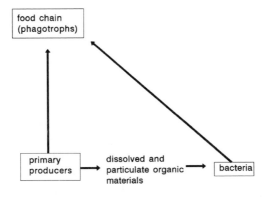

Figure 3.1 The "microbial loop"

and least in nutrient-rich upwelling systems and during the initial stages of algal blooms, when large algae dominate primary production. Although the term "microbial loop" was coined in the context of water-column ecology it does, in fact, describe almost all ecological systems equally well (e.g. terrestrial systems where only a fraction of the primary production is consumed by herbivores; most is degraded by micro-organisms which then enter food chains via soil protozoa and microfauna).

While the planktonic microbial loop is considered a relatively recent discovery, it is noteworthy that in 1934, Krogh[60] quantified dissolved organic matter in a few water samples, and concluded that it represented by far the largest pool of organic carbon in the sea. He also identified the most probable sources of this material, and reasoned that only bacteria utilise this resource and that they then enter food chains. While this concept had limited impact on biological oceanography at that time, it can be said that the discovery of the microbial loop is more than 60 years old.

In the following we discuss cyanobacterial primary production, biota of suspended heterotrophic bacteria, bacterial turnover of C, N and P and the role of bacteria associated with suspended particles. Stratified water columns with anoxic deep water are discussed separately in Section 6.4.

3.1 Prokaryotic primary producers

Large diatoms and dinoflagellates have traditionally been considered the most important primary producers in the water column. More recently, it has been found that a variety of smaller eukaryotic phototrophs are also important, especially under oligotrophic conditions.[84] One prokaryote, the oscillatorian *Trichodesmium*, qualifies among the larger and long-known marine plankton phototrophs. These filamentous organisms are interesting in that they are N_2 fixers and they occasionally form extensive blooms in N-limited but P-rich waters;[17,18] blooms of filamentous cyanobacterium *Nodularia* occur seasonally in the Baltic Sea.

Strains of the (unicellular) chroococcian genus *Synechoccus* are ubiquitous in the sea and in lake water. These $\sim$ 1 µm ovoid cells are easily identified on black filters under the fluorescence microscope due to their relatively large size, red chlorphyll *a* fluorescence in blue light, and bright fluorescence in green light due to phycoerythrin (an accessory pigment in cyanobacteria). There are typically around 10^5 cells per millilitre (representing about 10% of all bacteria present). They are especially important in oligotrophic waters since very dilute mineral nutrient concentrations favour small cell sizes (cf. Section 1.3.1); it has been estimated that under some conditions they can be responsible for 30–70% of the primary production.[47,61,77,98]

Most recently it has been found that prochlorophytes (a group of prokaryotes with oxygenic photosynthesis, but which seem not to be closely related to the cyanobacteria) occur in marine plankton. They are represented by the genus *Prochlorococcus*.[20,36] These organisms have unusual chlorophylls (divinyl chlorophylls *a* and *b*) and they lack phycobilins. Their small size is typical of planktonic bacteria in general and, due to the small amounts of photosynthetic pigments per cell, the chlorophyll autofluorescence is difficult to see; it is impossible to visualise in preparations stained with fluorochromes (used to visualise heterotrophs). They have so far been quantified only with flow cytometry. Their distribution is still incompletely known, and it is likely that they have been counted as heterotrophic bacteria in many studies of planktonic bacteria. So far they have been recorded only from oligotrophic, oceanic water samples (the Atlantic and Pacific oceans, the Red Sea) where they appear to play an important role as primary producers. Some studies have found them distributed in fairly constant numbers throughout the photic zone, whereas others found them mainly in the deepest part of the photic zone at > 100 m and beneath a maximum layer of *Synechoccus* cells.[14,37,62,69,101] The small size and high cellular concentration of photosynthetic pigments allow for a high blue light absorption efficiency and this would explain their competitive advantage in the deepest part of the photic zone.[75]

3.2   Water-column bacteria and mineralisation

3.2.1   Characteristics of the bacterial biota

Extensive counting of bacteria (based on cells that have been fixed and stained by a fluorochrome, collected on black membrane filters and observed in the fluorescence microscope)[43] indicates that there are about 10^6 cells per millilitre of water and counts are almost always within the range of 2×10^5 to 5×10^6 per millilitre. This applies to all sorts of aquatic systems, from eutrophic lakes to oligotrophic ocean sites. Since < 1% of the counted cells grow on nutrient agar plates, since planktonic bacteria are typically small and since ambient concentrations of dissolved organic nutrients are low, it has been suggested that total counts do not reflect metabolically active, dividing bacteria, but rather starving cells, dormant stages or even non-viable cells.[58,76]

Starving, dormant and non-viable cells undoubtedly occur in plankton. It is also possible that direct counts are inflated for other reasons. Viral particles which have been retained on filters may be mistaken for small bacteria. The fluorochromes used may stain substances other than DNA and RNA so that certain micrometre-sized detrital particles or bacterial "ghosts" (bacterial cell walls without contents) may be counted along with intact bacteria (such ghosts may be partly digested bacteria egested from protozoa or result from viral lysis). However, estimates of rates of growth and metabolism constrain the minimum number of metabolically active bacteria that can be present. Thus, observations on the increase in bacterial numbers in water samples, from which bacterial grazers (protozoa) have been removed, as well as indirect methods for estimating growth rates, show that bacterial populations on average grow with generation times that are typically within the range of 0.5 to a couple of days. Indirect methods for estimating growth rates include incorporation of radioactive thymidine (measuring DNA synthesis) or leucine (measuring protein synthesis), or quantification of the frequency of dividing cells in fixed samples (which is a measure of division rate, since the time used to pass a defined stage during cell fission is invariant except with respect to temperature).[19,30,39]

Measurements of *in situ* uptake rates of amino acids or glucose and the demonstration (using autoradiography) that most cells actively assimilate labelled compounds[31,45,49,50] show that the majority of the cells present are metabolically active. Finally, quantification of protozoan grazing on bacteria[4,26] and the frequency of viral infections[12,13] (although the quantitative significance of the latter is still not well established) show that bacterial populations have turnover times that are compatible with estimated division rates. Given that protozoan grazers effectively clear the entire water column of bacteria every 0.5–5 days (depending on time and site) it is inconceivable that large populations of non-dividing cells could be maintained.

The fact that bacterial densities remain relatively constant irrespective of the productivity of the water column, reflects control of numbers by grazing. Thus, the observed densities of bacteria correspond to those necessary to sustain their grazers. Conversely, the low ambient concentrations of readily available low molecular weight substrates (monosaccharides, amino acids) indicate continuous substrate consumption with rapid turnover. In fact, the concentrations of such substrates (typically within the range of 10–20 nM or about 1 μg organic C per litre; see Section 3.2.2) are not incompatible with theoretical considerations and with growth kinetics of laboratory strains of bacteria; see Section 1.3.1).

Plankton bacteria seem to share certain special properties although some aspects are speculative or poorly understood. For example, suspended free planktonic bacteria are small, typically with a cell volume within the range of $0.02-0.1\,\mu m^3$ or about 10 times lower than that of most attached bacteria in aquatic habitats (including bacteria associated with suspended detrital material in the water column) or in soils. Small size can be considered an adaptation to dilute substrate concentrations since it increases substrate affinity (see Section 1.3.1). However, it remains controversial whether small sizes reflect a special specific composition of the bacterial biota or whether planktonic bacteria are "normal" species that have phenotypically adapted to low nutrient conditions, but would grow larger in the presence of higher substrate concentrations. It has been claimed that planktonic bacteria have unusually high contents of C and N. Thus, values of $> 0.3\,pg\,C\,\mu m^{-3}$ have been reported; this would seem unlikely for actively metabolising cells and more recent results suggest that the elemental composition of plankton bacteria is not different from normal, laboratory grown species: $0.06-0.1\,pg\,C\,\mu m^{-3}$ and a C:N ratio of 4–6.[25] The reason for this discrepancy is undoubtedly related to the difficulty of estimating dimensions of these small cells (and thus cell volume) with sufficient precision.

Growth yields of planktonic bacteria have been estimated by various methods in incubations with natural sea or lake water. Values ranging between 10 and 60% have been reported; the latter value would be expected for an aerobic cell using monosaccharides or amino acids (see Section 1.2.1). In some cases lower values could be explained by nitrogen limitation; it has also been found that growth efficiency falls when bacteria depend on more recalcitrant organic material.[59,72]

The fact that only a small fraction of plankton bacteria grow on agar plates or in normal liquid bacterial media has led to the suggestion that the majority of planktonic bacteria are "obligatory oligotrophs", meaning organisms that are intolerant of high substrate concentrations.[85] Attempts to isolate planktonic bacteria by dilution cultures or chemostats based on sterile, filtered seawater have been successful; such isolates have in some cases been unable to grow in rich media, but in other cases this ability could be induced. The question whether isolates (based on traditional media or on dilution cultures with seawater) are actually representative of the quantitatively dominant types of bacteria has still not been fully answered.

Recently, preparation of molecular probes from colony-forming seawater isolates has been applied to study this problem.[83,88] Using rRNA sequences or whole DNA genome hybridisation it was found that the isolates were indeed representative of the planktonic biota and accounted for the majority of bacteria present. Furthermore, it was found that bacterial diversity was low at any point in time, but that there were characteristic seasonal succession patterns. Thus, it was concluded that planktonic bacteria are, in general, culturable by classical methods, but that plating efficiency is low. The reason for this low efficiency is not quite understood, but may be due to activation of virus in infected bacteria suddenly exposed to the nutrient-rich culture conditions. These studies indicate that the phenotypic diversity of planktonic bacteria is open to studies and that planktonic bacteria are less "exotic" than previously proposed.

Another approach has been simply to sequence rRNA genes from DNA extracted from seawater. This revealed the presence of bacteria of which some belong to established taxa, but also to sequences that have so far not been found in culture collections. These new sequences belong to a cluster of forms within the α-group of proteobacteria; these apparently unknown forms seem to be widely distributed in the sea. The presence of unknown types of archaebacteria has also been detected.[32-34] These results (knowing only 16S rRNA sequences) do not in themselves yield any information on the phenotypical properties of the bacteria or on the abundance of the carriers of particular sequences. Furthermore, the fact that the rRNA sequences cannot be found in available libraries does not mean that the carriers are "unculturable" or indeed that they have not already been cultured. On the other hand, it cannot be excluded that hitherto undiscovered and interesting phenotypes of bacteria do occur in plankton.

3.2.2 Organic matter: composition, origin and turnover

The weight ratios between dissolved organics, particulate organics and the living biota (bacteria, phototrophic plankton, protozoa and zooplankton) in the water column are approximately 100:10:2. Next to kerogen of sedimentary rocks, dissolved organic matter constitutes the largest pool of organic matter on Earth; it is usually estimated that the oceans together contain about 10^{18} g of dissolved organics.

Dissolved organic matter is defined operationally as material that passes a 0.2- or 0.1-µm filter. This is not entirely without problems: smaller colloidal particles may absorb to filters and conversely > 0.1 µm aggregates may disintegrate on filters so that the resulting smaller particles pass through them. Furthermore, living cells may lyse on the filter thus contributing to the solutes in the filtrate. The dissolved fraction may further be characterised by size or molecular weight using ultrafiltration.[15] Particles > 1 nm (molecular weight $> 10\,000$) are considered colloidal; most studies have found that colloids constitute 10–40% of the total pool of dissolved organic matter (the upper size limit of colloids is often defined as 1 µm, in which case they overlap with particulate matter as defined above).

The chemical analysis of dissolved organic matter in seawater is difficult and the bulk of the dissolved matter is only incompletely characterised. Humic acids (which are operationally defined by particular extraction methods) constitute 40–80% of the dissolved organic matter of freshwater, but only 5–25% in seawater. Humic acids are mainly refractile remains of plant polymers. Terrestrial and limnic humic acids derive mainly from the lignin of vascular plant tissue and they contain more aromatic groups than do humic acids extracted from seawater.[74] Combined carbohydrates, which to a large extent are resistant to hydrolysis, constitute about 10% and combined amino acids (peptides, proteins) about 1% of the dissolved organic matter. Bacteria can directly assimilate only low molecular weight compounds such as monosaccharides and free amino acids, and these compounds are therefore of special interest. They can now be quantified with high sensitivity by liquid chromatography. The concentrations of individual amino acids and monosaccharides are relatively constant within the range of 10–50 nM. Glucose is quantitatively dominant among the monosaccharides.[9,31,49,50,80,89] Concentration ranges of various constituents of dissolved organic matter (in terms of moles C per litre) are shown in Fig. 3.2.

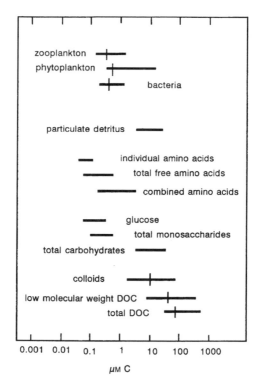

Figure 3.2   Concentration ranges for different constituents of dissolved organic matter (DOC) expressed as micromoles of C per litre. Vertical bars represent typical values for offshore seawater. Compiled from various sources

The origin of dissolved organic matter is diverse. Photosynthate excretion by algal cells is an important source of low molecular weight substances. It was discovered in the 1960s that growing algal cells excrete some of their photosynthate in the form of dissolved organics.[29] Estimates of the fraction of photosynthate thus lost to the cells are within a range of 5–40% and it probably constitutes the single most important source of low molecular weight substrates for bacteria in the photic zone of the water column. It was earlier believed that photo-oxidation leading to glycollate excretion is a likely mechanism, but this is no longer considered to be important. It is likely that healthy algal cells excrete dissolved photosynthate as a result of unavoidable leakage from small cells with large surface:volume ratios.[10] This is supported by the fact that small heterotrophic protozoa also excrete, for example, amino acids;[5] the topic is reviewed in Williams.[104] Viral lysis of bacteria may also provide some dissolved organic matter. Various modelling attempts have indicated that the "recycling" of organic matter excreted from bacterivorous protozoa and bacteria undergoing viral lysis seems to be of some significance.[5,11] Other sources of easily degradable dissolved organics include lysis of senescent cells and "sloppy feeding", meaning that protists or small zooplankton organisms leak dissolved organics when they are ingested by predators. Many diatoms, protists and bacteria secrete mucous substances which add to the pool of dissolved organics. Particulate organic material is also added in the form of copepod faecal pellets and abandoned larvacean mucous houses. The organic contents of such particles and aggregates formed from dissolved organics (colloids, mucus) are more or less completely hydrolysed by attached bacteria before they sink down to the sediment; aggregates are discussed in more detail in Section 3.2.3. In coastal areas mucous substances (including carbohydrates, alginate, etc.) deriving from fragmenting macroalgae and kelp constitute an important source of dissolved organic material.[64,66] Lakes and coastal waters also receive allochtonous organic matter via terrestrial run-off; the bulk of this material consists of recalcitrant molecules such as humic acids.

A general and schematic presentation of the role of bacteria in the mineralisation of organic carbon is shown in Fig. 3.3. Bacteria can directly assimilate low molecular weight organics which are first excreted from algal cells. Polymers and organic particles are hydrolysed by membrane-bound hydrolytic enzymes. A large number of studies, carried out at different sites and at different seasons, suggest that the amount of organic carbon channelled through the bacterial populations corresponds to 10–60% of the primary production.

Not all degradation of dissolved organic matter is due to biological (bacterial) processes. Some of the most recalcitrant components (including humic acids) seem primarily to be mineralised through photolysis.[38,73] Exposure of such otherwise undegradable organic compounds to short

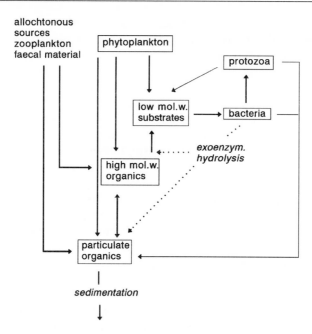

Figure 3.3   Microbial carbon flow in the water column; phytoplankton and protozoa are chanelled into zooplankton food chains and sedimenting particulate organic matter constitutes the basis for benthic food chains

wave light may also release molecules that are available to enzymatic hydrolysis and thus enhance bacterial growth.[63] It has been shown, however, that some bacterial degradation of humic acids takes place without previous non-biological degradation.[74] Conversely, it has also been shown that easily degradable proteins added to water samples may become refractile due to complexation with other dissolved organic components[51] and chelation with metals may have a similar effect.

The turnover time of different constituents of dissolved organic matter varies enormously: glucose and amino acids may have turnover times < 1 h while the most recalcitrant molecules may have turnover times that are measured in centuries. The radiocarbon age of deep-sea dissolved organic matter is about 6000 years. It is, however, assumed that this figure is due in part to fossil sedimentary organic matter which leaks to the water and that the real turnover time is shorter, but still of the order of several centuries.[73]

Notwithstanding that there is a continuum in the turnover time and degradability of the different constituents of dissolved organics, it may be convenient to make the distinction between a "superlabile pool", a "labile pool" and a "refractory pool". (In one long-term study of the mineralisation of organic carbon deriving from dead diatoms, four pools with different mineralisation rate constants were recognised.[82])

The superlabile fraction consists of amino acids and monosaccharides (especially glucose). Extensive investigations using radiolabelled substrates show that these pools turn over every few hours (or < 1 h in oligotrophic waters).[31,49,50] During photosynthesis, these pools are replenished mainly by algal exudates. Consequently, a daily rhythm of the bacterial populations can be followed in the photic zone: substrate uptake is highest during the middle of the day, and the bacteria divide mainly during the afternoon. Since the bacteria are constantly grazed by small protozoa, the number of bacteria also fluctuates on a daily basis.[21,105] Such fluctuations are also evident on a seasonal basis in that bacterial activity and population sizes peak during algal blooms.[70,72] Such patterns, however, are often disguised by protozoan predation which can lead to pronounced predator–prey cycles.

Pools with somewhat longer turnover times (the "labile pool" probably including polysaccharides, peptides and proteins, which are relatively easily hydrolysed) can be studied in water samples in which bacterial predators have been removed (by filtration) or in particle-free water samples together with an inoculum of the native bacterial biota. Mineralisation can then be followed over time as loss of dissolved organic matter, as increase in bacterial numbers or (most accurately) as O_2 consumption or CO_2 production. (If one of the two former plus one of the two latter measurements are made simultaneously, it is also possible to estimate bacterial growth efficiency.)

Early studies[78] showed that mineralisation rates systematically decreased during the first couple of days; it was also found that the average molecular weight of the residual dissolved organic matter increased. It has been generalised that degradation rates decrease with increasing substrate molecular weight; as discussed below this is partly, but not universally, true. More recent systematic studies[95,96] showed that mineralisation became quite slow after 4–7 days of incubation and the amount of material that had been mineralised prior to this stage was defined as the labile pool. The size of this pool correlated with the size of the total pool of dissolved organics; on average it constituted about 10% (including limnic and marine samples), but it was consistently lower in blackwater streams. There was, however, a considerable temporal variation; as could be expected the labile pool was relatively higher during algal blooms and lowest during winter.

When only polymers are available, bacteria must have exoenzymatic activity (proteases, carbohydrases). Such enzyme activity can conveniently be quantified in water samples by adding suitable substrates which release a fluorescent dye when hydrolysed. Most enzymatic activity is associated with the particulate fraction (bacterial cells). Some activity can also be detected in cell-free filtrates; this is probably caused by lysed bacteria.[87] It is also possible that bacteria to a limited extent excrete hydrolytic enzymes into the water, but it is unlikely to be an adaptive trait because it would also benefit other bacteria present in the surroundings. The activity level of different bacterial ectoenzymes may show a clear seasonality: thus in lakes hydrolytic enzyme activity increases after algal blooms when monomers and the more labile compounds are no longer available. Increase in hydrolytic activity correlates with an increasing fraction of the bacteria that are attached to suspended detrital matter.[71,72]

An example of a carbon budget for the microbial loop of the North Sea is shown in Fig. 3.4. It represents an annual average and shows that a large fraction of the bacterial substrates requires enzymatic hydrolysis. The entire carbon flow through bacteria was estimated to constitute 57% of the primary production.[9]

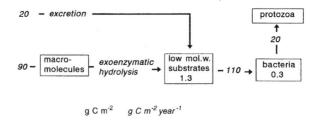

Figure 3.4   Microbial carbon budget for the water column of the North Sea; pool sizes and rates represent annual averages[9]

It stands to reason that polymers in general and very large molecules in particular are degraded more slowly than monomers and small molecules, and this is also basically true (although one study[44] reports that proteins are degraded as quickly as free amino acids). The reason for this is two-fold. The synthesis of hydrolytic enzymes requires energy and the hydrolysis of polymers may in itself be a rate-limiting step. Secondly, the diffusion coefficient of molecules is inversely related to molecular weight. Thus, according to the principles discussed in Section 1.3.1, the affinity for particular molecules must be inversely related to molecular size.

It has been observed, however, that larger colloidal particles may degrade relatively quickly.[3,71] It is probable that above a certain size (> 1–2 μm) particles may again become more accessible to bacteria. Transport of particles that measure < 1–2 μm is entirely dominated by diffusion (Brownian motion) and diffusivity decreases with increasing particle size. When particles become larger than a few micrometres, however, transport is also affected by turbulent shear and the probability of colliding with other particles increases. Colloids may thus form larger aggregates that are colonised by bacteria and this circumvents diffusional limitation.[52]

3.2.3   Suspended particles, their microbial biota and coupling between plankton and sediments

About 10% of the organic matter in the water column is in the form of non-living particles. The distinction between dissolved and particulate matter is, as already mentioned, not very sharp. Traditional methods probably also underestimate particulate matter since some aggregates are transparent and some types disintegrate very easily during sampling. Suspended particles are important because they create conditions for the mineralisation of organic matter which differ from those presented by dissolved organics. Particles are also important because they tend to sink at rates that vary according to specific density and size. The formation of particles is caused by biological and physical processes in the upper photic (and frequently turbulent) part of the water column. The organic contents are to a larger or smaller degree mineralised during their descent to the bottom, but what remains constitutes the basis for benthic food chains. The downward flux of particles therefore represents the coupling between the photic (productive) surface layers and sediments of the deep sea. The formation of aggregates from dissolved colloids plus suspended particles or from dissolved organics alone (with air bubbles constituting the necessary nucleation for aggregate formation) was first described by Riley,[91] who also stressed the significance of organic aggregates for biological processes in the sea.

The formation of particles, their downwards flux and mineralisation in the oceanic water column has recently drawn considerable interest. This is because organic particles that arrive in the deep sea essentially represent a (temporary) fossilation of carbon and the process is therefore a sink for atmospheric CO_2 (see Chapter 9). Beneath the turbulent (mixed) zone of the water column it is possible to quantify the vertical particle flux using sediment traps. The flux of organic particulate matter decreases strongly with depth in oceanic water columns, reflecting microbial mineralisation. Thus, in one case it has been found that 75–80% was lost during passage from 100 to 2000 m, and in another study only 9% of the organic material remained at 3100 m.[52,100] The amount of organic particulate matter that leaves the photic zone (the "new production") in turn represents only a very small fraction of the primary production in oceanic waters. In more productive coastal and estuarine waters and in lakes a larger part of the production is lost from the photic zone in the form of particulate matter through sedimentation: a value of about 20% has been estimated for the Kattegat (inner Danish waters).[79] In such shallow waters a large fraction of this material probably reaches the bottom to become mineralised in or on the sediment. Not all types of particles are equally likely to reach the sediment; copepod faecal pellets and large aggregations of diatom cells sink relatively quickly while fragile colloid aggregates probably tend to remain longer in suspension and to have higher rates of mineralisation.

There are several sources of particulate matter. Growth and multiplication of bacteria and phytoplankton imply the production of particulate matter from dissolved substances and the resulting cells, or their dead remains, may become a component of aggregates. Some organisms produce larger (aggregate) particles from smaller ones (faecal pellets and protozoan egesta). Individual particles are more or less sticky because many living organisms secrete mucus, or dead particles adsorb dissolved colloids or mucous flocks. When particles collide they will stick to each other with a certain probability and so aggregates tend to grow. Colloidal particles may also floculate and form mucous aggregates together with or without other types of particles.[53,91,102] Aggregates that consist only of flocculated mucous matter have recently been referred to as "transparent exopolymer particles".[2,81] When diatoms reach high densities during the end of the spring bloom they may form large aggregates and sink; these aggregates contain living diatoms and the adaptive significance of this phenomenon is still debated.[56] Suspended aggregates appear as white flocks in the water column and the phenomenon is referred to as "marine snow".[99] Most studies have found that the largest proportion of sedimenting particles consists of amorphous and microscopically unidentifiable material while copepod faecal pellets and diatom aggregates, for example, usually play a smaller quantitative role.[67]

The mechanism of aggregate formation is described by coagulation theory.[53,56] The probability that two particles will stick together depends on the probability of collision and on the probability of adhesion following a collision (the latter property is referred to as stickiness). Submicron particles depend exclusively on Brownian movement for collisions. Other mechanisms are more important and effective for larger particles. Turbulent shear is most important and episodic wind exposure of the water surface may therefore lead to rapid flocculation and subsequent sedimentation. Differential sinking velocity (which increases with the square of the diameter) is another mechanism by which particles may collide. Thus, large sinking particles tend to scavenge smaller ones on their way down through the water column.

Aggregate growth tends to accelerate with aggregate size. As the aggregates increase beyond a certain size, however, they also become more vulnerable to fragmentation by shear forces so some steady-state size will be reached depending on turbulence and stickiness.

Colloids tend to concentrate at air–water interfaces and this applies to the calm water surface as well as to the surface of air bubbles. Colloids adhering to bursting bubbles thus form a nucleus for further flocculation.[91] This is the basic principle of an established and efficient method of stripping seawater for dissolved organics (used in seawater aquaria). Sufficiently high densities of bacteria (which often excrete mucus) may also form flocs together with dissolved colloids.[8]

Marine snow particles do not consist only of inorganic or dead organic matter; they also host microbial communities including bacteria, various protozoa and algal cells. Bacteria undoubtedly grow and divide on suspended particles, but the initial colonisation is a coagulation process much as described above. When an object such as a microscopic slide is submerged in natural water, bacteria adhere to the glass surface within minutes. It is believed that this initial adherence is based on van der Waal's forces and the tendency to adhere varies among bacteria depending on the hydrophobicity of the cell envelope.[23,92] This type of adherence is reversible, but some bacteria may later cement themselves to surfaces with mucous secretions. A general treatment on bacterial adhesion is found in Marshall.[68]

Bacterial biota associated with suspended particles and aggregates seem to differ from the community of freely suspended bacteria. Attached bacteria are generally larger, but this might reflect more favourable conditions for growth. A difference in the composition of these two bacterial communities has also been demonstrated with molecular methods (rRNA sequences),[24] but this does not exclude the possibility that the adhering bacteria represent a subset of the community of freely suspended bacteria. Selective forces must include differential probability of adhesion in contact with surfaces and the ability to hydrolyse and exploit the organic matter in the aggregates.

Bacterial concentrations are 10^2- to 10^3-fold higher in aggregates (i.e. 10^8–10^9 cm^{-3}) as compared to the surrounding water. Most studies have found that only 10–15% of the bacteria in the water column are attached to particles at any time. This, however, does not allow for the conclusion that attached bacteria are unimportant. First of all, these estimates are mainly based on microscopic observations (fluorescence microscopy on filtered water samples). Filtration is likely to result in the disintegration of some types of aggregates and so their associated bacteria cannot be identified as having been associated with aggregates. It is also possible that attached bacteria have higher metabolic activity than suspended ones. Particles and aggregations are sites where not only particulate matter, but also adsorbed colloids, are mineralised. This is because large molecules in (dilute) solution can only be exploited inefficiently due to their low diffusivity, but this limitation does not apply when the dissolved colloids adsorb on flocculated material. A related phenomenon is the increased concentrations of macromoecules and bacteria in the microlayer of the sea surface.[22,23] Finally, the microbial biota on sedimenting particles are important because they control the amount of organic C and N that is transported to ocean sediments and the deeper parts of the water column.

The importance of flocculation for the bacterial exploitation of macromolecules can easily be demonstrated experimentally. When seawater samples are vigorously bubbled, a pronounced increase in microbial respiration follows. The effect is due to the adsorption of macromolucules to the surface of the bubbles and subsequent nucleation for floc formation. Bacteria are also adsorbed to the thus-formed aggregates which are subsequently easily degraded.[54,55]

There is direct evidence to show that bacteria attached to suspended particulate matter produce hydrolytic enzymes and that the bacteria grow at the expense of the organic particles on which they are attached.[46,94] Another aspect merits closer examination: in principle when bacteria are attached to sufficiently large particles they could benefit from advective water flow (due to sinking of the particles and to turbulence) by enhanced uptake of dissolved substances from the surrounding water;[65] see also Section 1.3.3. Bacterial metabolism, turnover and regeneration of nutrients may also be affected by the presence of eukaryotic micro-organisms (bacterivorous protozoa, phototrophs) associated with aggregates; this aspect has so far received limited attention, but protozoan grazing is known to enhance bacterial metabolic activity in terms of degradation of particulate detritus.[27]

Suspended aggregates may reach sizes exceeding several millimetres. This opens the possibility that the innermost part of such particles becomes anaerobic (see Section 1.3.2) and consequently that anaerobic metabolic processes may take place in the otherwise oxic water column. This is, of course, likely to be more important when the ambient p_{O_2} is low, such as in the oxygen minimum layers of the Pacific and Indian oceans. Some types of aggregates are very porous and this allows for internal advective water flow through them. But other types (e.g. faecal pellets) are quite compact and solute transport within these aggregates is exclusively diffusive. Some studies have observed anoxic or microaerobic conditions inside such particles and the presence of sulphate reduction and methanogenesis within the particles.[1,7,93] This possibly explains the fact that seawater is usually supersaturated with methane.

3.2.4   Bacteria and cycling of N and P

Heterotrophic bacteria are by definition mineralisers of organic material; in this chaper we have, so far, only considered this in terms of carbon. Bacteria, of course, also mineralise organic N and P. However, mineralisation of N and P may be decoupled from C mineralisation to the extent that the bacteria become net consumers of the inorganic forms of these elements in order grow on N- or P-depleted substrates. Such a decoupling occurs if the C/N or C/P ratios of the phototroph production (i.e. bacterial substrates) exceed the C/N or C/P ratios of bacterial cells to a certain degree (depending on bacterial growth yield; see Section 2.1.1). This effect is pronounced in terrestrial systems: the dominant primary producers (vascular plants) have very high C/N and C/P ratios. Bacteria that decompose such tissue must have a net assimilation of inorganic N and P in order to grow. These nutrients are then eventually mineralised by bacterial grazers.

Bacteria contain somewhat higher concentrations of N and P than do unicellular algae, but much higher concentrations than vascular plants. The decoupling between C and N and P cycling is therefore not likely to be so pronounced for heterotrophic planktonic bacteria. It is, however, easy to reproduce such a situation by adding large amounts of carbohydrates to a water sample. Available inorganic N and P will rapidly be removed from the water, and bacterial biomass will then no longer increase. If bacterial grazers are present, remineralisation will reach a steady state. If bacterial grazers are absent, bacterial metabolism and the rate of carbohydrate mineralisation will eventually slow. The addition of mineral N and P will accelerate biological activity again. The addition of bacterial grazers to the purely bacterial system will have a similar effect. In the 1960s such experiments[48] drew attention to the problem as did the somewhat earlier observation that bacteria are responsible for a large share of inorganic phosphate uptake in lake water.[90] In particular, it was discussed whether algae and bacteria compete for mineral nutrients, but this has not always been stated in a very succinct or meaningful way. While the smaller bacteria will be more efficient than the larger algal cells in terms of assimilating phosphate at very low concentrations, it is at the same time clear that the bacterial activity is limited by algal production. So, unless the system is loaded with large amounts of allochtonous, nutrient-deplete organic matter, it is not a simple case of resource competition.

The "net uptake versus net mineralisation of N and P problem" for water-column bacteria has recently been reviewed.[57] Extensive studies show that bacteria typically are responsible for more than 50% of the entire uptake of inorganic phosphate in lakes and in sea water. In the sea, bacteria may be responsible for between 10 and 75% of the ammonia uptake in the water column. These figures do not *per se* give any information on the simultaneous release of inorganic P and N. Lower ammonium uptake occurs in estuaries and coastal water than in oligotrophic oceanic waters. This accords with the observation that amino acids constitute a substantial fraction of the bacterial substrate in the former localities, whereas glucose is relatively more important in the latter type of habitats.

Another way of approaching this problem is to quantify the release of inorganic N (NH_4^+) for differently sized fractions of plankton. These studies have not always discriminated between bacterial and microprotozoal fractions. In one study[40] it was found that 39% of the NH_4^+ production in the water column was due to bacteria ($< 1\,\mu m$ fraction) and 50% due to organisms within the size range $1–35\,\mu m$; another study[35] found that most N remineralisation was due to the $< 10\,\mu m$ fraction. While the question is not entirely settled, it seems likely that some decoupling of N and C mineralisation takes place, but that bacteria, to a large extent, depend on organic sources of N and are responsible for some net mineralisation of N in the water column. In oligotrophic waters, however, it is likely that most

N and P mineralisation is due to bacterial grazers while the bacteria are net consumers of inorganic N (see also Caron and Goldman[16]).

Another type of decoupling between N and C has been observed for sinking particulate matter.[79] It was shown that the C/N ratio of such particles increases with age. The implication of this is that components with a higher N content are mineralised more rapidly and that this happens before the particles have left the photic zone. Consequently, there is no simple stoichiometry between external supply of N and new production (of reduced C) which is exported from the photic zone since N is partly recycled (for photosynthesis) before it is finally exported to below the photic zone. The actual release of NH_4^+ from the particles is probably due to the concerted activity of bacteria and their protozoan grazers.

References

1. Alldredge AL, Cohen Y (1987) Can microscale chemical patches persist in the sea? Microelectrode study of marine snow, fecal pellets. *Science* **235**: 689–691.
2. Alldredge AL, Passow U, Logan BE (1993) The abundance and significance of a class of large, transparent organic particles in the ocean. *Deep Sea Res* **40**: 1131–1140.
3. Amon RMW, Benner R (1994) Rapid cycling of high-molecular-weight dissolved organic matter in the ocean. *Nature* **369**: 549–552.
4. Andersen P, Fenchel T (1985) Bacterivory by microheterotrophic flagellates in seawater samples. *Limnol Oceanogr* **30**: 198–202.
5. Andersson A, Lee C, Azam F, Hagström Å (1985) Release of amino acids and inorganic nutrients by heterotrophic marine microflagellates. *Mar Ecol Prog Ser* **23**: 99–106.
6. Azam F, Fenchel T, Field JG, Gray JS, Meyer-Reil L, Thingstad F (1983) The ecological role of water column microbes in the sea. *Mar Ecol Prog Ser* **10**: 257–263.
7. Bianchi M, Marty D, Teyssié J-L, Fowler SW (1992) Strictly aerobic and anaerobic bacteria associated with sinking particulate matter and zooplankton fecal pellets. *Mar Ecol Prog Ser* **88**: 55–60.
8. Biddanda BA (1985) Microbial synthesis of macroparticulate matter. *Mar Ecol Prog Ser* **20**: 241–251.
9. Billen G, Joiris C, Meyer-Reil L, Lindeboom H (1990) Role of bacteria in the North Sea ecosystem. *Neth J Sea Res* **26**: 265–293.
10. Bjørnsen PK (1988) Phytoplankton exudation of organic matter: *why* do healthy cells do it? *Limnol Oceanogr* **33**: 151–154.
11. Blackburn N, Zweifel UL, Hagström Å (1996) Cycling of marine dissolved organic matter. II. A model analysis. *Aquat Microb Ecol* **11**: 79–90.
12. Bratbak G, Heldal M, Thingstad F, Rieman B, Haslund OH (1992) Incorporation of viruses into the budget of microbial C-transfer. A first approach. *Mar Ecol Prog Ser* **83**: 273–280.

13. Bratbak G Thingstad F, Heldal M (1994) Viruses and the microbial loop. *Microb Ecol* **28**: 209–221.

14. Campbell L, Nolla HA (1994) The importance of *Prochlorococcus* to community structure in the central North Pacific Ocean. *Limnol Oceanogr* **39**: 954–961.

15. Carlson DJ, Brann ML, Maque TH, Mayer LM (1985) Molecular weight distribution of dissolved organic material in seawater determined by ultrafiltration: a reexamination. *Mar Chem* **16**: 155–171.

16. Caron AC, Goldman JC (1990) Protozoan nutrient regeneration. In: Capriulo GM (ed.) *Ecology of Marine Protozoa*. Oxford University Press, New York, pp. 283–306.

17. Carpenter EJ (1982) Physiology and ecology of marine planktonic Oscillatoria (*Trichodesmium*). *Mar Biol Lett* **4**: 69–85.

18. Carpenter EJ, Romans K (1991) Major role of the cyanobacterium *Trichodesmium* in nutrient cycling in the North Atlantic Ocean. *Science* **254**: 1356–1358.

19. Chin-Leo G, Kirchman DL (1988) Estimating bacterial production in marine waters from simultaneous incorporation of thymidine and leucine. *Appl Environ Microbiol* **54**: 1934–1939.

20. Chisholm SW, Frankel S, Goerike R, Olson R, Palenik B, Urbach E, Waterbury J, Zettler E (1992) *Prochlorococcus marinus* nov gen nov sp: an oxytrophic marine prokaryote containing divinyl chlorophyll *a* and *b*. *Arch Microbiol* **157**: 297–300.

21. Coffin RB, Conolly JP, Harris PS (1993) Availability of dissolved organic carbon to bacterioplankton examined by oxygen utilization. *Mar Ecol Prog Ser* **101**: 9–22.

22. Dahlbäck B, Gunnarsson LÅH, Hermansson M, Kjelleberg S (1982) Microbial investigations of surface microlayers, water column, ice and sediment in the Arctic Ocean. *Mar Ecol Prog Ser* **9**: 101–109.

23. Dahlbäck B, Hermansson M, Kjelleberg S, Norkrans B (1981) The hydrophobicity of bacteria — an important factor in their initial adhesion at the air–water interface. *Arch Microbiol* **128**: 267–270.

24. DeLong EF, Franks DG, Alldredge AL (1993) Phylogenetic diversity of aggregate-attached vs free-living marine bacterial assemblages. *Limnol Oceanogr* **38**: 924–934.

25. Fagerbakke KM, Heldal M, Norland S (1996) Content of carbon, nitrogen, oxygen, sulfur and phosphorus in native aquatic and cultured bacteria. *Aquat Microb Ecol* **10**: 15–27.

26. Fenchel T (1982) Ecology of heterotrophic microflagellates. IV. Quantitative occurrence and importance as bacterial consumers. *Mar Ecol Prog Ser* **9**: 35–42.

27. Fenchel T, Harrison P (1976) The significance of bacterial grazing and mineral cycling for the decomposition of particulate detritus. In: Anderson JM, MacFadyen A (eds) *The Role of Terrestrial and Aquatic Organisms in Decomposition Processes*. Blackwell Scientific Publishers, Oxford, pp. 285–299.

28. Ferguson RL, Rublee P (1976) Contribution of bacteria to standing crop of coastal plankton. *Limnol Oceanogr* **21**: 141–145.

29. Fogg GE (1966) The extracellular products of algae. *Oceanogr Mar Biol Ann Rev* **4**: 195–212.

30. Fuhrman JA, Azam F (1982) Thymidine incorporation as a measure of heterotrophic bacterioplankton production in marine surface waters: evaluation and field results. *Mar Biol* **66**: 109–120.

31. Fuhrman JA, Ferguson RL (1986) Nanomolar concentrations and rapid turnover of dissolved free amino acids in seawater: agreement between chemical and microbiological measurements. *Mar Ecol Prog Ser* **33**: 237–242.

32. Fuhrman JA, Lee SH, Masuchi Y, Davis AA, Wilcox RM (1994) Characterization of marine prokaryotic communities via DNA and RNA. *Microb Ecol* **28**: 133–145.

33. Fuhrman JA, McCallum K, Davis AA (1992) Novel major archaebacterial group from marine plankton. *Nature* **356**: 148–149.

34. Giovannoni SJ, Britschgi TB, Moyer CL, Field KG (1990) Genetic diversity in Sargasso Sea bacterioplankton. *Nature* **345**: 60–63.

35. Glibert PM (1982) Regional studies of daily, seasonal and size fraction variability in ammonium remineralization. *Mar Biol* **70**: 209–222.

36. Goericke R, Repeta DJ (1992) The pigments of *Prochlorococcus marinus*: the presence of divinyl chlorophyll *a* and *b* in a marine prochlorophyte. *Limnol Oceanogr* **37**: 425–433.

37. Goericke R, Welschmeyer NA (1993) The marine prochlorophyte *Prochlorococcus* contributes significantly to phytoplankton biomass and primary productivity in the Sargasso Sea. *Deep Sea Res* **40**: 2283–2294.

38. Granéli W, Lindell M, Tranvik L (1996) Photo-oxidative production of dissolved inorganic carbon in lakes of different humic content. *Limnol Oceanogr* **41**: 698–706.

39. Hagström Å, Larsson U, Hörstedt P, Normark S (1979) Frequency of dividing cells, a new approach to the determination of bacterial growth rates in aquatic environments. *Appl Environ Microbiol* **37**: 805–812.

40. Harrison WG (1978) Experimental measurements of nitrogen mineralization in coastal waters. *Limnol Oceanogr* **23**: 684–694.

41. Hobbie JE (1967) Glucose and acetate in freshwater: concentrations and turnover rates. In: Golterman HL, Clymo RS (eds) *Chemical Environment in Aquatic Habitats*. North Holland Publishing Company, Amsterdam, pp. 245–251.

42. Hobbie JE, Crawford CC (1969) Respiration corrections for bacterial uptake of dissolved organic compounds in natural water. *Limnol Oceanogr* **14**: 528–532.

43. Hobbie JE, Daley RJ, Jasper S (1977) Use of nuclepore filters for counting bacteria by fluorescence microscopy. *Appl Environ Microbiol* **33**: 1225–1228.

44. Hollibaugh JT, Azam F (1983) Microbial degradation of dissolved proteins in seawater. *Limnol Oceanogr* **28**: 1104–1116.

45. Hoppe H-G (1976) Determination of actively metabolizing heterotrophic bacteria in the sea, investigated by means of microautography. *Mar Biol* **36**: 291–302.

46. Hoppe H-G, Ducklow H, Karrash B (1993) Evidence for dependency of bacterial growth on enzymatic hydrolysis of particulate organic matter in the mesopelagic ocean. *Mar Ecol Prog Ser* **93**: 277–283.

47. Ittunga R, Mitchell BG (1986) Chroococcoid cyanobacteria: a significant component in the food web dynamics of the open ocean. *Mar Ecol Prog Ser* **28**: 291–297.

48. Johannes RE (1964) Uptake and release of dissolved organic phosphorus by representatives of a coastal marine ecosystem. *Limnol Oceanogr* **9**: 224–234.

49. Jørgensen NOG, Kroer N, Coffin RB, Yang X-H, Lee C (1993) Dissolved free amino acids, combined amino acids, and DNA as sources of carbon and nitrogen to marine bacteria. *Mar Ecol Prog Ser* **98**: 135–148.

50. Jørgensen NOG, Jensen RE (1994) Microbial fluxes of free monosaccharides and total carbohydrates in freshwater determined by PAD–HPLC. *FEMS Microbiol Ecol* **14** : 79–94.

51. Keil RG, Kirchman DL (1994) Abiotic transformation of labile protein to refractory protein in seawater. *Mar Chem* **45**: 187–196.

52. Karl DM, Knauer GA, Martin JH (1988) Downward flux of particulate organic matter in the ocean: a particle decomposition paradox. *Nature* **332**: 438–441.

53. Kepkay PE (1994) Particle aggregation and the biological reactivity of colloids. *Mar Ecol Prog Ser* **109**: 293–304.

54. Kepkay PE, Johnson BD (1988) Microbial response to organic particle generation by surface coagulation in seawater. *Mar Ecol Prog Ser* **48**: 193–198.

55. Kepkay PE, Johnson BD (1989) Coagulation on bubbles allows the microbial respiration of oceanic carbon. *Nature* **338**: 63–65.

56. Kiørboe T, Hansen JLS (1993) Phytoplankton aggregate formation: observations of patterns and mechanisms of cell sticking and the significance of exopolymeric material. *J Plankton Res* **15**: 993–1018.

57. Kirchman DL (1994) The uptake of inorganic nutrients by heterotrophic bacteria. *Microb Ecol* **28**: 255–271.

58. Kjelleberg S, Hermansson M, Mårdén P, Jones GW (1987) The transient phase between growth and nongrowth of heterotrophic bacteria, with emphasis on the marine environment. *Ann Rev Microbiol* **41**: 25–49.

59. Kroer N (1993) Bacterial growth efficiency on natural dissolved organic matter. *Limnol Oceanogr* **38**: 1282–1290.

60. Krogh A (1934) Conditions of existence of aquatic animals. *Ecol Monogr* **4**: 420–429.

61. Kuylenstierna M, Karson B (1994) Seasonality and composition of pico- and nanoplanktonic cyanobacteria and protists in the Skagerrak. *Bot Mar* **37**: 17–33.

62. Li WKW (1994) Primary production of prochlorophytes, cyanobacteria, and eukaryotic ultraphytoplankton: measurements from flow cytometric sorting. *Limnol Oceanogr* **39**: 169–175.

63. Lindell M, Granéli W, Tranvik L (1995) Enhanced bacterial growth in response to photochemical transformation of dissolved organic matter. *Limnol Oceanogr* **40**: 195-199.

64. Linley EAS, Nedwell RC, Bosma SA (1981) Heterotrophic utilisation of mucilage released during fragmentation of kelp (*Ecklonia maxima* and *Laminaria pallida*). I. Development of microbial communities associated with the degradation of kelp mucilage. *Mar Ecol Prog Ser* **4**: 31-41.

65. Logan BE, Kirchman DL (1991) Uptake of dissolved organics by marine bacteria as a function of fluid motion. *Mar Biol* **111**: 175-181.

66. Lucas MI, Newell RC, Velimirov B (1981) Heterotrophic utilisation of mucilage released during fragmentation of kelp (*Ecklonia maxima* and *Laminaria pallida*). II. Differential utilisation of dissolved organic components from kelp mucilage. *Mar Ecol Prog Ser* **4**: 43-55.

67. Lundsgaard C, Olesen M (1997) The origin of sedimenting detrital matter in a coastal system. *Limnol Oceanogr* **42** (in press).

68. Marshall KC (ed.) (1986) *Microbial Adhesion and Aggregation*. Springer, Berlin.

69. McManus GB, Dawson R (1994) Phytoplankton pigments in the deep chlorophyll maximum of the Carribbean Sea and the western tropical Atlantic Ocean. *Mar Ecol Prog Ser* **113**: 199-206.

70. Middelboe M, Søndergaard M (1993) Bacterioplankton growth yield: seasonal variations and coupling to substrate lability and ß-glucosidase activity. *Appl Environ Microbiol* **59**: 3916-3921.

71. Middelboe M, Søndergaard M (1995) Concentration and bacterial utilization of submicron particles and dissolved organic carbon in lakes and a coastal area. *Arch Hydrobiol* **133**: 129-147.

72. Middelboe M, Søndergaard M, Letark Y, Borch NH (1995) Attached and free-living bacteria: production and polymer hydrolysis during a diatom bloom. *Microb Ecol* **29**: 231-248.

73. Mopper K, Zhou XL, Kieber RJ, Sikorski RJ, Jones RD (1991) Photochemical degradation of dissolved organic carbon and its impact on the oceanic carbon cycle. *Nature* **353**: 60-62.

74. Moran MA, Hodson RE (1994) Support of bacterioplankton production by dissolved humic substances from three marine environments. *Mar Ecol Prog Ser* **110**: 241-247.

75. Morel A, Ahn Y-H, Partinsky P, Vaulot D, Claustre H (1993) *Prochlorococcus* and *Synechococcus*: a comparative study of their optical properties in relation to their size and pigmentation. *J Mar Res* **51**: 617-649.

76. Morita RY (1985) Starvation and miniaturisation of heterotrophs with special emphasis on maintenance of the starved viable state. In: Fletcher M, Floodgate GD (eds) *Bacteria in their Natural Environment*. Academic Press, New York, pp. 111-131.

77. Never S (1992) Growth dynamics of marine *Synechoccus* spp in the Gulf of Alaska. *Mar Ecol Prog Ser* **83**: 251-262.

78. Ogura N (1975) Further studies on decomposition of dissolved organic matter in coastal seawater. *Mar Biol* **31**: 101–111.

79. Olesen M, Lundsgaard C (1995) Seasonal sedimentation of autochthonous material from the euphotic zone of a coastal system. *Estuarine Coast Shelf Sci* **41**: 475–490.

80. Pakulski JD, Benner R (1994) Abundance and distribution of carbohydrates in the ocean. *Limnol Oceanogr* **39**: 930–940.

81. Passow U, Alldredge AL, Logan BE (1994) The role of particulate carbohydrates in the flocculation of diatom blooms. *Deep Sea Res* **41**: 335–357.

82. Pelt RJ (1989) Kinetics of microbial mineralization of organic carbon from detrital *Skeletonema costatum* cells. *Mar Ecol Prog Ser* **52**: 123–128.

83. Pinhassi J, Zweifel UL, Hagström Å (1997) Dominant bacterioplankton found among colony forming species using a new species density protocol. *Appl Environ Microbiol* **63**: 3359–3366.

84. Platt T, Li WKW (eds) (1986) Photosynthetic picoplankton. *Can Bull Fish Aquat* **214**: 1–583.

85. Pointdexter JS (1987) Bacterial responses to nutrient limitation. In: Fletcher M, Gray TRG, Jones JG (eds) *Ecology of Microbial Communities*. Cambridge University Press, Cambridge, pp. 283–317.

86. Pomeroy LR, Johannes RE (1968) Occurrence and respiration of ultraplankton in the upper 500 meters of the ocean. *Deep Sea Res* **15**: 381–391.

87. Rego JV, Billen G, Fontigny A, Somville M (1985) Free and attached proteolytic activity in water environments. *Mar Ecol Prog Ser* **21**: 245–249.

88. Rehnstam A-S, Bäckman S, Smith DC, Azam F, Hagström Å (1993) Blooms of sequence-specific culturable bacteria in the sea. *FEMS Microbiol Ecol* **102**: 161–166.

89. Rich JH, Ducklow HW, Kirchman DL (1996) Concentration and uptake of neutral monosaccharides along 140°W in the equatorial Pacific: contribution of glucose to heterotrophic bacterial activity and the DOM flux. *Limnol Oceanogr* **41**: 595–604.

90. Rigler FH (1956) A tracer study of the phosphorus cycle in lake water. *Ecology* **37**: 550–562.

91. Riley GA (1963) Organic aggregates in seawater and the dynamics of their formation and utilization. *Limnol Oceanogr* **8**: 372–381.

92. Rosenberg M, Kjelleberg S (1986) Hydrophobic interactions: role in bacterial adhesion. *Adv Microb Ecol* **9**: 353–393.

93. Shanks AL, Reeder ML (1993) Reducing microzones and sulfide production in marine snow. *Mar Ecol Prog Ser* **96**: 43–47.

94. Simon M, Alldredge AL, Azam F (1990) Bacterial carbon dynamics on marine snow. *Mar Ecol Prog Ser* **65**: 205–211.

95. Søndergaard M, Hansen B, Markager S (1995) Dynamics of dissolved organic carbon lability in a eutrophic lake. *Limnol Oceanogr* **40**: 46–54.

96. Søndergaard M, Middelboe M (1995) A cross-system analysis of labile dissolved organic carbon. *Mar Ecol Prog Ser* **118**: 283–294.

97. Sorokin Yu I (1977) The heterotrophic phase of plankton succession in the Japan Sea. *Mar Biol* **41**: 107–117.

98. Stockner JG, Antia NJ (1986) Algal picoplankton from marine and freshwater ecosystems: a multidisciplinary perspective. *Can J Fish Aquat Sci* **43**: 2472–2503.

99. Suzuki N, Kato S (1953) Studies of suspended materials. Marine snow in the sea. I. Sources of marine snow. *Bull Fac Fish Hokkaido Univ* **4**: 132–135.

100. Turley CM, Mackie PJ (1994) Biogeochemical significance of attached and free-living bacteria and the flux of particles in the NE Atlantic Ocean. *Mar Ecol Prog Ser* **115**: 191–203.

101. Veldhuis MJW, Kaay GW (1994) Cell abundance and fluorescence of picoplankton in relation to growth, irradiance and nitrogen availability in the Red Sea. *Neth J Sea Res* **31**: 135–145.

102. Wells ML, Goldberg ED (1993) Colloid aggregation in seawater. *Mar Chem* **41**: 353–358.

103. Williams PJ leB (1981) Incorporation of microheterotrophic processes into the classical paradigm of the planktonic food web. *Kieler Meeresforsch, Sonderh* **5**: 1–28.

104. Williams PJ leB (1990) The importance of losses during microbial growth: commentary on the physiology, measurement and ecology of the release of dissolved organic material. *Mar Microb Food Webs* **4**: 175–206.

105. Zweifel UL, Norman B, Hagström Å (1993) Consumption of dissolved organic carbon by marine bacteria and demand for inorganic nutrients. *Mar Ecol Prog Ser* **101**: 23–32.

4

Biogeochemical cycling in soils

Although soils cover $< 25\%$ of Earth's surface, they cycle elements intensively and have a major global impact. When expressed on an areal basis, rates for many terrestrial biogeochemical processes substantially exceed rates for the same processes in aquatic systems. For example, areal rates of microbial respiration in soils exceed respiration rates in most marine sediments and the water column.[50,57] This is obviously a consequence of high soil organic carbon contents and higher rates of terrestrial than marine primary production per unit area. Of course, it is equally true that certain biogeochemical transformations, for example dissimilatory sulphate reduction, occur much more rapidly in aquatic systems, usually as a result of some unique characteristics (e.g. the prevalence of anoxia and sulphate in marine sediments).

We explore here some basic features of the soil environment that contribute to differences in rates and modes of biogeochemical cycling between terrestrial and aquatic systems. We propose that the basic structure of biogeochemical pathways and the functional types of organisms that operate them do not differ substantially between terrestrial and aquatic systems. Rather, the differences are often in the relative importance of particular pathways and, in some cases, the responsible phylogenetic groups. For instance, the terrestrial sulphur cycle is dominated by vascular plant sulphate assimilation and microbial organosulphur mineralisation;[71] in contrast, the marine sulphur cycle is dominated by dissimilatory sulphate reduction (see Chapters 1 and 5). However, all of the basic sulphur transformations that occur in one system can be found to a greater or lesser extent in others. Thus, it is not our intent simply to reiterate with examples from soils pathways that are described elsewhere (e.g. Chapters 1–3). We also stress that our purpose is not to conduct an exhaustive analysis of soil microbiology, ecology or biogeochemistry, for which other references are more appropriate.[14,57,71] Likewise for a detailed description of soil genesis, composition and physical–chemical characteristics other sources should be consulted.[24,66]

Instead, we structure our analysis around two basic premises. First, we view soils as an end-member in a continuum of particle and water distribution that may be defined on the basis of volumetric water content. For soils, water contents are often $< 50\%$ (cm^3 water cm^{-3} soil) and time variable with minima of perhaps 10% or so; soils are seldom water saturated and contain a gas phase with a composition more or less similar to the atmosphere. Sediments (excluding those composed primarily of sand) typically contain > 50% to 90 + % water, exhibit little temporal variability and generally remain water saturated; the composition of dissolved gases may differ dramatically from equilibrium with air, and the occasional quasi-stable sediment gas phase (i.e. bubbles) often consists of methane–CO_2–nitrogen mixes. Obviously, the water column represents a second end-member with water content > 99% by volume and temporal variability limited to events such as sediment resuspension. Dissolved gases in the water column may differ significantly from equilibrium with air, and there is no stable gas phase *per se*.

We stress water content as a defining feature of soils for a number of reasons. At regional to global scales, it is evident that water regimes, measured as evapotranspiration, are correlated with litter decomposition rates. Primary production, which provides the substrate for decomposition, is correlated with precipitation.[38,39,57] Obviously, water *per se* is not a factor in aquatic decomposition or primary production. In addition, the distribution and activity of bacteria depend on the availability of liquid water. Ignoring for a moment the physical–chemical concept of water availability (discussed subsequently), bacterial activity requires immersion by at least a thin film of water.[9] While this requirement is clearly fulfilled in the water column and in sediments, it is at times a constraint in soils. More generally, the nature of water distribution among soil particles determines a number of interdependent parameters (e.g. Fig. 4.1) or processes that regulate microbial activity: (i) the ability of bacteria to migrate from one microhabitat to another; (ii) transport mechanisms for solutes (usually diffusion, occasionally mass flow during precipitation or soil flooding); and (iii) transport mechanisms for gases (often diffusion but occasionally advection).

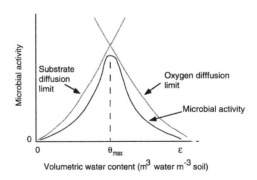

Figure 4.1   Conceptualisation of the relationships between microbial activity, oxygen and substrate diffusion, and soil water content[62]

Soil water content also acts as a strong selective force that determines the composition and activity of the soil biota (including below-ground plant tissues). For example, fungi appear much more important as heterotrophs in soils than in sediments,[28,71] largely due to the fact that fungal growth and activity are not strictly confined to an aqueous medium. As a result, soils represent a niche in which fungi have an advantage over bacteria. In contrast, sediments and the water column may represent niches where bacterial dominance is favoured by more rapid intrinsic growth rates.

In addition, temporal variability in water content affects both the composition and activity of the soil biota as well as soil nutrient status.[14] Nutrient leaching is well known in laterite soils of tropical rain forests and in the soils of other regions with heavy precipitation.[7] The potential for nutrient leaching associated with certain precipitation regimes determines in part plant community composition, and more generally patterns of biogeochemical cycling.

Our second basic premise is that the structure of microbial biogeochemical cycles in soils is determined to a large extent by the evolutionary responses of primary producers to the terrestrial environment. Obviously, microbial processes in virtually all ecosystems depend on interactions with primary production, and in this regard soils are no different from algal mats, the water column or sediments. However, certain details of primary production differ substantially among systems, and these differences contribute to the structure of microbial biogeochemical cycles.

For example, terrestrial plant C:N ratios are considerably higher than algal ratios (> 20 and < 10, respectively;[53,71] and see Chapter 2). Consequently, nitrogen limits microbial biosynthesis to a greater extent during decomposition of terrestrial plants. High plant tissue C:N ratios may also favour fungal heterotrophs since higher C:N ratios in fungi than in bacteria (> 6 versus 4) facilitate greater fungal biomass production per unit of available nitrogen. Though some bacteria use atmospheric nitrogen if fixed nitrogen is otherwise unavailable;[71] (see Chapter 1), this trait may not greatly affect bacterial–fungal competition since nitrogen fixation is energetically expensive and not ubiquitous. Nonetheless, plant C:N ratios may affect microbial community composition by selecting for taxa with nitrogenase genes.

Other important responses of plants to the terrestrial environment include the production and dynamics of roots (which affect the distribution of organic matter and mediate competition for nutrients); transpiration (which affects soil water regimes); production of large amounts of very high molecular weight structural polysaccharides (e.g. cellulose and hemicelluloses which require complex hydrolytic enzymes for degradation, and which usually require some form of physical disaggregation for bacterial utilisation); production of lignins, tannins and other aromatics, and waxes, all of which are more or less resistant to degradation. Each of these responses optimises the "success" of terrestrial plants, but impacts the nature and dynamics of microbial biogeochemistry.

4.1 Soil water as a master variable

4.1.1 Physical and chemical principles

Microbiologists and others often refer to soil water status using various weight- or volume-based measures.[71] These measures can provide useful insights, especially since they are often directly related to parameters such as gas diffusion and advection.[6,11] Certainly, the temporal patterns in aerobic and anaerobic biogeochemical transformations, and the distribution of the soil fauna, are related at least in part to the volumetric water content of soils. Nonetheless, weight- and volume-based measurements shed little light on the physiological responses of micro-organisms to soil water status, and provide an incomplete basis for comparisons among systems.

Soil scientists, agronomists and plant physiological ecologists have long recognised that soil water status can be more completely specified using physical–chemical terms.[9,28,44] For example, the availability of soil water to plants is predicted not by water content but by "soil suction", which is related to the energetic status of soil water and measured with lysimeters. Food microbiologists and some microbial physiologists and ecologists have likewise recognised the utility of physical–chemical descriptions,[9] often referring to water "activity" when measuring microbial activity in aqueous solutions. Water activity is another energy-related parameter.

There is a simple rationale for using a physical–chemical description of water in lieu of or in addition to volumetric measures (Brown[9] and Griffin[28] are highly recommended for lucid discussions). The movement of water across cell membranes is paramount in importance for all organisms, microbial or otherwise. The *direction* of water movement cannot be predicted on the basis of weight or volumetric measures of water content on either side of a membrane (e.g. internal to a soil microbe, external in the soil matrix). However, vectorial water fluxes can be predicted using measures of the relative "abundance" of water in solutions on both sides of a membrane, where abundance is defined more precisely below.

One measure of relative abundance is derived from the mole fraction of water in a solution:

$$N_w = n_w/(n_w + n_i)$$

where n_w is the number of moles of water per kilogram of solvent (about 55.51 $mol\,kg^{-1}$) and n_i is the moles of solute per kilogram of solvent. (Note: this definition requires that solution concentrations are expressed on a *molal*, not molar, basis; a 1 molal aqueous solution is thus 1 mol of an ideal solute per kilogram of water.) Since typical solutions are non-ideal in thermodynamic terms, a correction is applied giving rise to a definition for water activity, a term commonly encountered in the microbiological literature:

$$a_w = \gamma N_w$$

where γ is an activity coefficient specific for a given solute. Water activity has proven useful in defining the ability of various microbes to tolerate or grow under a variety of circumstances. However, a_w does not necessarily predict directions of water flow, and is an inadequate measure for a complex system such as soil.

A second measure of abundance is embodied in the concept of water potential. Abundance in this case is not simply a measure of water and solute concentrations *per se*. Water potential is defined in energetic terms as the partial molal free energy of a solution of water under specified conditions of solute composition, temperature, pressure and gravitational potential:

$$\mu_w = (\partial G / \partial n_w)_{n_i, T, P, h}$$

where G is Gibbs free energy, n_i is solute concentration, P is pressure, and h is height (usually ignored for applications in microbiology). From this a working expression for the chemical potential of water can be derived:

$$\mu_w = \mu_w{}^0 + RT \ln(a_w) + V_w P$$

where $\mu_w{}^0$ is the chemical potential of water in a standard reference state; R, T (in Kelvin) and P as usual, and V_w the partial molal volume of water (about $1.8 \times 10^{-5} \, m^3 \, mol^{-1}$ at $25\,°C$). Rearranging yields:

$$(\mu_w - \mu_w{}^0)/V_w = RT \ln(a_w)/V_w + P$$

where the left-hand expression is designated water potential and symbolised as

$$\psi = RT \ln(a_w)/V_w + P$$

which indicates that water potential in a solution can be subdivided into a pressure term (taken as a departure from 1 atm) and a solute-dependent term. As applied to soils, total water potential is typically distributed among three terms:

$$\Psi = \psi_s + \psi_p + \psi_m$$

where ψ_s, ψ_p, ψ_m are the potentials due to solutes, pressure and the soil matrix, respectively. Total water potential is < 0 and expressed in units of bars or pascals ($N\,m^{-2}$). Unlike water activity or any other measure of water status, water potential provides a complete description that can be compared among diverse aqueous systems and used to predict the direction of water flows. With respect to the latter, water always moves from a high to a low potential (more positive to more negative).

Since the matric potential term, ψ_m, is especially relevant in soils, it deserves at least limited comment here. The potential that arises as a result of the interaction of water at surfaces in a porous matrix has been described by analogy to the behaviour of water in a capillary tube immersed in pure water. The force associated with the rise of water a distance h in a capillary is related to the matric potential within the capillary $(= h \rho g$, where ρ is water density [kg m^{-3}] and g is the gravitational constant [m s^{-2}]); the height of capillary rise is inversely proportional to the capillary radius, r. To a first approximation, soil can be considered as a porous matrix in which the matric potential is related to pore size (i.e. pore radius) and the distribution of water among pores (a function of water content). When all pores are filled (i.e. at saturation), the matric potential is zero. Upon desaturation, the matric potential decreases due to the loss of water from the larger pores and retention in pores of smaller radii. Continued water loss confines the remaining water to ever smaller pores at progressively lower potentials.

Because matric potential depends on pore size and pore-size variation depends on soil composition (e.g. sand, silt and clay), water potentials vary significantly among soils for a given water content (Fig. 4.2). Water content alone is therefore an incomplete comparative index at best, and for some purposes only marginal in value. Water potential also varies for a given soil as a function of its wetting history. As indicated above, matric potentials are determined during drying by progressive water loss from large to small pores; however, when a soil is wetted, the larger pores fill first with subsequent equilibration of the smaller pores. This results in a hysteresis: at a given water content, water potential is lower during a period of decreasing rather than increasing moisture. A further complication arises from the shrinking and swelling of clay colloids that occurs at low water contents and that varies in significance as a function of clay type.

The relationship between water potential and soil pore sizes has a number of important consequences. With decreasing water content and matric potential, solute diffusion becomes increasingly constrained by the loss of continuity among pores, which clearly limits substrate availability. However, in contrast to solutes, gas transport increases with decreasing water potential;[61,62] fluctuations in air-filled (or water-filled) pore spaces have profound effects on a number of anaerobic processes in soil (e.g. denitrification) as well as soil–atmosphere gas exchange (see below). The movement of soil microbes (excluding hyphal extension by fungi) and indeed the general habitability of soils also depend on the diameters of water-filled pores.[56] Water potentials lower than about -0.5 MPa typically inhibit many bacterial activities due to physical constraints on substrate transport, cell movement and the thickness of films available for bacterial immersion. The soil fauna is likewise sensitive to water potential and the distribution of water-filled pore spaces;[14] some taxa (e.g. bactivorous nematodes) are much

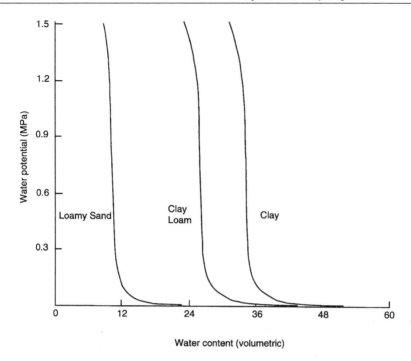

Figure 4.2 Water potential versus water content (cm³ water cm⁻³ soil) for three different soil types[57]

less tolerant of low potentials than others (e.g. the microarthropods). For these reasons, some workers[11,19,38] have recommended the use of water-filled pore space as a parameter for expressing the impact of soil water content on processes such as gas transport. Water-filled pore space (*WFPS*) is calculated from gravimetric water content (GC = g water [g dry soil]$^{-1}$), bulk density (BD = g dry soil cm^{-3} soil), and particle density (PD; typically 2.65 g dry weight cm^{-3} dry soil) according to:

$$\%WFPS = 100\% \times GC \times BD/(1 - BD/PD)$$

Alternatively, *WFPS* can be calculated from the quotient of volumetric water content (g water cm^{-3} soil) and total soil pore space (cm^{-3} pore space cm^{-3} soil).

4.1.2 Water stress physiology

In addition to its impact on substrate availability and habitable pore space, water potential has a profound impact on microbial physiology in soils.[9] This is due to the fact that water moves freely and spontaneously from high to low potential across cell membranes, while most solutes are membrane impermeable. As a result, the total intracellular water potentials (Ψ^i) of bacteria and other microbes equilibrate rapidly with total extracellular potentials (Ψ^e). In order to achieve a positive turgor pressure for cell wall extension and growth, bacteria typically maintain $\psi_s^i < \Psi^e$, thereby creating an inward water flow that pressurises the cell wall sacculus to an extent determined by the intra- and extracellular potential difference. This is usually accomplished by accumulating intracellular solutes to relatively high levels that vary as a function of Ψ^e.

However, the ability of cells to respond to decreases in Ψ^e, which are accompanied by losses of water and turgor, is limited. In particular, only certain substrates (compatible solutes or osmolytes; see Brown[9]) can be accumulated without deleterious effects on enzymatic activity or metabolic pathways. Solute accumulation can also be energetically expensive, especially if solutes must be synthesised *de novo*. Moreover, the demand for solutes at low matric potentials may exceed local solute supply in soils with relatively low soil water ionic strengths, thus severely taxing biosynthesis. In such cases, cells cannot maintain turgor and senesce. In contrast, increases in Ψ^e result in stresses of the cell wall and some cytoplasmic dilution due to water influx, the extent of which depends on the magnitude and rate of Ψ^e increase. Large and rapid changes lead to cell lysis if cell wall elasticity is exceeded. More gradual changes can be accommodated simply by solute export. Clearly, the soil microbiota must be reasonably adaptable to survive changes in matric and solute potentials that occur on hourly, daily and seasonal time scales, and on both large and small spatial scales due to variability in soil composition.

Based on their responses to, or tolerance of, water stress a number of microbes have been assigned to one of five groups defined loosely by optimum and minimum water potentials (ψ_s) for growth.[28] Three of these groups are most relevant for soils:

Group 1 Optimum -0.1 MPa; minimum about -2 MPa. This group contains some fungi and a variety of Gram-negative bacteria, including taxa involved in trace gas dynamics (see below).

Group 2 Optimum about -1 MPa; minimum about -5 MPa. This group contains many phycomycete fungi, actinomycetes and Gram-negative bacteria.

Group 3 Optimum about -1 MPa; minimum -10 to -15 MPa. This group contains a variety of ascomycete and basidiomycete fungi, actinomycetes and Gram-positive bacteria.

Though hardly complete, and lacking assignments for specific functional categories (e.g. nitrogen fixers, cellulose degraders, etc.), the preceding organisation provides a basis for understanding the role of water potential as a structuring agent in soil microbial communities and in patterns of biogeochemical cycling.

4.1.3 Interactions among soil water content, water potential and biogeochemistry

The diverse processes associated with the terrestrial nitrogen cycle illustrate well the complex interactions among soil water regimes, biogeochemical transformations and terrestrial ecosystem structure. Two especially relevant processes, ammonia oxidation (which requires oxygen) and denitrification (which requires anoxia), are linked in that the latter provides substrate for the former (see Chapters 1 and 2). Nonetheless, water regimes that favour ammonia oxidation by enhancing ammonium diffusion and oxygen availability essentially inhibit denitrification (Fig. 4.3), thus complicating an otherwise simple relationship based on nitrate supply. As a result of their divergent responses to oxygen, ammonia oxidation and denitrification might be expected to exhibit significant spatial and temporal differentiation as has indeed been observed (e.g. Williams *et al.*[79]; see also Tiedje[72]).

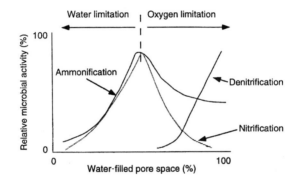

Figure 4.3   Relative rates of ammonification, respiration, nitrification and denitrification as a function of water-filled pore space[38]

However, soil water regimes do not regulate either ammonia oxidation or denitrification entirely (Fig. 4.4). A number of factors confound correlations between the activity of each process and water content.[79] For instance, soil texture and aggregate size contribute to substantial physical heterogeneity and process microzonation. Parkin[46] documented the point well by showing that 85% of the active denitrification in a soil column could be attributed to < 0.1% of the soil mass (Fig. 4.5). The variability in denitrification derives in part from the coexistence at millimetre scales of oxic and anoxic conditions (Fig. 4.6) facilitating both ammonia oxidation and denitrification.[60] The high degree of variability in texture and aggregate distribution at cm to > m scales and among soils of differing type clearly obviates the stratification in ammonia oxidation and denitrification reported for sediments.[3,54]

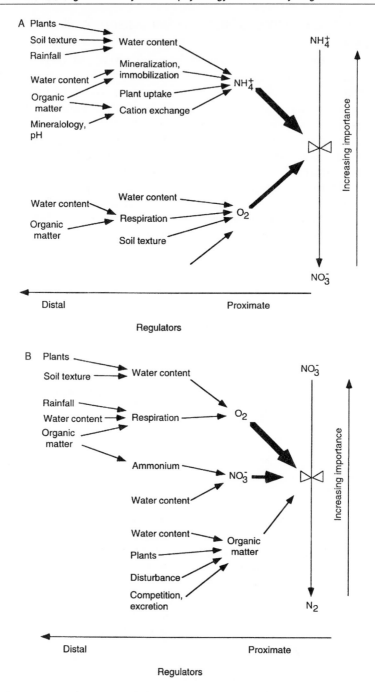

Figure 4.4   Conceptual model of controls of nitrification (A) and denitrification (B); proximal controls act at a cellular level[19,30,72,79]

Microzonation of denitrification

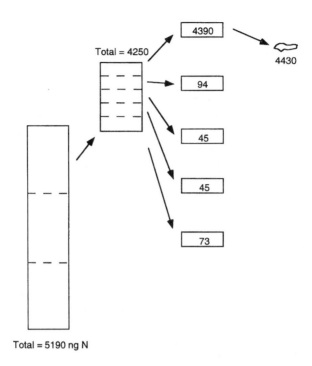

Total = 5190 ng N

Figure 4.5   Rates of denitrification (ng N day^{-1}) in a segmented core; core segments assayed for activity are indicated by dashes. Rates of denitrification are indicated within or adjacent to segments. More than 80% of the activity in the intact core occured within the upper third of the soil. Subsequent segmentation revealed that an 80 mg subsample near the core surface was responsible for most of the activity.[46]

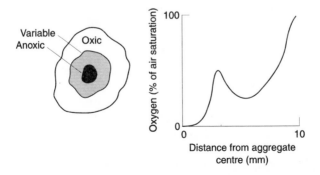

Figure 4.6   Isopleths of oxygen distribution in soil aggregates at a submillimetre scale measured using oxygen microelectrodes[60]

The availability of ammonia also affects temporal and spatial patterns of activity. Although organic nitrogen mineralisation increases ammonium availability, competition among plants, ammonia-oxidising bacteria and heterotrophs growing on high C:N (> 20) substrates decreases nitrification (see also Chapter 2). While ammonia oxidisers can compete effectively with plants for ammonium,[20] plant roots promote substantial below-ground heterogeneity, and affect both the distribution and dynamics of ammonia oxidation and denitrification (e.g. Christensen *et al.*[12] and Fig. 4.7).

A great deal is known about root–microbe interactions in soils, especially for agricultural plants and taxa that support nitrogen-fixing symbioses.[49,50,72] However, much remains to be learned about interactions among plant species composition, successional change and the dynamics of microbial nitrogen transformations within specific ecosystems. These interactions are particularly important in the context of rising atmospheric CO_2, temperature and nitrogen eutrophication, all of which can be expected to result in changes in the relative abundance of plants, if not species composition. Whether such changes will induce corresponding changes in microbial populations is a matter of conjecture at present.

The physiology of denitrifiers also impacts the spatial distribution of denitrification. Tiedje[72] has suggested that the *modus vivendi* for these bacteria is primarily as aerobic heterotrophs. As such, they can colonise virtually the entire soil volume, switching to anaerobic respiration when oxygen is absent but nitrate is present. The physiological versatility of denitrifiers and their impressive phylogenetic diversity undoubtedly account for the ubiquity of denitrification, and contrast markedly with the more limited physiology, phylogeny and distribution of ammonia-oxidizing bacteria.[1]

Owing to the complex controls of both ammonia oxidation and denitrification, the relationship between rates of activity and soil water content has varied among sites, and seasonally within a site.[18,30,79] However, a much more consistent relationship has been observed for two extremely important byproducts of these processes, $NO_x (= NO + NO_2)$ and nitrous oxide. NO_x, which affects atmospheric hydroxyl radical, tropospheric ozone and acid deposition, is both produced and consumed by soils.[15,79] Nitrous oxide, which is a potent greenhouse gas that affects stratospheric ozone,[76] can also be produced and consumed by soils although net production tends to be typically observed.[4,5,8,42,43,81] Ammonia oxidisers and denitrifiers predominate in the dynamics of these gases, with the ammonia oxidisers tending to produce more NO_x under fully oxic conditions while denitrifiers produce nitrous oxide under fully reduced conditions; ammonia oxidisers produce nitrous oxide under hypoxic (oxygen-limiting) conditions.[19] Thus, the nitrous oxide/NO_x ratio reflects soil oxygen status, and indirectly the status of soil water content which determines oxygen regimes. Accordingly, Davidson and Schimel[19] have reported that nitrous oxide/NO_x increases from < 1 to > 6 with increasing soil water content (Figs 4.8 and 4.9).

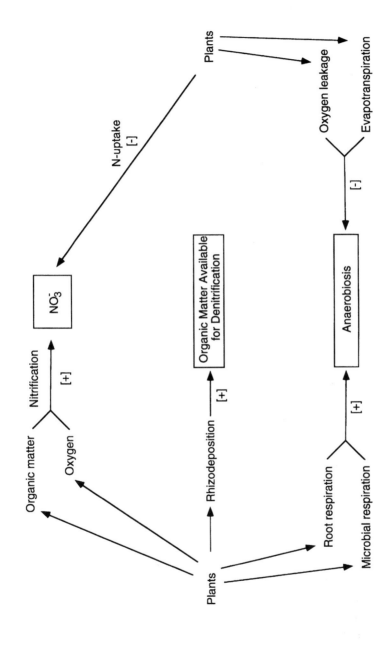

Figure 4.7 Theoretical interactions between plants and denitrification[12]

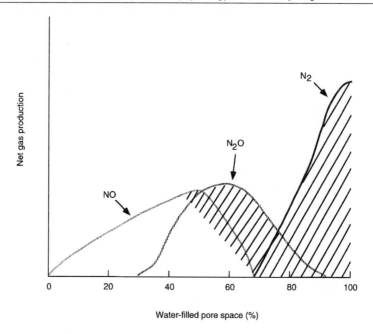

Figure 4.8   Relationship between NO, N_2O and N_2 production as a function of water-filled pore space[17]

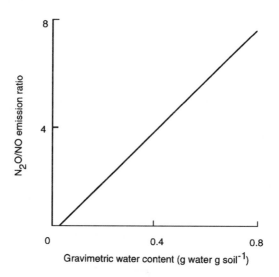

Figure 4.9   Ratio of N_2O/NO emissions versus soil water content for a variety of tropical to temperate soils[19]

NO_x and nitrous oxide consumption are also sensitive to soil water regimes, but the relationships are more complex than for the nitrous oxide/NO_x ratio. This is due to the fact that NO consumption occurs under oxic or anoxic conditions, while nitrous oxide consumption occurs primarily under anoxic conditions.[15,79] Chemodenitrification, an abiological process confined to acidic soils, also consumes NO; in addition, abiological NO production has been proposed as an explanation for the pulsed production of NO that occurs upon wetting dried soils.[17] This NO pulse has been attributed to a series of chemical transformations that occur in thin films of water surrounding clay minerals during drying (Fig. 4.10).

An additional complicating factor alluded to earlier (Fig. 4.3) arises when either gravimetric water content drops significantly below about 20% (< about 60% water-filled pore space), or water potential falls much lower than −0.5 MPa. Ammonia oxidation rates decline as soils dry due to constraints on diffusive transport of ammonium; ultimately, activity is reduced due to physiological limitations imposed by water stress[67] (Fig. 4.11). A similar pattern has been described recently for methane oxidation (see Chapter 9). Results also indicate that ammonia oxidation can tolerate substantial drying, and recover relatively quickly from severe water stress (−9 MPa) after wetting.[20] Tolerance of severe water stresses is possible since heterogeneity in the distribution of water at a microscale allows for immersion of some bacteria in a film of water; tolerance is also possible through the formation of inactive states during drying. The ability to recover from stress is especially important in systems that undergo periods of short- or long-term drying, since a slow recovery would likely mean greater competition for ammonium with heterotrophic bacteria and plants. However, it should be noted that it is not possible unambiguously to attribute the observed water stress responses to heterotrophs or chemolithotrophs due to the difficulty of distinguishing between their activities *in situ*.[15]

In contrast to ammonia oxidation, denitrification is probably not affected directly by water stress, since aeration would become inhibitory long before water potentials did. Indirect effects are possible though, if water regimes reduce denitrifier diversity. The extent of such a phenomenon has not been addressed experimentally.

Water stress is also highly unlikely to affect either ammonia oxidation or denitrification in wetlands or similar water-saturated systems. However, as with many sediments and unlike soils, oxygen availability is a critical factor for ammonia oxidation at virtually all times, limiting activity to either a millimetre to centimetre zone at the soil surface, or the oxic rhizosphere of rooted aquatic vegetation.[52] Denitrification is also highly constrained spatially, but mostly by nitrate availability; as a result it typically occurs in anoxic soils just beneath or adjacent to zones of ammonia oxidation.[65]

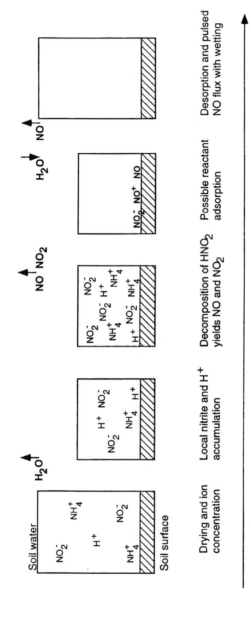

Figure 4.10 Conceptual model for production of NO by nitrification and chemodenitrification in soils undergoing wetting–drying cycles[17]

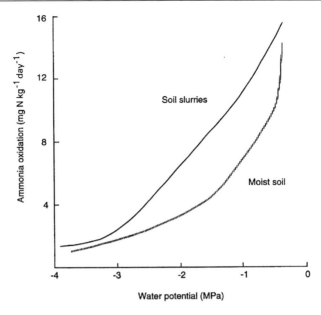

Figure 4.11   Nitrification (ammonia oxidation) as a function of water potential in soil slurries or moist soils amended with ammonia; potassium sulphate was used to adjust total water potential[67]

Though much remains to be learned about the interactions of soil water regimes and biogeochemical dynamics, substrate transport and water stress physiology provide a suitable framework for ordering observations and developing hypotheses. In contrast, the effects of water regimes on microbial diversity and community structure are understood much less. Not only is total microbial diversity in soils uncertain (though some estimates have been made),[73,74] it is not clear how diversity is related to function. For example, rates of processes and the relative importance of various bacterial functional groups vary from arid to permanently flooded soils, but to what extent are these variations associated with changes in microbial diversity?

Some broad generalisations can be offered in response to these questions. The relative importance of fungi in soil organic matter cycling seems inversely proportional to water content, and fungi are reportedly much more significant in soils than in sediments[29,71] Soil water regimes thus favour certain major microbial groups and their physiological and biochemical traits over others.

The selective capacity of soil water regimes is also evident from the fact that methanogens are largely absent from typically oxic soils. Though methane production in forest soils has been documented by several groups,[59] it appears unusual and atypical for methanogens. In addition, acetogenesis but not methanogenesis has been observed readily in grassland soils incubated for extended periods under anoxic conditions,[75,76] again suggesting that methanogens are rare or absent in most soils. Rice paddy (or wetland) soils that alternate between water saturation and desaturation are well known for copious methane production.[15] However, in these soils the combination of large methanogen populations during flooded periods, and desiccation and oxygen tolerance during dry periods, appears to allow cyclic activity and the perpetuation of populations even under adverse conditions.[25] Anoxia tolerance by aerobes (e.g. methanotrophs),[55] oxygen tolerance by anaerobes (e.g. sulphate reducers) or water stress tolerance and adaptations to variable nutrient regimes may also contribute more generally to the maintenance of microbial diversity during periodic shifts in soil water content.

The response of the soil fauna and rooted plants to water represents an additional but seldom-considered factor that may affect soil microbial diversity. Bactivores and fungivores (e.g. protozoa, nematodes, tardigrades, microarthropods, annelids, etc.) differ in their requirements for water and susceptibility to water stress.[14] Though little is known about the impact of selective feeding or digestion on soil microbial diversity, it seems likely that changes in grazing may have an impact. However, at present one cannot predict unequivocally whether grazing is a minor or major control of diversity.

Since rooted plants affect soil water content via transpiration, they may affect microbial diversity by altering conditions in the rhizosphere. However, the extent of any impact is likely limited by the fact that transpiration stops at the wilting point, which for many plants occurs at soil potentials around $-1.5\,MPa$.[45] This means that plants alone cannot reduce potentials sufficiently to induce substantial water stresses. The impact of plants on soil microbial diversity is therefore more likely manifest through various plant-rhizosphere-microbe interactions, including mycorrhizal and bacterial symbioses, and the timing and nature of organic inputs that control substrate availability.

Although the varied interactions among microbes and higher organisms undoubtedly affect microbial diversity, the relationship of diversity to biogeochemical cycling is uncertain. At a minimum, taxonomic and phylogenetic diversity in a given soil (or any other system) must correlate with metabolic and process diversity. However, it is unclear at present whether microbial diversity includes significant functional redundancy, whether diversity can be linked directly to rates of biogeochemical cycling, or what role diversity plays in system stability. The application of traditional microbiological methodologies and the several powerful molecular approaches for determining microbial diversity represent a promising area for experimental effort and theoretical development.

4.2   Responses to plant organic matter

As indicated earlier, we suggest that the details, if not the basic structure, of microbial biogeochemical cycling in soils differ from those of aquatic systems in part due to the nature of plant organic inputs. Soils can be distinguished from other systems (excepting wetlands and lentic systems) to a large extent by the fact that organic matter enters from both external (above-ground) and internal (belowground) sources in macroparticulate form, either as leaves and various woody tissues, or as roots.[14] In many aquatic systems (e.g. large lakes, the oceans) sedimentary organic matter is derived largely from microparticulates, for example phytodetritus (perhaps repackaged in faecal pellets), although in some instances (small lakes with root macrophytes, streams) macroorganics are important.[57]

The size of organic inputs determines, in part, susceptibility to microbial degradation. The rate of polymer hydrolysis per unit mass or particulate organic matter is a function of the surface area that can be colonised by bacteria or accessed by hydrolytic enzymes, many of which are "exoenzymes" in soils[10,71] (see Chapter 2). Clearly, for equal masses hydrolytic rate is potentially much greater for microparticulates than for macroparticulates. Although bacteria cannot perceive particle size *per se*, they do perceive the availability of hydrolysis products at the cell surface, and express this through growth and metabolism. Thus, differences among systems in the size distribution of organic inputs can have a profound effect on microbial activity.

The size distribution of particulate organic matter also affects higher organisms, determining feeding strategies (e.g. active foraging or passive acquisition) and mechanisms for ingestion and digestion. The feeding behaviours of higher organisms often involve direct interactions with microbes (e.g. bactivory; hindgut organic processing by microbes in various insects)[14] and a variety of indirect interactions (e.g. disaggregation of leaf litter, redistribution of microbial biomass by soil fauna;[14] formation of organic aggregates by microbial polysaccharides).[71] Most of these interactions enhance rates of biogeochemical cycling, and in some cases affect patterns of cycles. For instance, bioturbation by earthworms in soil and various invertebrates in sediments plays a major role in oxygenation and carbon mineralization.[14,36]

The composition of organic inputs is as important as size in determining rates of microbial activity and differences in cycling among systems. A complex microcrystalline cellulose base with various intercalated and distinct hemicelluloses, pectins and lignins, all of which contain very little organic nitrogen, forms the bulk of terrestrial plant matter. In contrast, a more simple system of polysaccharides characterises the cell walls of most phytoplankton and some aquatic macrophytes (e.g. macroalgae); in addition, phytoplankton biomass typically contains a greater proportion of nitrogen (lower C:N ratio). The wall composition of terrestrial plants obviously reflects mechanical constraints imposed by the terrestrial environment that are absent or differ substantially in aquatic systems.

There are several ramifications of the major differences in composition of aquatic micro- and terrestrial macrophytes. First, the relatively more complex terrestrial organics likely require a more diverse array of hydrolytic enzymes for degradation, and perhaps a more diverse array of hydrolytic bacteria as well. Unfortunately, the relationship between polymer diversity and the diversity of polymer degraders is largely unknown and in need of focused analysis. However, a greater diversity of hydrolytic microbes may exist in soils than in aquatic systems, since both fungi and bacteria are important hydrolytic agents in the former, while bacteria dominate the latter.

Second, typically high C:N ratios (> 60) for terrestrial plant organic matter have a major impact on the nitrogen cycle. A number of empirical and theoretical analyses have established a strong linkage between nitrogen mineralisation, assimilation and organic matter decomposition.[71] In particular, C:N ratios > 30 decrease mineralisation and increase assimilation instead, with the balance between the two processes dependent on the nitrogen content of microbial biomass (Fig. 4.12). The latter parameter sets the minimum nitrogen requirement for biosynthesis per unit amount of substrate metabolised. Non-symbiotic nitrogen fixation can ameliorate

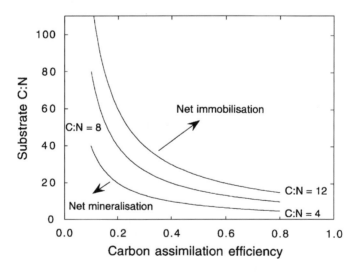

Figure 4.12   Relationship between net mineralisation and immobilisation of nitrogen as a function of substrate C:N ratio and microbial assimilation efficiency (F) for three different biomass nitrogen levels equivalent to C:N ratios of 12–6 for $f_n = 0.4$–0.8[63]

nitrogen limitation, and to some extent high C:N ratios may be a determinant of soil microbial diversity, even though non-symbiotic nitrogen fixation is generally viewed as but a fraction of total nitrogen fixation.[71]

Aside from any impacts on decomposition rates, the balance between mineralisation and assimilation obviously determines the availability of nitrogen for ammonia oxidation, and therefore denitrification. Thus, to a large extent the basic biochemical composition of plants plays an important role in regulating the flow of nitrogen through pathways that result in losses or transport out of terrestrial systems. In contrast, relatively low C:N ratios for phytoplankton and aquatic detritus promote flows though mineralisation, nitrification and denitrification. This can be viewed as a feature that stabilises aquatic systems overall; for example, ammonia fluxes from sediments (referred to as benthic–pelagic coupling) are often critical in the nitrogen budgets and productivity of shallow marine and freshwater systems.

Of course, it must be emphasised that net nitrogen assimilation during decomposition of terrestrial organic matter does not mean nitrogen retention in microbial biomass. The turnover of microbial biomass through grazing or cell death ultimately transfers nitrogen from the mineralised fractions of organic inputs to inorganic pools. It is simply the timing and complexity of these transfers that differ between aquatic and terrestrial systems.

A third important ramification of the differences in composition between terrestrial macrophytes and aquatic microphytes arises from the presence of lignin in the former. Lignin, a high molecular weight complex aromatic structure randomly polymerised from various oxyphenylpropane units, occurs in all vascular plants, including the aquatic angiosperms.[16] It does not play an important role in plant cell wall structure, but appears instead to serve as a defensive compound. Lignin residues are abundant in Carboniferous coal deposits, and indeed lignin is considered an important coal precursor. Lignin likely predates the Carboniferous period, however, as lignin-like compounds (possible lignin precursors) occur in certain phylogenetically ancient algae, in which they are thought to function as antimicrobials.[21] Since lignin and similar compounds are highly resistant to microbial decomposition, it is tempting to speculate that they evolved in response to fungal parasitism of algae, for which there is microfossil evidence from the Devonian.[70]

Irrespective of the selective pressures leading to its accumulation as a major component of vascular plant biomass, lignin has had a major impact on terrestrial carbon cycling since at least Carboniferous times. It is conceivable that the advent of lignin precipitated significant changes in patterns of elemental cycling. The significance of lignin derives from the fact that it is not readily degraded and that it is closely woven into the fabric of plant cell walls. The resistance of lignin to degradation arises not only from its aromaticity, but that unlike other polymers, for example cellulose, it has no regular structure to serve as a target for the evolution of hydrolytic enzymes.

Considerable empirical evidence documents the importance of lignin to terrestrial elemental cycling. For example, Wessman *et al.*[77] found that nitrogen mineralisation in Wisconsin forests was strongly and inversely correlated with leaf lignin content (Fig. 4.13). Melillo *et al.*[41] showed that litter weight loss after 1 year of decomposition *in situ* was strongly and inversely correlated with initial litter lignin : nitrogen ratios (Fig. 4.14). In fact, it appears that lignin content provides a better predictor of both nitrogen mineralisation and decomposition rate than do C:N ratios. Lignin is therefore a "keystone" molecule to which the terrestrial microbiota have had to adapt specifically.

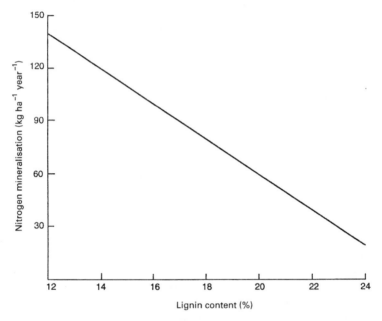

Figure 4.13   Estimated nitrogen mineralisation as a function of % lignin in leaves of Wisconsin forest canopies[77]

Benner *et al.*[2] and many others have reported that celluloses and hemicelluloses are degraded much more rapidly than lignin (Fig. 4.15). Lignin degradation in soils appears largely the result of fungal and actinomycete activity, though some bacteria may also play a role.[16,32] Though considerable quantities of lignin pass through the guts of the soil fauna, little evidence exists to suggest that they are significant agents of degradation. Lignin degradation appears to require a variety of non-specific

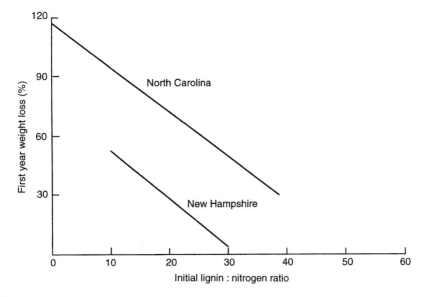

Figure 4.14   Relationship between initial litter loss (decomposition) and lignin/nitrogen ratios for fresh litterfall from North Carolina and New Hampshire[40]

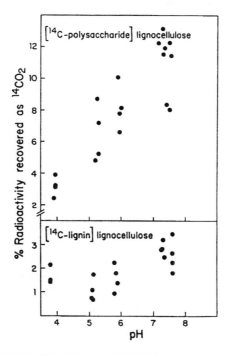

Figure 4.15   Degradation of ^{14}C-labelled lignocellulose as a function of pH; note higher degradation of polysaccharide than lignin components at all pH values[2]

enzymes, including peroxidases (ligninase and manganese peroxidases), more or less specific mixed-function oxygenases, phenol oxidases, such as laccase, and oxidases that produce peroxide.[32] Most of these enzymes are known primarily from fungi, and may provide a rationale for the dominance of fungi in lignin degradation. Lignin degradation appears to proceed by limited release of small aromatic subunits (oligolignols) and by local dearomatisation of the molecule with subsequent mineralisation of the aliphatic components. The molecule is in essence eroded by exo- or surface-bound enzymes that require molecular oxygen.

For a number of fungi, for example *Phanerochaete chrysosporium*, lignin degradation requires co-substrates, such as cellulose or glucose.[16] Whether this is true *in situ* is unknown. In addition, some of the enzymes for lignin degradation by *P. chrysosporium* are induced by nitrogen starvation (high C:N substrate ratios). This would appear to contradict field results that indicate greater litter mineralisation at lower rather than higher lignin : nitrogen ratios (Fig. 4.14). However, it is not clear to what extent litter lignin : nitrogen ratios can be used to predict nitrogen availability for a specific organism or group of organisms, nor do the weight loss data indicate lignin degradation explicitly. It is likely that most of the lignin degradation occurs after the primary loss of other polymers, at which point nitrogen availability could be much lower than it is initially. Laboratory results also indicate that consortia of fungi and bacteria enhance lignin degradation relative to that for fungi alone, even if the bacteria are not lignin degraders *per se*.[16] Stimulation may be due to symbioses involving vitamins or other co-factors. Such interactions are highly likely *in situ*.

Laboratory analyses have also shown that fungal lignin degradation depends strictly on the availability of molecular oxygen at relatively high levels.[16,22,32] Thus, lignin degradation in soils may depend in part on soil water content, as do processes such as ammonia and methane oxidation. Anaerobic lignin degradation has been reported for both freshwater and marine sediments.[2,13] However, rates appear much slower than in oxic soils, as is evident from the association of coal formation with water-saturated, anoxic systems that preserve lignin, and some argue that true lignin degradation does not occur under anoxic conditions at all.[32]

Since a variety of anaerobes degrade monomeric lignin precursors via several ring cleavage mechanisms,[13,47] biochemical limitations would not seem to account for slow lignin degradation under anoxic conditions. A key factor may be that anaerobic bacterial aromatic degradation requires uptake of substrates into the periplasm or across the cell membrane; this would preclude metabolism of all but some lignin side chains or low molecular weight hydrolysis products. In contrast, substantial oxic degradation appears to proceed with exoenzymes. In any case, it is evident that water content, as it affects oxygen distribution, plays a major role in the cycling of an important reservoir of organic carbon and its associated elements.

In addition to lignin, a number of other plant polymers are relatively resistant to degradation, especially under anoxic conditions. These include long-chain hydrocarbon waxes in the cuticle of leaves, and wall compounds such as sporopollenin. However, in contrast to lignin and cellulose, few of these compounds account for a large fraction of organic inputs to soils or aquatic sediments. As a consequence, they appear to have little impact on patterns of biogeochemical cycling or differences among systems.

4.3   Responses of soil biogeochemistry to disturbance and change

In considering the biogeochemistry of soils, we have emphasised soil water and terrestrial plant polymers as agents that structure cycles. While these parameters are profoundly important, a number of other parameters play important roles as determinants of rates and patterns of elemental transformations. Temperature, pH and nutrient regimes, the geology and chemistry of soil formation, meio- and macrofaunal activity, land use, and natural and anthropogenic disturbances contribute to substantial differences in cycling rates and patterns among systems, even if water contents and polymer inputs are comparable. For example, in acidic soils supporting conifer growth, chemolithotrophic ammonia oxidation appears very limited, and unable to respond to ammonium additions,[8] presumably as a result of pH stresses (but see Stark and Hart[68] for results suggesting that rapid microbial assimilation of nitrate has led to underestimates of nitrification). In contrast, ammonium additions to more alkaline soils are readily nitrified, irrespective of the dominant plant community present. As an additional example, the soil fauna, which plays a critical role in elemental cycling,[35] can vary markedly in activity, diversity and density as a function of relatively small changes in temperature, water-filled pore space and organic input.[14]

Because of the pervasive and rapid changes resulting from anthropogenic activity, the multiple interactions among parameters that affect rates and patterns of biogeochemical cycling in soils have become the focus of considerable attention. This is due in part to the realisation that soils represent one of the largest reservoirs of organic carbon on Earth, and that changes in pools of soil carbon and associated elements can either amplify or damp trends in global change. For instance, it appears at present that soils of temperate forests function as a net sink for atmospheric carbon, and as such provide a negative feedback on climate change.[31,33] In contrast, the well-documented inhibition of atmospheric methane consumption by soils and increased emission of NO_x and nitrous oxide from soils amplify climate change (see Chapter 9).

Since all but a cursory analysis is beyond the scope of this book, we discuss here only briefly a few of the interactions and topics that are the focus of current global change research. Chief among these are responses of soils to warming and changes in water content, drying in particular. Obviously, depending on the magnitude of change, not only are specific microbial processes subject to perturbation,[26,58,78,80] but state changes in whole ecosystems can ensue, due to destabilisation of mineral cycling pathways, and changes in populations of key macro- and meiofaunal species that regulate microbial activity.[35] For instance, a decline in earthworm populations due to increased desiccation could substantially reduce soil aeration and the availability of litter organics, ultimately reducing nutrient recycling, at least temporarily. While the effects of altered temperature and water regimes on specific microbial processes or animal populations are being addressed in field and laboratory studies, additional synoptic, "whole system" analyses are needed if we are to develop any reliable predictive capabilities.

Responses to nitrogen eutrophication also require continued attention. Nitrogen inputs to soils, as both nitrate and ammonium, have been increasing and may become problematic at a global scale. Depending on soil water status at local to regional scales, these inputs have several effects: (i) increased primary production (potential negative feedback on climate warming), enhanced litter quality (lower C:N ratios), shifts in plant species composition; (ii) increased emission of nitrogen oxides through nitrification or denitrification; and (iii) enhanced degradation of high C:N soil organic matter. However, nitrogen eutrophication can also inhibit some processes (e.g. atmospheric methane consumption) directly, and perhaps affect others indirectly by decreasing soil pH in the case of nitrate inputs. In at least one study, nitrogen eutrophication has been reported to decrease microbial biomass, a response that could obviously elicit major changes at an ecosystem level.[64] Sulphur eutrophication through acid rain deposition can likewise result in deleterious effects, not only through direct responses to lower pH, but also through nutrient leaching and increased abundance of toxic metals.

Responses of the soil microbiota and biogeochemical cycling to successional changes in plant or animal communities have not been investigated extensively, although it is evident that nitrogen-fixers and other symbionts are dependent in part on the abundance of plant hosts. Because of the nearly ubiquitous and often highly specific associations between microbes and higher organisms, successional changes can precipitate changes in microbial diversity, if not overall activity. Conversely, the diversity of microbes at a given locale may partially regulate the nature and rate of succession by determining the competitive capabilities of individual plant species.

Plant–animal–microbe interactions, including successional changes, are subject to a number of perturbations. Temperature, soil water content, and eutrophication are but a few of the important parameters. Elevated atmospheric CO_2 has a variety of direct and indirect effects. In addition to contributing to higher temperatures, elevated CO_2 increases plant water and nitrogen use efficiency and possibly plant production,[23,41,48] both of which have far-reaching ramifications for ecosystem dynamics and biogeochemical cycling. Changes in the susceptibility of plants to herbivory and microbial decomposition have also been suggested as consequences of elevated CO_2, although data supporting such changes are limited.[41,51] Any such changes would directly affect the supply of organic matter to soils, and therefore rates of cycling.

However, significant indirect effects may also arise from changes in herbivory. For example, Gehring and Whitham[27] have shown that colonisation of pinyon pine (*Pinus edulis*) roots by ectomycorrhizal fungi is inhibited by herbivory, which doubtless decreases the supply of photosynthate to the fungi. Because of their importance in soil nutrient dynamics, organic matter mineralisation and soil food webs, changes (positive or negative) in ectomycorrhizal abundance can have a profound impact on biogeochemical cycles.

Finally, land-use change and other anthropogenic activities (e.g. pollution) represent major disturbances of biogeochemical cycles. Agriculture, for instance, has substantially depleted pools of soil organic matter, inhibited atmospheric methane consumption, and increased nitrogen oxide emissions among many other effects.[31,34,37] Deforestation has similar serious consequences.[44,69] Most anthropogenic activities, not just agriculture and deforestation, result in mobilisation and redistribution of elements. Both processes typically involve micro-organisms at some point (e.g. trace gas emission), and ultimately increase the flow of mass from continents to the oceans where both subtle and more substantial changes are initiated.

References

1. Bédard C, Knowles R (1989) Physiology, biochemistry, and specific inhibitors of CH_4, NH_4^+, and CO oxidation by methanotrophs and nitrifiers. *Microbiol Rev* **53**: 68–84.
2. Benner R, Maccubbin AE, Hodson RE (1984) Anaerobic biodegradation of the lignin and polysaccharide components of lignocellulose and synthetic lignin by sediment microflora. *Appl Environ Microbiol* **47**: 998–1004.
3. Blackburn TH, Blackburn ND (1992) Model of nitrification and denitrification in marine sediments. *FEMS Microbiol Lett* **100**: 517–522.
4. Blackmer AM, Bremner JM (1976) Potential of soil as a sink for atmospheric nitrous oxide. *Geophys Res Lett* **3**: 739–742.
5. Blackmer AM, Robbins SG, Bremner JM (1982) Diurnal variability in rate of emission of nitrous oxide from soils. *Soil Sci Soc Am Proc* **46**: 937–942.
6. Born M, Dörr H, Ingeborg L (1990) Methane consumption in aerated soils of the temperate zone. *Tellus* **42(B)**: 2–8.

7. Boul SW, Hole FD, McCracken RJ (1980) *Soil Genesis and Classification*. University of Iowa Press, Ames.

8. Bowden RD, Melillo JM, Steudler PA, Aber JA (1991) Effects of nitrogen additions on annual nitrous oxide fluxes from temperate forest soils in the northeastern United States. *J Geophys Res* **96(D)**: 9321-9328.

9. Brown, AD (1990) *Microbial Water Stress Physiology: Principles and Perspectives*. Wiley, New York.

10. Burns RG (1982) Enzyme activity in soil: location and a possible role in microbial ecology. *Soil Biol Biochem* **14**: 423-427.

11. Castro MS, Steudler PA, Bowden RD (1995) Factors controlling atmospheric methane consumption by temperate forest soils. *Glob Biogeochem Cyc* **9**: 1-10.

12. Christensen S, Groffman P, Mosier A, Zak DR (1990) Rhizosphere denitrification: a minor process but indicator of decomposition activity. In: Revsbech NP, Sørensen J (eds) *Denitrification in Soil and Sediment*. Plenum Press, New York, pp. 199-211.

13. Colberg PJ (1988) Anaerobic microbial degradation of cellulose, lignin, oligolignols, and monoaromatic lignin derivatives. In: Zehnder AJB (ed.) *Biology of Anaerobic Micro-organisms*. Wiley, New York, pp. 333-372.

14. Coleman DC, Crossley DA, Jr (1996) *Fundamentals of Soil Ecology*. Academic Press, New York.

15. Conrad R (1996) Soil micro-organisms as controllers of atmospheric trace gases (H_2, CO_2, CH_4, OCS, N_2O, NO). *Microbiol Rev* **60**: 609-640.

16. Crawford RL (1981) *Lignin Biodegradation and Transformation*. Wiley, New York.

17. Davidson EA (1992) Sources of nitric oxide and nitrous oxide following wetting of dry soil. *Soil Sci Soc Am J* **56**: 95-102.

18. Davidson EA, Matson PA, Vitousek PM, Riley R, Dunkin K, Méndez-García G, Maass JM (1993) Processes regulating soil emissions of NO and N_2O in a seasonally dry tropical forest. *Ecology* **74**: 130-139.

19. Davidson EA, Schimel JP (1995) Microbial processes of production and consumption of nitric oxide, nitrous oxide and methane. In: Matson PA, Harriss RC (eds) *Biogenic Trace Gases: Measuring Emissions from Soil and Water*. Blackwell Science, Oxford, pp. 327-357.

20. Davidson EA, Stark JM, Firestone MK (1990) Microbial production and consumption of nitrate in an annual grassland. *Ecology* **71**: 1968-1975.

21. Delwiche CF, Graham LE, Thomson N (1989) Lignin-like compounds and sporopollenin in Coleochate, an algal model for land plant ancestry. *Science* **245**: 399-401.

22. Dobbie KE, Smith KA, Priemé A, Christensen S, Degorska A, Orlanski P (1996) Effect of land use on the rate of methane uptake by surface soils in Northern Europe. *Atmos Environ* **30**: 1005-1011.

23. Drake BG (1992) The impact of rising CO_2 on ecosystem production. *Wat Air Soil Pollu* **64**: 25-44.

24. Fanning DS, Fanning MCB (1989) *Soil Morphology, Genesis and Classification*. Wiley, New York.

25. Fetzer S, Bak F, Conrad R (1993) Sensitivity of methanogenic bacteria from paddy soil to oxygen and desiccation. *FEMS Microbiol Ecol* **12**: 107–112.
26. Gallardo A, Schlesinger WH (1994) Factors limiting microbial biomass in the mineral soil and forest floor of a warm-temperate forest. *Soil Biol Biochem* **26**: 1409–1415.
27. Gehring CA, Whitham TG (1991) Herbivore-driven mycorrhizal mutualism in insect-susceptible pinyon pine. *Nature (Lond)* **353**: 556–557.
28. Griffin DM (1981) Water and microbial stress. *Adv Microb Ecol* **5**: 91–136.
29. Griffin DM (1985) A comparison of the roles of bacteria and fungi. In: Leadbetter ER, Poindexter JS (eds) *Bacteria in Nature*, Vol. 1. Plenum, New York, pp. 221–255.
30. Groffman PM, Tiedje JM (1989) Denitrification in north temperate forest soils: relationships between denitrification and environmental factors at the landscape scale. *Soil Biol Biochem* **5**: 621–626.
31. Harrison KG, Broecker WS, Bonani G (1993) The effect of changing land use on soil radiocarbon. *Science* **262**: 725–726.
32. Kirk TK, Farrell RL (1987) Enzymatic "combustion": the microbial degradation of lignin. *Ann Rev Microbiol* **41**: 465–505.
33. Kirshbaum MUF (1993) A modelling study of the effects of changes in atmospheric CO_2 concentration, temperature and atmospheric nitrogen input on soil organic carbon storage. *Tellus* **45(B)**: 321–334.
34. Kruse CW, Iversen N (1995) Effect of plant succession, ploughing, and fertilization on the microbiological oxidation of atmospheric methane in a heathland soil. *FEMS Microbiol Ecol* **18**: 121–128.
35. Lavelle P (1997) Faunal activities and soil processes: adaptive strategies that determine ecosystem function. *Adv Ecol Res* **27**: 93–132.
36. Levinton JS (1982) *Marine Ecology*. Prentice Hall, New York.
37. Li C, Narayanan V, Harriss RC (1996) Model estimates of nitrous oxide emissions from agricultural lands in the United States. *Glob Biogeochem Cyc* **10**: 297–306.
38. Linn DM, Doran JW (1984) Effect of water-filled pore space on carbon dioxide and nitrous oxide production in tilled and nontilled soils. *Soil Sci Soc Am J* **48**: 1267–1272.
39. Meentemeyer V (1978) Macroclimate and lignin control of litter decomposition rates. *Ecology* **59**: 465–472.
40. Melillo JM, Aber JD, Muratore JF (1982) Nitrogen and lignin control of hardwood litter decomposition dynamics. *Ecology* **63**: 621–626.
41. Melillo JM, Callaghan TV, Woodward FI, Salati E, Sinha SK (1990) Effects on ecosystems. In: Houghton JT Jenkins GJ, Ephraums JJ (eds) *Climate Change: The IPCC Scientific Assessment*. Cambridge University Press, Cambridge, pp. 287–310.
42. Mosier AR, Parton WJ, Valentine DW, Ojima DS, Schimel DS, Delgado JA (1996) CH_4 and N_2O fluxes in the Colorado shortgrass steep, I, Impact of landscape and nitrogen addition. *Glob Biogeochem Cyc* **10**: 387–400.

43. Mosier AR, Parton WJ, Valentine DW, Ojima DS, Schimel DS, Heinemeyer O (1997) CH_4 and N_2O fluxes in the Colorado shortgrass steep 2. Long-term impact of land use change. *Glob Biogeochem Cyc* **11**: 29-42.

44. Niemelä S, Sundman V (1977) Effects of clear-cutting on the composition of bacterial populations of northern spruce forest soil. *Can J Microbiol* **23**: 131-138.

45. Nobel PS (1991) *Physiochemical and Environmental Plant Physiology*. Academic Press, New York.

46. Parkin T (1990) Characterizing the variability of soil denitrification. In: Revsbech NP, Sørensen J. (eds) *Denitrification in Soils and Sediment*. Plenum Press, New York, pp. 213-228.

47. Phelps CD, Young LY (1997) Microbial metabolism of the plant phenolic compounds ferulic and syringic acids under three anaerobic conditions. *Microb Ecol* **33**: 206-215.

48. Polley HW, Johnson HB, Marino BD, Mayeux HS (1993) Increase in C3 plant water-use efficiency and biomass over Glacial to present CO_2 concentrations. *Nature (Lond)* **361**: 61-64.

49. Rabatin SC, Stinner BR (1991) *Vesicular-Arbuscular Mycorrhizae, Plant and Invertebrate Interactions in Soil*. Wiley, New York.

50. Raich JW, Potter CS (1985) Global patterns of carbon dioxide emissions from soils. *Glob Biogeochem Cyc* **9**: 23-36.

51. Randlett DL, Zak DR, Pregitzer KS, Curtis PS (1996) Elevated atmospheric carbon dioxide and leaf litter chemistry: influences on microbial respiration and net nitrogen mineralization. *Soil Sci Soc Am J* **60**: 1571-1577.

52. Reddy KR, Patrick WH Jr, Lindau CW (1989) Nitrification-denitrification at the plant-root-sediment interface in wetlands. *Limnol Oceanogr* **34**: 1004-1013.

53. Redfield AC, Ketchum BH, Richards FA (1963) The influence of organisms on the composition of seawater. In: Hill MN (ed.) *The Sea*. Wiley Interscience, New York, pp. 26-77.

54. Revsbech NP, Sørensen J (1990) Combined use of the acetylene inhibition technique and microsensors for quantification of denitrification in sediments and biofilms. In: Revsbech NP, Sørensen J (eds) *Denitrification in Soil and Sediment*. Plenum, New York, pp. 259-275.

55. Roslev P, King GM (1994) Survival and recovery of methanotrophic bacteria under oxic and anoxic conditions. *Appl Environ Microbiol* **60**: 2602-2608.

56. Rutherford PM, Juma NG (1992) Influence of texture on habitable pore space and bacterial-protozoan populations in soil. *Biol Fertil Soils* **12**: 221-227.

57. Schlesinger WH (1991) *Biogeochemistry: An Analysis of Global Change*. Academic Press, New York.

58. Schuler S, Conrad R (1991) Hydrogen oxidation activities in soil as influenced by pH, temperature, moisture, and season. *Biol Fertil Soils* **12**: 127-130.

59. Sexstone AJ, Mains CN (1990) Production of methane and ethylene in organic horizons of spruce forest soils. *Soil Biol Biochem* **22**: 135-139.

60. Sexstone AJ, Revsbech NP, Parkin TB, Tiedje JM (1985) Direct measurement of oxygen profiles and denitrification rates in soil aggregates. *Soil Sci Soc Am Proc* **49**: 645-651.

61. Skopp J (1985) Oxygen uptake and transport in soils: analysis of the air-water interfacial area. *Soil Sci Soc Am J* **49**: 1327-1331.

62. Skopp J, Jawson MD, Doran JW (1990) Steady-state aerobic microbial activity as a function of soil water content. *Soil Sci Soc Am J* **54**: 1619-1625.

63. Söderlund R, Rosswall T (1982) The nitrogen cycles. In: Hutziner O (ed.) *The Handbook of Environmental Chemistry*, Vol. 1/Part B. Springer, Berlin, pp. 60-81.

64. Söderström B, Bååth E, Lundgren B (1983) Decrease in soil microbial activity and bimasses owing to nitrogen amendments. *Can J Microbiol* **29**: 1500-1506.

65. Sørensen J, Revsbech NP (1990) Denitrification in stream biofilm and sediment: *in situ* variation and control factors. In: Revsbech NP, Sørensen J (eds) *Denitrification in Soil and Sediment*. Plenum, New York, pp. 277-289.

66. Sposito G (1989) *The Chemistry of Soils*. Oxford University Press, New York.

67. Stark JM, Firestone MK (1995) Mechanisms for soil moisture effects on activity of nitrifying bacteria. *Appl Environ Microbiol* **61**: 218-221.

68. Stark JM, Hart SC (1997) High rates of nitrification and nitrate turnover in undisturbed coniferous forests. *Nature (Lond)* **385**: 61-64.

69. Steudler PA, Melillo JM, Feigl BJ, Neill C, Piccolo MC, Cerri CC (1996) Consequence of forest-to-pasture conversion on CH_4 fluxes in the Brazilian Amazon Basin. *J Geophys Res* **101(D)**: 18547-18554.

70. Stewart WN, Rothwell GW (1993) *Paleobotany and the Evolution of Plants* (2nd edn). Cambridge University Press, New York.

71. Tate RL (1995) *Soil Microbiology*. Wiley, New York.

72. Tiedje JM (1988) Ecology of denitrification and dissimilatory nitrate reduction to ammonium. In: Zehnder AJB (ed.) *Biology of Anaerobic Micro-organisms*. Wiley, New York, pp. 179-244.

73. Torsvik V, Goksøyr J, Daae FL (1990a) High diversity in DNA in soil bacteria. *Appl Environ Microbiol* **56**: 782-787.

74. Torsvik V, Salte K, Sørheim R, Goksøyr J (1990b) Comparison of phenotypic diversity and DNA heterogeneity in a population of soil bacteria. *Appl Environ Microbiol* **56**: 776-781.

75. Wagner C, Grießhammer A, Drake HL (1996) Acetogenic capacities and the anaerobic turnover of carbon in a Kansas prairie soil. *Appl Environ Microbiol* **62**: 494-500.

76. Watson RT, Rodhe H, Oeschger H, Siegenthaler U (1990) Greenhouse gases and aerosols. In: Houghton JT, Jenkins GJ, Ephraums JJ (eds) *Climate Change: The IPCC Scientific Assessment*. Cambridge University Press, Cambridge, pp. 287-310.

77. Wessman CA, Aber JD, Peterson DL, Melillo JM (1988) Remote sensing of canopy chemistry and nitrogen cycling in temperate forest ecosystems. *Nature (Lond)* **335**: 154-156.

78. Whalen SC, Reeburgh WS (1996) Moisture and temperature sensitivity of CH_4 oxidation in boreal soils. *Soil Biol Biochem* **28**: 1271–1281.
79. Williams EJ, Hutchinson GL, Fehsenfeld FC (1992) NO_x and N_2O emissions from soil. *Glob Biogeochem Cyc* **6**: 351–388.
80. Winkler JP, Cherry RS, Schlesinger WH (1996) The Q_{10} relationship of microbial respiration in a temperate forest soil. *Soil Biol Biochem* **28**: 1067–1072.
81. Zak DR, Grigal DF (1991) Nitrogen mineralization, nitrification and denitrification in upland and wetland ecosystems. *Oecologia* **88**: 189–196.

5

Aquatic sediments

The task of reviewing the immense literature on marine and freshwater sediments is beyond the scope of this book. We have, therefore, decided to generalise as much as possible and to elucidate the general principles that control the biology of sediments, rather than attempt case by case comparisons. The similarities and differences between the two sediment types are reviewed below, but it is important to stress that the main characteristic of sediments and soils is that they are essentially heterotrophic systems in which the products of primary production are mineralised and recycled again to the photoautotrophs. Because of this primary heterotrophic function, the quantity of particulate organic matter (POM) input is the most important factor controlling the processes which occur in sediments. Without organic input there is no sediment biology. This dependence on the input of organic detritus, which is common to both sediment types, makes it possible to model detrital degradation and emphasise similarities between the two systems. Modelling is a very convenient way to examine a process as a function of a variable, for example quantity or quality of detrital input. Another, and equally important advantage of modelling is that processes can be examined at a spatial and temporal resolution unachievable by conventional techniques. Not only is it impossible to assay most reaction rates at a resolution of $< 1\,mm$, but it is also difficult to measure concentration profiles. Exceptions to this generalisation are the measurement by microsensors of O_2,[35] HS^-,[11,22] NO_3^-,[20,25] and CH_4.[18]

We believe that our slightly unconventional treatment of mineralisation in aquatic sediments, with an emphasis on POM hydrolysis and modelling, gives the most logical framework for understanding the interaction of the various element cycles. Unfortunately, we have had to cite much of our own work to illustrate these principles, for which we ask the reader's understanding.

5.1 Comparison of freshwater and marine sediments

The areas occupied by marine sediments are vastly greater than those of freshwater sediments. This results directly from the 70% of the Earth's surface occupied by the oceans and the relatively small area of land covered by lakes and rivers. Although we are principally interested in the organic content of sediments, it is obvious that the majority of sediments, in both lakes and oceans, contain much mineral material. This mineral content is composed of silt or sand, depending on the hydrodynamic regime; active water movement results in sands and pebbles, while fine-grained silts are deposited in regions of little water disturbance. Mineral particles are derived from the disintegration of rocks by weathering, physical abrasion and biological activity. Some sediment addition to oceans and lakes is by deposition from the air, but most sediment particles are carried to lakes and oceans by rivers. Many oceanic sediments contain silicate- and carbonate-containing skeletons of planktonic algae. The edges of lakes and oceans are disturbed by wind-induced waves, both have currents of differing intensity, but tidal disturbance is largely confined to the oceans. However, the two types of water bodies are more similar than different, with respect to sediment source and hydrodynamic characteristics. Similarities are not confined to the physical characteristics. The same degradation of organic matter occurs in both sediment types, dominated by microbes but assisted by sediment fauna. There is, however, a difference in the composition of the organic detritus reaching the sediment in the two different situations. In general, there is a greater input of vascular plant detritus to lake sediments, partially due to the extensive shore line of most lakes. The detritus is often in the form of fallen leaves, but can also consist of larger plant fragments. These fragments contain much cellulose, hemicelluloses and lignins. Except in near-shore environments, these polymers seldom enter marine sediments. The main organic detritus reaching marine sediments originates from pelagic algal cells, either directly or as faecal pellets of zooplankton, which have eaten the algae. But pelagic algae and faecal pellets can also be a major source of POM to lake sediments.

The similarities and differences between lake and marine sediments may be assessed with reference to Fig. 5.1. There is little input of coarse particulate organic matter (CPOM) to marine sediments. The disintegration of CPOM in lake sediments is performed by a number of organisms of different sizes. Decapods and insects are of particular importance in the physical shredding of plant material. It is interesting that insects play no role in marine sediments. In marine shore lines, CPOM (from salt marshes, mangroves and seagrass beds) is disintegrated by, for example, arthropods, polychaetes and molluscs, and the physical shredding is accompanied by enzymatic hydrolytic attack within the digestive system of the animal. The retention of particulate material relative to the passage of fluids through the animal's alimentary tract promotes hydrolysis (see Chapter 7). The hydrolytic enzymes may be produced by a microbial flora within the animal, the latter acting as a mini-ruminant. The plant material is frequently subjected to preconditioning before it is ingested by the animal. This conditioning, which makes the plant material more susceptible to gastric digestion, may be due to ascomycetes and chytrids in both lake

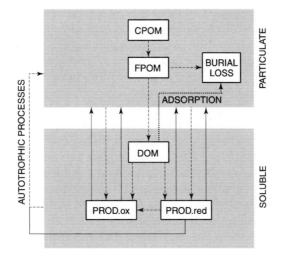

Figure 5.1   Fate of coarse particulate organic matter (CPOM) and fine particulate organic matter (FPOM) in sediments. Hydrolysis gives soluble organic matter (DOM) which is metabolised by micro-organisms into cells (also FPOM) plus oxidised or reduced products. Some of the latter inorganic molecules are oxidised by chemoautotrophs. Inorganic nutrients are cycled back to the primary producers (photoautotrophs)

and marine sediments, at sites where plant litter is present. It is presumed that hydrolytic bacteria have a similar function, but they may be less effective than the ascomycetes whose hyphae can penetrate into the plant fibrous material very efficiently. This conditioning may be very important in lignocellulose breakdown, particularly where subsequent degradation in the animal alimentary tract occurs under anaerobic conditions, because lignin is not broken down readily by anaerobes. Predigestion of the lignin could facilitate the subsequent hydrolysis of the plant material in the anaerobic digestive system.

The end products of macrofaunal ingestion of CPOM are fine particulate organic material (FPOM), minerals and macrofaunal biomass. The latter may have important implications with regard to the amount of POM that is permanently lost by burial. Food webs, which include soft animal biomass, may possibly result in less burial loss than webs that contain only bacteria.[4] Macrofaunal ingestion may result in the defaecation of FPOM at a lower depth in the sediment than that in which the CPOM was eaten. Subsequent degradation of this FPOM occurs anaerobically, but not necessarily fermentatively. The absence of O_2 not only restricts the degradation of lignin, but may also result in the retardation of other hydrolyses in marine sediments[23] and presumably also in lake sediments. FPOM (which includes micro-organisms) is also processed by animals. In marine and lake sediments, some representatives of the meiofauna are probably active in eating FPOM. This meiofaunal population is most active in the upper sediment strata, but is not entirely restricted to an aerobic life style. Meiofaunal numbers are controlled by the availability of food, growth rate and predation. Little is known about most of these control factors and the importance of the meiofauna in FPOM degradation has not been measured with precision. Fungi are presumed to be of less importance in the degradation of FPOM compared to that of CPOM, but their role cannot be assessed. Similarly, both aerobic and anaerobic bacteria with hydrolytic exoenzymes must be of great significance, but few isolations have been made. The product of this hydrolytic exoenzyme attack is dissolved organic matter (DOM) which is available not only to the micro-organism producing the enzyme, but also to any neighbouring cells. Some of the DOM may become adsorbed to mineral particles and be rendered resistant to enzymatic attack. FPOM particles themselves may be resistant to degradation and therefore lost by burial in the lower sediment layers. Macrofauna, meiofauna and micro-organisms all re-enter the POM pool and become substrates for attack (Fig. 5.1). More food (carbon) is processed by bacteria and fungi than by the meio- and macrofauna. The efficiency of carbon incorporation into biomass, relative to the amount processed for energy, will depend on the quality of the food resource and on the presence or absence of electron acceptors.

The single most important difference between lake and marine sediments is the presence of high sulphate concentrations in sea water, and thus in the sediments. Sulphate is the most important electron acceptor in the oxidation of carbon in marine sediments with the coupled production of sulphide. Sulphate reduction can also be an important process in lake sediments, accounting for 30% of the carbon oxidation, even though the SO_4^{2-} concentrations are low.[38] High sulphate concentrations can occur in small meromictic lakes, where evaporation is high in relation to inflow.

Both lakes and seas are subject to stratification. In seas, this can be due to salinity or temperature differences, but only thermal stratification can occur in freshwater lakes. Thermal stratification is a summer phenomenon and occurs when wind energy is low. In shallow near-shore waters, stratification often results in O_2 depletion in the underlying water accompanied by anoxia in the sediments (see Section 6.4). Although stratification and anoxia can occur in both fresh- and saltwater sediments, the production of sulphide (HS^-) in the latter has a number of undesirable side effects, which are discussed later. The end product of carbon degradation in lake sediments (in addition to CO_2) is methane (CH_4). CH_4, unlike HS^-, is not toxic and has a low solubility in water. This results in CH_4 bubbling out of lake sediments which have received a high organic input. This loss of CH_4 before it can be oxidised by O_2, results in an apparently low C:N ratio in the mineral products of POM degradation.[20] However, CH_4 loss and HS^- immobilisation have a similar effect on the nitrogen cycle in both sediment types. Because carbon is mineralised in the absence of O_2 and neither CH_4 nor FeS is oxidised immediately by O_2, there is a greater tendency for NH_4^+ to be oxidised by O_2, as competition from CH_4 and FeS is reduced. Modelling studies, based on these factors, predict similarly high rates of coupled nitrification/denitrification in lake and marine sediments.

POM as it sinks through the water column is degraded. The length of time in reaching the sediment is proportional to the amount of POM that is mineralised in the water column.[39] Not much POM reaches sediments underlying deep water, unless the particles are large or dense. In addition, preferential losses of P and N occur relative to C, as POM sinks through the water column. These factors related to the sinking of POM are common to both lakes and seas, but in general lakes are not as deep as seas. This results in a higher proportion of primary productivity reaching the sediment in freshwater systems thus less loss of P and N from POM. As a consequence of this, there is a tendency to associate low C:N ratios with fresh, readily degradable POM. This is not always true, however, as old resistant POM can have quite a low C:N ratio.

The accumulation (burial or preservation) of organic material occurs in both lakes and seas. A high rate of organic addition (organic loading) results in a high rate of organic burial, up to 78%, which may be associated with sulphate reduction in marine sediments.[13] However, for normal sediment to accumulate, there must also be an addition of inorganic matter, as discussed above.[4] Much has been written about pelagic–benthic coupling, but in essence the process is very simple. POM produced in the pelagic photic zone sinks to the sediment. The quantity of POM arriving at the sediment depends on water depth. The quantity of mineral products leaving the sediment depends on the amount of POM that has entered the sediment as well as on the amount buried. All unburied POM is mineralised and is potentially available again to the primary producers. This relationship is clearly illustrated in Fig. 5.1. The water depth and degree of stratification determine when the mineral products reach the photic zone. If the overlying water is deep or stratified, the coupling will be poor. In shallow coastal waters, coupling will be tight. However, it is a mistake to view sediments, in lake or sea, as being a new source of nutrients to the primary producers. Sediments can only degrade and mineralise POM which they have received from the overlying water column. It is equally important to remember that a large amount of secondary POM synthesis occurs in sediments, producing CPOM in the form of macrofauna and FPOM from micro-organisms.

Both lakes and seas are fringed by rooted macrophytes, which are additional sources of POM to these limited, but ecologically very important, zones. In addition to being sources of lignocellulose polymers, these macrophytes have a very profound influence on the sediments inhabited by their roots. They can transfer gases from the sediment to the atmosphere[17] and oxygen from the atmosphere to the rhizosphere.[36] What is most interesting is the addition of DOM to the sediments via the roots. The dissolved organic carbon (DOC) forms a transient pool in sediments, dependent on recent photosynthesis. In marine sediments, oxidation of this DOC is coupled to sulphate reduction and other biogeochemical processes, which are discussed below.

The importance of hydrolytic control on the heterotrophic processes in both types of sediment cannot be overemphasised. Once the insoluble carbon polymers have been hydrolysed and made soluble, the degradation of the these soluble products is inevitable, with the exception of the small amount lost through burial. The heterotrophic carbon cycle is thus inextricably bound to hydrolysis, and the other cycles are equally firmly bound to the carbon cycle. If there were no carbon cycle, there would not be any N, S or P cycles.

5.2 The carbon cycle

5.2.1. Heterotrophy

The carbon cycle, in both lake and sea sediments, is predominantly heterotrophic, but there are some aspects of autotrophy, which will also be discussed. The DOM produced from hydrolysis of FPOM is processed by a number of oxidative and fermentative processes in aqueous sediments. The oxidants are O_2, NO_3^-, Mn^{4+}, Fe^{3+}, SO_4^{2-} and CO_2 in sequence from the top of the sediment downward. The DOC component of DOM has limited possibilities: it can be oxidised by one of the listed oxidants or it can leave the sediment unoxidised. This is an obvious conclusion, but it has some interesting connotations, for example the proportion of C oxidised by O_2, etc. and the determination of the factors influencing this proportion. Clearly, the quantity of DOC will determine the depth of O_2 penetration and the extent to which O_2 can participate in C oxidation. An equally important factor is the depth at which DOC is produced by hydrolysis of POM: the deeper the site of POM hydrolysis, the more likely will be the anoxic processing of the soluble products. Another very important factor is the degree to which HS^- is free to diffuse in marine sediments. If HS^- can diffuse to the sediment surface and react with O_2, the depth of O_2 penetration will be greatly reduced. Simulation modelling of the fate of DOC under different conditions emphasises these simple relationships in marine sediments.[8]

Figure 5.2 shows the rates at which DOC is stipulated to be produced by hydrolysis of POM at approximately 6, 36 and 60 mmol m^{-2} day^{-1} either at the surface of the sediment (TOP), or as a linear gradient from the surface down (LINEAR), or equally to all sediment layers down to 5 cm (MIX). The matrix in Fig. 5.2 represents sediments receiving increasing quantities of POM (arrow down) at increasing frequencies (arrow across). Both processes,

Frequency of POM arrival

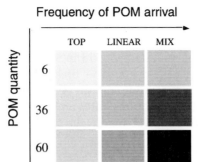

Figure 5.2   A representation of the addition of particulate organic matter (POM) to sediments. The amount of POM added increases from top to bottom. The frequency of POM addition increases from left to right. The amount of POM hydrolysed per day is 6, 36 or 60 mmol C m^{-2} day^{-1}. Infrequent addition of POM will not support macrofaunal development and will result in POM hydrolysis at the sediment surface (TOP distribution), whereas more frequent additions will result in increasing macrofaunal populations (LINEAR and MIX distributions) and greater sediment mixing. Dark shading represents the degree of bioturbation

within limits, will increase benthic faunal populations and bioturbation, but sediments receiving excessive organic loading will tend to become anoxic and devoid of macrofauna. This matrix represents the variety of sediment types that might be found in various parts of the world and under various water depth and productivity conditions. These hypothetical types are used as a proxy for actual sediment variety, but if the world's sediments could be categorised in terms of quantity, quality and frequency of POM input, simulations could predict all the important biogeochemical rates and nutrient profiles. Higher primary productivity or lower water depths could explain increased POM input. More continuous productivity (longer season) would result in a more continuous input than that found in ice-covered seas. The TOP distribution would represent a sediment subject to very little disturbance either by water movement or by bioturbation. At the lower DOC addition rates, this would be equivalent to a shelf sediment with no macrofauna. The absence of macrofauna could be attributed to a single pulse of POM per year, as might occur in an Arctic ocean. The highest DOC inputs to the sediment surface may be unrealistic, as macrofauna might be expected to arise in response to this amount of organic input, unless the input occurs as a single pulse. The LINEAR distribution might be found in sediments receiving multiple POM pulses, supporting a moderately active macrofaunal population, which transports a small amount of the fresh POM to 5 cm depth. The MIX distribution is equivalent to a very active macrofaunal population or to very efficient mixing by waves or tides.

The main characteristics of the simulation model are seen in Fig. 5.3 and some parameters are given in the associated caption. It should be emphasised that this is a reaction–diffusion model, which describes the flows from one pool to another within any sediment layer and the diffusional flows between layers. The breakdown of DOC controls the breakdown of DON, depending on the C:N ratio of the DOM, which equalled 6. In the following figures, rates are sometimes related to C (DOC) and sometimes to N (DON), but are always connected by the C:N ratio of 6.

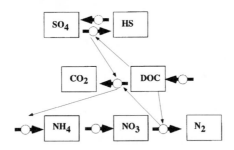

Figure 5.3   A system dynamic diagram of the main components of the model. The heavy arrows show the flows in the C, N and S cycles. The circles represent the controls on the mass flow between the pools (agents). Oxygen is not shown, but influences positively or negatively all reactions except the inflow of NH_4^+ and dissolved organic carbon (DOC). The thin arrows show other interdependencies, emphasising the central role of DOC oxidation. Dissolved organic nitrogen (DON) (not shown) breakdown to ammonium is linked to DOC by a C:N ratio of 6. In these simulations, the reaction–diffusion model was run under steady-state conditions with a time step of 4 s. There were 12 sediment layers ranging in thickness from 0.05 cm at the top to 1.6 cm at 5.3 cm depth. Reactions (flows between pools in a single sediment layer) were controlled by Michaelis–Menten kinetics. The rate constant (day^{-1}) was 30 for nitrification, denitrification, DOC oxidation by O_2 and HS^- oxidation by O_2; the rate constant for sulphate reduction was 5. Oxygen completely inhibited denitrification and sulphate reduction at 30 nmol cm^{-3}. Oxygen-dependent processes were at V_{max} when O_2 was greater than 10 nmol cm^{-3}. Diffusion between pools in different sediment layers was controlled by an equation taking account of porosity and tortuosity. Diffusion coefficients were in the range 1.64–1.81 cm^2 day^{-1}, except for DOC which was 0.5 cm^2 day^{-1}, and HS^- which was zero, on the assumption that all sulphide reacted with iron and did not diffuse

In these simulations (Fig. 5.4), HS^- is not allowed to diffuse, but FeS reacts with any O_2 that reaches it in the lower sediment layers. Sulphate reduction in these models includes oxidation of DOC by Fe and Mn, but without diffusion of reduced Mn or Fe. The most obvious feature of Fig. 5.4 is the exact equality of the C budgets at the three levels of DOC addition, irrespective of the distribution, stressing that once POM is mineralised a

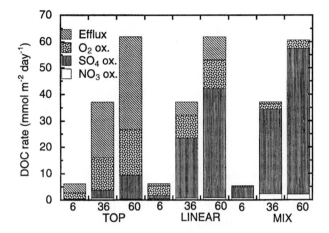

Figure 5.4 The fate of dissolved organic carbon (DOC) added to sediment. DOC is added at 6, 36 or 60 mmol m^{-2} day^{-1} to sediment in three different modes of distribution: TOP, where the addition is to the top surface sediment; LINEAR, where the addition is in a linear gradient decreasing from the surface to 5 cm; and MIX, where the addition is uniformly mixed down to 5 cm. DOC either effluxes from the sediment or is oxidised by O$_2$, NO$_3^-$ or SO$_4^{2-}$[8]

strict C budget may be made. What is of more interest is the fate of DOC at the different addition rates and distributions. All NO$_3^-$ in these simulations arises from sediment nitrification. There is no external source of NO$_3^-$.

The relative effluxes of DOC and the proportions of DOC oxidised by O$_2$, NO$_3^-$ and SO$_4^{2-}$ are seen in Table 5.1. It is predicted that DOC efflux

Table 5.1 Fate of dissolved organic carbon (DOC) produced at different rates and at different sites in a marine sediment*

	Rate (mmol m^{-2} day^{-1})	Efflux (%)	O$_2$ (%)	NO$_3^-$ (%)	SO$_4^{2-}$ (%)
TOP	6	55	37	5	4
	36	57	33	2	8
	60	57	28	1	14
LINEAR	6	12	60	8	20
	36	13	23	2	61
	60	14	18	1	67
MIX	6	2	9	14	75
	36	2	5	6	87
	60	0	5	3	91

*DOC is either lost by efflux or is oxidised by O$_2$, NO$_3^-$ or SO$_4^{2-}$. The data are taken from a simulation model.[8]

from the sediment is greatest (55–57%) for the TOP distribution and lowest (0–2%) for the MIX distribution. Because of the large efflux losses of DOC with the TOP distribution, the relatively small amount of remaining DOC is mostly oxidised by O_2. Only DOC that diffuses down into the sediment has an opportunity to be oxidised by O_2 (37–28%) with a small proportion oxidised by NO_3^- (5–1%) or SO_4^{2-} (4–14%). The LINEAR distribution has the highest proportion of oxidation by O_2 at the lowest addition rate of DOC (60% at 6 mmol m^{-2} day^{-1}), decreasing with organic loading (18% at 60 mmol m^{-2} day^{-1}). The same trend occurs with NO_3^- oxidation (8–1%), due to the dependence of NO_3^- on nitrification. An opposite trend is noted for sulphate reduction which increases from 20 to 67% with higher organic loading. In the MIX distribution, anoxic processes dominate with only 9–5% oxidation of DOC by O_2, and 75–91% oxidation by SO_4^{2-}. A higher proportion of DOC is oxidised by NO_3^- (14, 6 and 3%) in MIX than in the other distributions. At first sight, this oxic generation of NO_3^- from NH_4^+ seems to be at variance with the high degree of anoxia related to sulphate reduction. However, nitrification in these simulations is associated with the relatively deep penetration of O_2 in the MIX distribution, at additions of DOC of 36 and 60 mmol m^{-2} day^{-1} (Fig. 5.5). This point is discussed in relation to the nitrogen cycle (see Section 5.3).

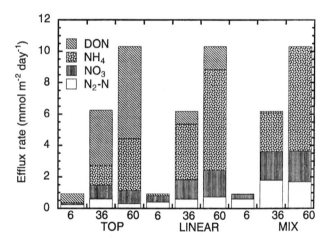

Figure 5.5   The fate of dissolved organic nitrogen (DON) added to sediment. DON is added (at 1/6 the dissolved organic carbon (DOC) addition rate of 6, 36 or 60 mmol m^{-2} day^{-1}) to sediment in three different modes of distribution: TOP, where the addition is to the top surface sediment; LINEAR, where the addition is in a linear gradient decreasing from the surface to 5 cm; and MIX, where the addition is uniformly mixed down to 5 cm. Nitrogen effluxes from the sediment either directly as DON, or as inorganic nitrogen in the form of NH_4^+, NO_3^- or N_2[8]

It is not possible to consider all sediment types underlying multiple water depths in various geographical locations. It is hoped that the simulations, representing different POM loading (different primary productivities or water depths) reaching sediments in single or multiple pulses and supporting different macrofaunal densities (degree of mixing), are representative of medium low to relatively high marine sediment activity. Similar simulations of freshwater sediments are not presented, but it is likely that lake sediments function in much the same way. As already discussed, the escape of CH_4 from freshwater sediments results in a deeper penetration of O_2 than would occur if all CH_4 were oxidised. This will have an important influence on N cycling and will also be associated with the small proportion of DOC that is oxidised by O_2. In the mixed marine sediment (MIX), this oxidation was predicted to be only 5–9% of the DOC mineralised. Experimental data from marine sediments bordering Denmark and Sweden, with O_2 uptake rates of 10–16 mmol m^{-2} day^{-1}, indicated that probably 4–17% of carbon was oxidised by O_2.[15] Most oxygen was used to reoxidise reduced Mn, Fe and HS^-, which were the end products of anoxic carbon respiration. In a different simulation, in which DOC mineralisation is 18.6 mmol m^{-2} day^{-1}, 8.4% DOC is oxidised by O_2 when HS^- diffusion is restricted, compared with only 1.8% when HS^- diffuses freely.[7] Most O_2 is consumed by HS^- (or Mn and Fe) when diffusion is not restricted. As discussed earlier (see Chapter 2), the non-diffusion of HS^-, and the other reduced products of anoxic respiration, results in an uncoupling of C oxidation and O_2 uptake; CO_2/O_2 ratios can be very high. In the simulation above, where DOC mineralisation is 18.6 mmol m^{-2} day^{-1} and there is no HS^- diffusion, the ratio CO_2/O_2 is 2.6, similar to measured values.[19]

5.2.2 Autotrophy/heterotrophy

All heterotrophic activity in sediments is obviously dependent on photoautotrophy, but the coupling can be quite weak, as when algal detritus reaches sediment underlying very deep water. However, coupling can be very tight when the primary producer grows on the sediment surface or has roots that penetrate into the lower strata.

Organic matter can also enter sediments from vascular plants growing at the interface between water and land. The organic input can be in the form of detrital POM from leaves and dead tissues, but it can also enter sediments as soluble secretions from the roots. The following observations were made for mangrove[26] and semi-tropical seagrass,[12] but probably apply to rooted plants on the fringes of many seas and lakes:

- DOC is released into the rhizosphere.
- DOC release is coupled to photosynthesis.
- DOC pool is short lived and turns over several times per day.
- DOC is oxidised by sulphate reduction (marine).

- HS^- is oxidised by O_2 from the roots.
- Ratios of CO_2/O_2 can be $5/1$.

In marine sediments, the short-lived DOC is oxidised by sulphate reduction and the resulting HS^- is reoxidised by O_2, also from the roots. This addition of O_2 to the sediment is not included in measured fluxes across the sediment–water interface; the result is very high measured ratios of CO_2 efflux to O_2 influx. Presumably similar events occur in lake sediments, but CH_4 is produced instead of HS^-, and is oxidised by O_2 from the root. There is evidence from marine sediments with macrophyte roots that the sediments are more reduced (contain more HS^-) during sunlight hours, as a result of DOC secretions and sulphate reduction.

Benthic algae can have quite high rates of photosynthesis in shallow water and algal mats.[34] They are very efficient in taking up nutrients at the sediment surface. The cells, especially in marine algal mats, provide substrate for very high rates of sulphate reduction.[33] This is analogous to the rooted vascular plants, but on a different scale.

Chemoautotrophic lithotrophic bacteria represent a type of autotrophy, but are ultimately dependent on photoautotrophy to furnish them, via heterotrophic breakdown, with the reduced compounds NH_4^+, Mn^{2+}, Fe^{2+}, HS^-, H_2 and CH_4. The oxidation of these inorganic compounds yields varying amounts of energy for cell growth: NH_4^+ oxidation yields little energy, the importance of Mn oxidation is unknown, at normal pH values Fe oxidation occurs spontaneously with O_2 as does the oxidation of HS^-. H_2 is not usually found together with O_2, but there is quite a lot of free energy to be gained from HS^- oxidation and this may result in the production of a significant cell biomass at marine sediment surfaces. Methane is not spontaneously oxidised (unless ignited) and is oxidised biologically at oxic–anoxic interfaces in lake sediment. The oxidation of CH_4 yields much energy and potentially a significant microbial cell biomass.

5.3   The nitrogen cycle

5.3.1   Hydrolysis and ammonium production

The production of dissolved organic nitrogen (DON) occurs in conjunction with the production of DOC, by hydrolytic enzyme attack on POM. N-containing polymers (protein, polynucleotides) are more easily hydrolysed than the mainly C-containing structural cell components (cellulose, chitin, bacterial cell wall). There is, therefore, a preferential mineralisation of N compared to C.

The hydrolysis of proteins and polynucleotides in POM yields oligopeptides, amino acids, oligonucleotides, nucleotides and nucleosides, all compounds having a low C:N ratio. If this preferential hydrolysis of low C:N polymers in POM occurs at the sediment surface, there is a strong possibility that the dissolved hydrolytic products may escape into the overlying water. As a consequence of this loss of DON, the residual POM has an increased C:N ratio. There are very few reported measurements of DON loss from sediments, but losses of 68% of probable PON input occurred from the surface of Arctic sediments, which were not actively bioturbated and received POM infrequently.[10] The high C:N ratio of the residual POM resulted in a very restricted N cycle in these sediments, with very little escape of dissolved inorganic nitrogen (DIN) to the overlying water. It is very likely that this is not an unique situation: DON loss from non-bioturbated sediments is to be expected in many locations, associated with a correspondingly low DIN efflux.

Preferential PON hydrolysis occurs within the sediment, in addition to being important at the water column and at the sediment surface. This results in the hydrolysis of POM with a low C:N relatively close to the sediment surface where fresh detritus has been deposited, whereas POM with a high C:N (20) is mineralised at depth in the sediment.[2] This situation is explained by the mixing downward of POM which has been preferentially stripped of N, but is still degraded more rapidly than the bulk sediment POM with a lower C:N ratio (10). Similarly, an incubation of sediment cores indicated that POM with a C:N = 4 was first hydrolysed; further incubation resulted in a gradual shift to POM of C:N > 20 after 60 days.[24]

Ammonium cannot be liberated without a parallel attack on the organic carbon of the DOM, yielding CO_2 if there is an electron acceptor present, or producing a mixture of CO_2 and CH_4 in a completely fermentative system (e.g. below the SO_4^{2-} zone in marine sediments or below the oxic zone in lake sediments). Ammonium production is thus bound to, and dependent on, organic C mineralisation. Once DOM is liberated, the C:N ratio of NH_4^+ to $CO_2 + CH_4$ is determined by the ratio in the DOM, which is often lower than the C:N ratio in the original POM. There is thus an inevitability about the gross quantity of NH_4^+ production, once hydrolysis of POM has occurred. Gross production will not necessarily equal net production, as NH_4^+ incorporation into cells and later POM detritus will probably occur, as discussed in Section 5.3.4. Unless there is a permanent loss of organic N ("burial" in Fig. 5.1), all inorganic nitrogen is liberated into the overlying water for phytoplankton production or is utilised more directly by rooted macrophytes or by benthic algae. The inorganic nitrogen is composed mainly of NH_4^+, NO_3^- and N_2 whose sum will often be equal to the gross NH_4^+ production (if DIN is not incorporated into cells and buried). The ratios of the three components will vary depending on various factors within the sediment, but the total will remain unchanged irrespective of the oxidation–reduction characteristics of the sediment. These characteristics have a very important influence on the oxidation of NH_4^+ to NO_3^- (see Section 5.3.2). Accompanying the mineralisation of carbon (Fig. 5.4), in the simulation matrix of increasing quantity and frequency of POM addition (Fig. 5.2), N mineralisation is expressed as the rates of nitrogen effluxes (Fig. 5.5). The high efflux of DON from the TOP distribution is proportional to the DOC efflux. The sum of the effluxes is the actual rate of N mineralisation at DOC mineralisation rates of 6, 36 and 60 mmol m^{-2} day^{-1}. The sum of the effluxes of $NO_3^- + N_2$ is the rate of nitrification.

5.3.2 Nitrification

Two main factors regulate nitrification: the availability of NH_4^+ and the availability of O_2. The availability of NH_4^+ depends on the C:N ratio of the DOM being mineralised. A high C:N ratio results not only in a small production of NH_4^+, but also in the large amount of carbon that consumes O_2. Nitrification efficiency is highest when O_2 penetration into the sediment is greatest. Under these conditions, there is least diffusional loss of NH_4^+ which comes from the lower sediment layers to the oxygenated zone. A long diffusional path, from the base of this zone to the sediment surface, results in a lower net efflux of NH_4^+. Reference to these factors clarifies the reasons for the different rates (Fig. 5.5) and efficiencies (Table 5.2) of nitrification. The predicted rates in N cycling are expressed as percentages of the rate of NH_4^+ production in Table 5.2. The NH_4^+ production rates are used in preference to DON production as the latter effluxes and does not participate in sediment processes. In general, increasing POM loading results in higher rates of nitrification (Fig. 5.5). An exception to this is the lower rate $(\Sigma\ NO_3^- + N_2)$ for DOC loading at 60 compared with 36 mmol m^{-2} day^{-1} in the TOP distribution. This low rate of nitrification at the higher loading is associated with a large efflux rate for NH_4^+ and a large proportional

Table 5.2   Fate of dissolved inorganic nitrogen (DIN) molecules produced at different rates and at different sites in a marine sediment*

	DON production (mmol m^{-2} day^{-1})	NH_4^+ (%)	NO_3^- (%)	N_2 (%)	$N_2 + NO_3^-$ (%)
TOP	1	0	36	64	100
	6	46	32	22	54
	10	74	19	7	26
LINEAR	1	0	50	50	100
	6	66	23	11	34
	10	73	19	8	27
MIX	1	0	33	67	100
	6	40	30	30	60
	10	65	19	16	35

*The efflux of dissolved organic nitrogen (DON) is not shown and the other effluxes are expressed as a percentage of the DON mineralised in the sediment. The sum of NO_3^- efflux plus N_2-N efflux is the percentage of NH_4^+ that is nitrified.[8]

NH_4^+ efflux of 74%. This is consistent with a low penetration of O_2, which is confirmed in Fig. 5.6, where O_2 penetration is lowest for the TOP distribution. Similarly, the TOP distribution gave the lowest NO_3^- peak at the highest loading (Fig. 5.7). The location of the DOM source close to the sediment surface results in disproportionally high losses of both DON and NH_4^+. This decrease in nitrification rate at increased organic loadings for the TOP distribution, and partially in the MIX distribution, was predicted in an earlier simulation model and was seen in sediment incubations,[37] but there is little additional published evidence to confirm this hypothesis.

A striking feature of these simulations is the high rate of nitrification for the MIX distribution at the DOC loading of $36\,mmol\,m^{-2}\,day^{-1}$ (Fig. 5.5) and the high proportion of NH_4^+ that is nitrified (60%, Table 5.2). This high rate is attributed to the deep penetration of O_2 in the MIX distribution compared to TOP and LINEAR at the same loading (Fig. 5.6). The deep O_2 penetration is a result of most DOC being oxidised by SO_4^{2-}. Only a small amount of DOC is available in the upper sediment layers and there is also little HS^- because of restricted diffusion: this results in optimum conditions for the oxidation of NH_4^+ diffusing from below. A high NO_3^-, lower in the sediment than the other distributions at this loading, is consistent with the nitrification occurring at depth in the sediment. If HS^- is allowed to diffuse freely, O_2 penetration is small and nitrification is reduced from 3.11 to $0.57\,mmol\,m^{-2}\,day^{-1}$ under approximately the same MIX and loading conditions.[7] If NO_3^- can compete with O_2 for DOC at low O_2 concentrations, an external NO_3^- source could result in a deeper penetration of O_2 and a slightly higher rate of nitrification.[5] A higher rate of nitrification is also to be expected if there is NH_4^+ in the overlying water. This would seldom occur in normal ecosystems, but it could be an important stimulus of nitrification in polluted estuaries. Even a modest increase in NH_4^+ concentration in sediment–water incubations is predicted to have a marked effect on nitrification rate.[3,5]

The deep penetration of O_2 associated with the non-diffusion of HS^- in marine sediments is paralleled in lake sediments by the predicted deep O_2 penetration associated with CH_4 loss. This penetration is associated with very active nitrification.[9,21] The introduction of oxygenated water into the sediment by bioturbation and animal burrows tends to increase nitrification in both marine and freshwater sediments.[28–32] It is possible that O_2, introduced by plant roots into sediment, will stimulate nitrification. There is little published information relating to nitrification in the rhizosphere of aquatic plants, but it is likely that any O_2 will be used in the oxidation of DOC, reduced Fe, Mn and HS^- in preference to the limited amount of NH_4^+ in oligotrophic rooted environments.

In conclusion, one may speculate as to the importance of nitrification. Would there be a major disruption of biogeochemical cycling if no nitrification occurred? The recycling of nitrogen back to the primary

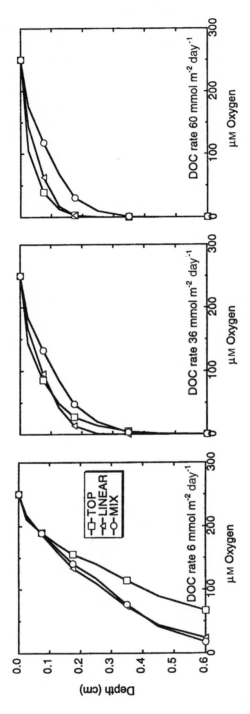

Figure 5.6 Oxygen profiles in sediment in which varying quantities of dissolved organic carbon (DOC), distributed as in Fig. 5.4, are mineralised[8]

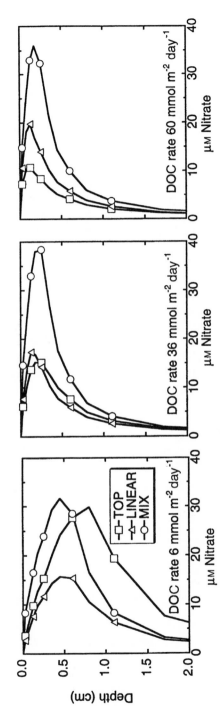

Figure 5.7 Nitrate profiles in sediment in which varying quantities of dissolved organic nitrogen (DON), distributed as in Fig. 5.5, are mineralised[8]

producers would not be affected in any obvious way, as most photoautotrophs prefer NH_4^+ as a nitrogen source. There might be some benefits, as NO_3^- is lost more readily from soils and sediments than NH_4^+, which is retained by ion exchange. However, if there were no nitrification, biological denitrification would not be possible either, and this would have important consequences. Inhibition of sediment nitrification would result in the efflux of only NH_4^+, rather than the efflux of variable ratios of NH_4^+, NO_3^- and N_2.

5.3.3 Denitrification

Denitrification is judged to be a beneficial process in relation to the removal of unwanted nitrogen from an environment in which it is causing too much and unwanted primary productivity, for example in coastal regions, principally in marine environments where nitrogen tends to be a limiting nutrient. In relation to agriculture, dinitrification of nitrate fertiliser is not a process that is viewed with joy by farmers. Denitrification in sediment can be at the expense of NO_3^- in the overlying water (Dw) or of NO_3^- produced in the sediment by nitrification (Dn). An elegant tracer method using ^{15}N-NO_3^- allows both types of denitrification to be measured.[27] The factors that regulate the rate of nitrification are also those that control Dn. This is not surprising since Dn is dependent on nitrification. Dn increases with organic loading up to a certain level (Fig. 5.8); increased loading ($60\,mmol\,m^{-2}\,day^{-1}$) usually results in a decrease in denitrification, as in nitrification. The coupling between nitrification and Dn changes with the level of organic loading; the ratio Dn/nitrification falls with increasing

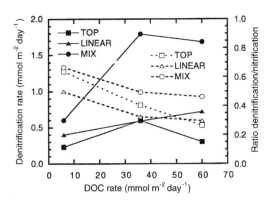

Figure 5.8   Changes in the rate of denitrification and in the ratio of denitrification/nitrification in response to dissolved organic carbon (DOC) added to sediment. DOC is added at 6, 36 or $60\,mmol\,m^{-2}\,day^{-1}$ to sediment in three different modes of distribution: TOP, where addition is to the top surface sediment; LINEAR, where the addition is in a linear gradient decreasing from the surface to 5 cm; and MIX, where addition is uniformly mixed down to 5 cm[8]

loading (Fig. 5.8). This is due to the movement of the zone of nitrification closer to the sediment surface, resulting in a greater efflux of NO_3^- and a proportional decrease in the efficiency of Dn.

The amount of sediment NO_3^- that is denitrified will depend on the activity of the nitrifying population, expressed as the V_{max} for denitrification. The degree of physical coupling between the two processes (closeness of the two zones) is a factor of almost equal importance. This is determined by the sensitivity of denitrification to O_2. If denitrification is relatively insensitive to O_2 (e.g. incomplete inhibition $< 10\,\mu M$) an overlap can occur between the nitrification and denitrification zones, allowing a tight coupling between the processes and much of the NO_3^- can be denitrified. If denitrification is very sensitive to inhibition by O_2 (e.g. complete inhibition at $0.01\,\mu M$), then no overlap can occur between zones and coupling is weak. This is because NO_3^- has a greater probability of diffusing into the water column, as the diffusional path to the zone of denitrification is long. There are few published data on the sensitivity of denitrifying populations to O_2, but there is some indirect modelling evidence for a freshwater sediment, showing that the denitrifying bacteria were insensitive to O_2 inhibition.[9] It was impossible to simulate the actual rate of Dn unless the population was incompletely inhibited by $\sim 10\,\mu M\,O_2$. The sensitivity of denitrification to O_2 is also an important factor in relation to Dw. In the same simulation[9] it was impossible to obtain a sufficiently high Dw rate unless the denitrification zone was close to the sediment surface. Sensitivity to O_2 forces denitrification to a lower depth in the sediment and the long diffusional distance prevents sufficient NO_3^- from reaching this depth.

In general, the other factors that stimulate nitrification (high O_2 concentration in the water, nitrification at depth in the sediment, availability of NH_4^+, low C:N ratio in DOM) also increase Dn.[6] Nitrification was predicted to be inhibited by diffusion of HS^- to the surface, and Dn is similarly affected. This has implications in defining a relationship between oxygen uptake and Dn. Oxygen uptake is often used as a proxy for POM degradation. This is invalid, if there is uncoupling of carbon mineralisation and O_2 uptake, as happens when HS^- is bound as FeS and cannot diffuse. Figure 5.9 illustrates where problems can arise. Simulations are presented for three levels of NH_4^+ production (3.1, 6.2 and 9.3 mmol m^{-2} day^{-1}), with and without HS^- diffusion. When HS^- does not diffuse there are low rates of O_2 uptake associated with high rates of Dn, increasing with organic loading (NH_4^+ availability). When HS^- diffuses, rates of O_2 uptake are predicted to be high and Dn rates are low, decreasing with organic loading. In natural sediments, intermediate degrees of coupling between C oxidation and O_2 uptake would be expected, leading to a large scatter in plots of O_2 uptake and Dn. Oxygen and oxic conditions have a contrary effect on Dw, which is stimulated by low O_2 penetration and by low O_2 concentration in the overlying water.

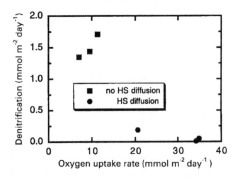

Figure 5.9   Changes in the rate of denitrification associated with changes in O_2 uptake rate, with and without HS^- diffusion, at three different levels of DOC addition[7]

There is the possibility, at high levels of organic loading, that instead of denitrification, NO_3^- will be reduced to NH_4^+. This process, unlike denitrification, is fermentative rather than respiratory. It is not likely to be of importance except in highly eutrophic environments.

Macrofaunal bioturbation, which stimulates nitrification, obviously also increases Dn. It might be expected that Dw would not be increased by bioturbation, but in the event of overlying water being pumped into a burrow intermittently so that oxygen depletion occurs during the stagnant period, there can be an increase in denitrification of external NO_3^-.

Aquatic macrophytes were judged to have little influence on nitrification and are, therefore, unlikely to affect Dn. It is also unlikely that they have much effect on Dw, as the rhizosphere influence probably does not have much effect on the surface sediment and thus on NO_3^- in the water. Phragmites' rhizosphere seems to have some effect on denitrification in the treatment of waste water, so it is possible that either Dw or Dn may be stimulated. There is insufficient evidence on which to base a reliable judgement as to the mechanism involved.

Denitrification is important as it removes NO_3^- from the biosphere, producing N_2 which can be assimilated by a very limited number of bacteria. Again, one may speculate about the result, should all denitrification be prevented. One result would be the efflux of only NH_4^+ and NO_3^- from sediment, both of which may be assimilated by phytoplankton. An increase in phytoplankton growth might be anticipated, depending on the time involved for the effluxed nitrogen to reach the photic zone. This time lag is mostly a function of water depth and stratification. An increase in phytoplankton productivity could be undesirable in some circumstances, but it might be too pessimistic to anticipate a world flooded by NO_3^-. It is probable that nitrogen fixation would be inhibited by an increase in NO_3^-, thus keeping nitrogen availability in balance. Dissimilative reduction is often thought to be the only possible fate for

NO_3^- in sediment, but there is evidence that assimilative reduction can also occur.

5.3.4 Inorganic nitrogen incorporation

It is reasonable to assume that most of the nitrogen requirements of bacteria in sediment are met by NH_4^+ uptake, since there is usually a high NH_4^+ concentration present. It is also assumed, perhaps with less justification, that the C:N of the substrate is that of algal cells (~ 6) or of sediment organic matter (~ 10). Under these conditions net uptake of NH_4^+ is not expected yet can occur, but presumably only for a short time before the microbial cells themselves are recycled.[2] Net assimilation of NO_3^- would not be expected in most sediments. However, in some sediments there is evidence of NO_3^- uptake: profiles of pore water NH_4^+ and NO_3^- from a station off the Arctic island, Svalbard, indicate that uptake had occurred.[10] The gradient of NO_3^- to the surface predicts a net flux to the water of $0.062\,mmol\,m^{-2}\,day^{-1}$, whereas the measured net flux was in the opposite direction ($0.03\,mmol\,m^{-2}\,day^{-1}$). This would indicate a net uptake of NO_3^- ($0.032\,mmol\,m^{-2}\,day^{-1}$) in the upper 10 mm of sediment. There was a predicted net flux of NO_3^- downward ($0.170\,mmol\,m^{-2}\,day^{-1}$), but the measured total rate of denitrification ($Dn + Dw$) was only $0.021\,mmol\,m^{-2}\,day^{-1}$, suggesting a net uptake of $0.149\,mmol\,m^{-2}\,day^{-1}$. There was probably insufficient NH_4^+ in the upper sediment to meet the nutritional requirement of the bacteria, as there was also a slight uptake of NH_4^+ from the water ($0.01\,mmol\,m^{-2}\,day^{-1}$), although the steep gradient would have indicated an efflux of NH_4^+. Similarly, there was evidence for NH_4^+ assimilation just below the zone of O_2 depletion at 20 mm depth. This coincided with the zone of NO_3^- uptake (Fig. 5.10). It is likely that microbial cells, synthesised in the two uptake zones, were hydrolysed and metabolised in the zone of NH_4^+ and NO_3^- production at ~ 15 mm depth. This was a pattern repeated for many

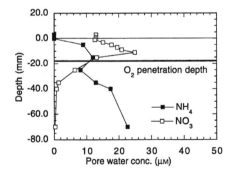

Figure 5.10   Nitrate and ammonium profiles for a sediment near Svalbard. The depth of oxygen penetration is indicated[10]

sediments around Svalbard. There is reason to believe that this uptake of NH_4^+ and NO_3^- was associated with the bacterial metabolism of very high C:N substrate.[10]

5.3.5 Nitrogen fixation

Nitrogen fixation by sediment bacteria is even less likely than NO_3^- assimilation, because of the presence of fixed nitrogen and the requirement for a plentiful energy source. The universal presence of dinitrogen fixing enzymes in anaerobic sulphate-reducing bacteria and in aerobic methane oxidisers is an unexplained mystery, particularly for the former, which normally are found in N-rich environments. It is likely, however, that sulphate-reducing bacteria are important in the indirect supply of fixed nitrogen to aquatic macrophytes. There is a very fast turnover of a small NH_4^+ pool, associated with sulphate reduction of DOC released by roots and nitrogen fixation in the rhizosphere of seagrasses[12] and mangroves.[26] It is suggested that N-starved plants secrete DOC, which then sets N fixation in motion. Breakdown of the bacterial cells releases fixed nitrogen, which the plants can utilise. This is probably a common process for aquatic macrophytes in oligotrophic environments.

5.4　The phosphorus cycle

Phosphate is a limiting nutrient in most freshwater environments and probably often in marine waters also. The cycling of P is thus of great importance ecologically, but it is of less interest biologically, due to the absence of oxidation–reduction states. P always occurs as phosphate, in its many inorganic and organic molecules. It occurs in cells mostly in polynucleotides, which are hydrolysed quite readily and the phosphate is easily mineralised from the hydrolytic products. Cell detritus loses P preferentially and POM reaching the sediment often has a high C:P ratio. Preferential P stripping continues in the sediment, but this does not automatically lead to a rapid efflux of P from sediment to water (and to phytoplankton). Phosphate is retained in oxidised ferric complexes at the sediment surface and is released in quantity only when the sediment becomes reduced or when phosphate complexes with organic molecules. The control of phosphate release is thus partially biological, as it is microbial O_2 uptake that creates the conditions for phosphate recycling to the photic zone.

5.5　Manganese and iron

Oxidised manganese and iron (Mn^{4+} and Fe^{3+}) are involved as electron acceptors in the oxidation of organic carbon compounds. Their importance is determined by the availabilty of the oxidised form, which is controlled by bioturbation.[14-16] Unoxidised manganese can precipitate at depth as $MnCO_3$.[1] Bioirrigation also stimulates the export of Mn^{2+} to the overlying water, where it precipitates as Mn oxides. Export of reduced Mn from sediments depends on the oxic status of the sediment and overlying water. It may be transported laterally until it is oxidised by O_2 and re-enters the sediment as a precipitate.[41] Manganese concentration can be quite high in O_2-depleted water overlying sediments in which manganese reduction has occurred.[40] However, significant manganese reoxidation occurs in a thin subsurface zone of sediments.[41] Oxidation by manganese can account for more than 90% of total carbon oxidation in some marine sediments,[1,16] but in general its importance is much less and can be zero.[41]

Mn is almost exclusively a marine oxidant, but Fe probably plays a role in freshwater sediments. Their chemistry is of considerable interest, but if they are not present it is likely that carbon oxidation will proceed via other electron acceptors or by fermentation. Mn and Fe are almost certainly not involved in the hydrolysis of POM, so at worst, in their absence, unoxidised DOC may efflux from the sediment. This will not interrupt the general mineralisation process, which may be displaced to a different site.

References

1. Aller RC (1990) Bioturbation and manganese cycling in hemipelagic sediments. *Phil Trans R Soc Lond A* **332**: 51-68.
2. Blackburn TH (1980) Seasonal variations in the rate of organic-N mineralization in anoxic marine sediments. In: *Biogéochimie de la Matière Organique à l'Interface Eau-Sédiment Marin*. Édition du CNRS, Paris, pp. 173-183.
3. Blackburn TH (1990) Denitrification model for marine sediment. In: Revsbech NP, Sørensen J (eds) *Denitrification in Soil and Sediment*. Wiley, Chichester, pp. 323-337.
4. Blackburn TH (1991) Accumulation and regeneration: processes at the benthic boundary layer. In: Mantoura RCF, Martin J-M, Wollast R (eds) *Ocean Margin Processes and Global Change*. Wiley, Chichester, pp. 181-195.
5. Blackburn TH (1995) The role and regulation of microbes in sediment nitrogen cycle. In: Joint I (ed.) *Molecular Ecology of Aquatic Microbes*. Springer, Berlin, pp. 55-71.
6. Blackburn TH (1996) Nitrogen gas flux from sediments: insights from simulation modelling. *Aquat Microb Ecol* **10**: 209-211.
7. Blackburn TH, Blackburn ND (1993) Coupling of cycles and global significance of sediment diagenesis. *Mar Geol* **113**: 101-110.
8. Blackburn TH, Blackburn ND (1993) Rates of microbial processes in sediments. *Phil Trans R Soc Lond A* **344**: 49-58.
9. Blackburn TH, Blackburn ND, Jensen K, Risgaard-Petersen N (1994) Simulation model for the coupling between nitrification and denitrification in a freshwater sediment. *Appl Environ Microbiol* **60**: 3089-3095.

10. Blackburn TH, Hall POJ, Hulth S, Landén A (1996) Organic-N loss by efflux and burial associated with a low efflux of inorganic-N and with nitrate assimilation in Arctic sediments (Svalbard, Norway). *Mar Ecol Prog Ser* **141**: 283–293.

11. Blackburn TH, Kleiber P, Fenchel T (1975) Photosynthetic sulfide oxidation in marine sediments. *Oikos* **26**: 103–108.

12. Blackburn TH, Nedwell DB, Wiebe WJ (1994) Active mineral cycling in a Jamaican seagrass sediment. *Mar Ecol Prog Ser* **110**: 233–239.

13. Canfield DE (1989) Sulfate reduction and oxic respiration in marine sediments: implications for organic carbon preservations in euxinic environments. *Deep-Sea Res* **36**: 121–138.

14. Canfield DE (ed.) (1993) *Organic Matter Oxidation in Marine Sediments. Interactions of C, N, P and S Biogeochemical Cycles and Global Change*. Springer, Berlin.

15. Canfield DE, Jørgensen B, Hansen JW, Nielsen LP, Hall POJ (1993) Pathways of organic carbon oxidation in three continental margin sediments. *Mar Geol* **113**: 27–40.

16. Canfield DE, Thamdrup B, Hansen JW (1993) The anaerobic degradation of organic matter in Danish coastal sediments: iron reduction, manganese reduction, and sulfate reduction. *Geochim Cosmochim Acta* **57**: 3867–3883.

17. Dacey JWH, Klug MJ (1979) Methane efflux from lake sediments through water lilies. *Science* **203**: 1253–1255.

18. Damgaard LR, Larsen LH, Revsbech NP (1995) Microscale biosensors for environmental monitoring. *HTRAC* **14**: 300–303.

19. Hargrave BT, Phillips GA (1981) Annual *in situ* carbon dioxide and oxygen fluxes across a subtidal marine sediment. *Estuar Coast Shelf Sci* **12**: 725–737.

20. Jensen K, Revsbech NP, Nielsen LP (1993) Microscale distribution of nitrification activity in sediment determined with a shielded microsensor for nitrate. *Appl Environ Microbiol* **59**: 3287–3296.

21. Jensen K, Sloth NP, Risgaard-Petersen N, Rysgaard S, Revsbech NP (1994) Estimation of nitrification and denitrification from nitrate and oxygen microprofiles in model sediment systems. *Appl Environ Microbiol* **60**: 2094–2100.

22. Jørgensen BB, Revsbech NP, Blackburn TH, Cohen Y (1979) Diurnal cycle of oxygen and sulfide microgradients and microbial photosynthesis in a cyanobacterial mat. *Appl Environ Microbiol* **38**: 46–58.

23. Kristensen E, Ahmed SI, Devol AH (1995) Aerobic and anaerobic decomposition of organic matter in a marine sediment: which is fastest? *Limnol Oceanogr* **40**: 1430–1437.

24. Kristensen E, Blackburn TH (1987) The fate of organic carbon and nitrogen in experimental marine sediment systems: influence of bioturbation and anoxia. *J Mar Res* **47**: 231–257.

25. Larsen LH, Revsbech NP, Binnerup SJ (1996) A microsensor for nitrate based on immobilised denitrifying bacteria. *Appl Environ Microbiol* **62**: 1248–1251.

26. Nedwell DB, Blackburn TH, Wiebe WJ (1994) Dynamic nature of organic carbon, nitrogen and sulfur in sediment of a Jamaican mangrove forest. *Mar Ecol Prog Ser* **110**: 223–231.

27. Nielsen LP (1992) Denitrification in sediment determined from nitrogen isotope pairing. *FEMS Microbiol Ecol* **86**: 357-362.

28. Pelegrí SP, Blackburn TH (1994) Bioturbation effects of the amphipod *Corophium volutator* on microbial nitrogen transformations in marine sediments. *Mar Biol* **121**: 253-258.

29. Pelegrí SP, Blackburn TH (1995) Effect of bioturbation by *Nereis* sp., *Mya arenaria* and *Cerastoderma* sp. on nitrification and denitrification in estuarine sediments. *Ophelia* **42**: 289-299.

30. Pelegrí SP, Blackburn TH (1995) Effects of *Tubifex tubifex* (Oligochaeta: Tubificidae) on N-mineralization in freshwater sediments measured with ^{15}N isotopes. *Aquat Microb Ecol* **9**: 296-303.

31. Pelegrí SP, Blackburn TH (1996) Nitrogen cycling in lake sediments bioturbated by *Chironimus plumosus* larvae, under different degrees of oxygenation. *Hydrobiologia* **325**: 231-238.

32. Pelegrí SP, Nielsen LP, Blackburn TH (1993) Denitrification in estuarine sediment stimulated by the irrigation activity of the amphipod *Corophium volutator*. *Mar Ecol Prog Ser* **105**: 285-290.

33. Revsbech NP, Jørgensen BB, Blackburn TH (1983) Microelectrode studies of the photosynthesis and O_2, H_2S and pH profiles of a microbial mat. *Limnol Oceanogr* **28**: 1062-1074.

34. Revsbech NP, Jørgensen BB (1981) Primary production of microalgae in sediments measured by oxygen microprofile, $H^{14}CO_3$, and oxygen exchange methods. *Limnol Oceanogr* **26**: 717-730.

35. Revsbech NP, Jørgensen BB, Blackburn TH (1980) Oxygen in the sea bottom measured with a microelectrode. *Science* **207**: 1355-1356.

36. Sand-Jensen K, Prahl C, Stokholm H (1982) Oxygen release from roots of submerged aquatic macrophytes. *Oikos* **38**: 349-354.

37. Sloth NP, Blackburn TH, Hansen LS, Risgaard-Petersen N, Lomstein BÅ (1995) Nitrogen cycling in sediments with different organic loading. *Mar Ecol Prog Ser* **116**: 163-170.

38. Smith RL, Klug MJ (1981) Reduction of sulfur compounds in the sediments of a eutrophic lake basin. *Appl Environ Microbiol* **41**: 1230-1237.

39. Suess E (1980) Particulate organic carbon flow in the oceans — surface productivity and oxygen utilization. *Nature* **288**: 260-262.

40. Thamdrup B, Canfield DE, Ferdelman TG, Glud RN, Gundersen JK (1996) A biogeochemical survey of the anoxic basin Golfo Dulce, Costa Rica. *Rev Biol Trop* **44** (Suppl 3): 19-33.

41. Thamdrup B, Fossing H, Jørgensen BB (1994) Manganese, iron, and sulfur cycling in a coastal marine sediment, Aarhus bay, Denmark. *Geochim Cosmochim Acta* **58**: 5115-5129.

6

Microbial mats and stratified water columns

Bacteria and bacterial processes dominate mat communities. Under suitable conditions, mats develop on solid surfaces or on sediments in aquatic or semi-aquatic habitats. In microbial mats dissolved nutrients and metabolites are transported by one-dimensional (vertical) molecular diffusion. The distinction between microbial mats and biofilms is not sharp; we use the following criteria for a definition (but see Hamilton[27]). Microbial mats are typically stratified vertically with respect to different functional types of bacteria. Microbial mats are thicker (often several millimetres) than biofilms. In microbial mats various types of filamentous prokaryotes are the most conspicuous part and they are responsible for the mechanical coherence of the mat. The mechanical stability of microbial mats is sometimes reinforced by the bacterial excretion of mucous polymers (exopolymers) producing a gelatinous matrix. In contrast, biofilms are much thinnner, usually consisting of only a single layer of bacteria (typically together with various eukaryotic micro-organisms) and they do not show a characteristic vertical zonation or a strong mechanical coherence. In microbial mats there is a high degree of mutual interaction and interdependency between different functional groups of bacteria; in biofilms all the organisms depend largely on dissolved nutrients in the overlying water or on the hydrolytic degradation of their solid substratum. Microbial mats originate as biofilms (i.e. the colonisation of surfaces by a single layer of cells belonging to one or few species of bacteria). Most biofilms (such as found on the surface of detrital particles, mineral grains or other submersed surfaces) do not develop into microbial mats due to energy limitation (in terms of access to light or reducing compounds) and due to protozoan and animal grazing. Microbial mats may be photosynthetic or chemotrophic. Some of the highest known metabolic rates have been recorded in certain mats.

Requirements for the development of microbial mats include a suffi-cient energy supply and conditions that more or less exclude eukaryotic activity, especially grazing and mechanical disturbance (bioturbation). In sediments, disturbances include mixing of surface sediment, bulk sediment ingestion, burrowing and generation of advective water transport within the sediment. Animals are usually abundant in and on aquatic sediments, in which case microbial mats do not develop. The absence of macrofauna may be brought about by periodic desiccation, hyperhaline conditions, periodic or permanent anaerobic or hypoxic conditions in the overlying water column, or extreme temperatures.

Mats that develop under hyperthermal conditions (such as in hot springs and around hydrothermal vents) differ in several respects from other mats in terms of organisms and functional diversity. When temperatures exceed 55–60 °C these mats are altogether devoid of eukaryotic organisms. Hydrothermal mats are treated separately as an example of life in "extreme environments" in Chapter 8. Other types of mats seem always to harbour some eukaryotes in addition to the dominant prokaryotic constituents: various phototrophic and phagotrophic protists and usually also representatives of the meiofauna (i.e. animals measuring from about 0.1 to a couple of millimetres) such as nematodes, harpacticoid copepods, oligochaetes and rotifers. Many of these organisms graze on mat-forming prokaryotes and it is an open question under which conditions mats can maintain their integrity in spite of some phagotrophic activity. The essential condition for mat formation seems to be the absence of macrofauna that are capable of mixing and burrowing into sediments and of producing large-scale advective water transport.

Microbial mats are widely distributed in spite of the somewhat special conditions required for their formation and integrity. In most places they are transient or seasonal phenomena of limited extension and they grow at most to a few millimetres in thickness. Under particular circumstances (mainly in hyperhaline lakes or lagoons) cyanobacterial mats may be more permanent structures and they then slowly (over many years) accumulate resilient organic material (and $CaCO_3$); the growth rates of some such mats have been estimated to be about 1 mm per year. Even in these thicker mats almost all biological activity takes place in the upper few millimetres. But below this superficial layer a thick deposit with a laminated structure is found. The lamina represent the remains of previous surface layers and reflect some sort of seasonality in growth or seasonal patterns of sedimentation. Such mats are referred to as stromatolithic mats.

Microbial mats (although known for a long time) have more recently drawn particular interest. This is because laminated sedimentary rocks (stromatolites) dating from about 3.5×10^9 years to the end of the Precambrian are now (with much confidence, but see Grotzinger and Rothman[25]) interpreted as the fossil remains of ancient cyanobacterial mats (see Chapter 10). Recent microbial mats are, therefore, considered to be analogues to the earliest known biotic communities on Earth as discussed in detail in Schopf and Klein.[51] Other important general treatments of the biology of microbial mats are listed.[8,46,56,58]

6.1   Mats based on colourless sulphur bacteria

Lithotrophic bacteria which oxidise reduced sulphur compounds (colourless sulphur bacteria) are ubiquitous in aquatic sediments. Under some circumstances they form cohesive mats on the sediment surface; macroscopically, such mats form conspicuous white patches or extensive white areas on the sea floor. The white colour is due to the intracellular accumulation of elemental sulphur. The colourless sulphur bacteria depend on simultaneous access to sulphide and O_2. They are generally microaerophiles (preferring a p_{O_2} of around 5% atmospheric saturation) and they do not tolerate very high sulphide concentrations. They are therefore "gradient organisms" which in most habitats are confined to the narrow 0.1–1-mm thick zone where low concentrations of HS^- and O_2 coexist (see Fig. 6.1; also Nelson[40] and Nelson *et al.*[41]).

The chemosensory behaviour of these organisms is an important property for understanding the structure of these mats. Sulphide oxidisers apparently respond only to p_{O_2} and not to sulphide. *Beggiatoa* filaments incessantly glide between aerobic and anaerobic conditions, changing direction whenever the environmental O_2 tension becomes too high or too low. The unicellular *Thiovulum* can reverse its swimming direction by 180° when it encounters adverse conditions. In this way the organisms optimise their position within sufficiently steep O_2 gradients, and maintain clearly delimited bacterial layers which measure only a fraction of a millimetre. Conversely, the metabolic activity of the mat organisms maintains the steep chemical gradients and thus the integrity of the habitat.[16,39]

Mats of colourless sulphur bacteria may form under somewhat different circumstances. Transient mats appear above accumulations of degrading organic material (seaweeds, dead animals) buried in or lying on sediments, because this creates a high local production of sulphide based on sulphate reduction and to some extent fermentation of S-containing proteins. Such white patches are often seen on shallow-water sediments which are protected from strong wave action. In marine shallow-water sites (and presumably elsewhere) such biota initially undergo characteristic successional changes during the first few days. Following active sulphide production, the patches are first colonised by unicellular colourless sulphur bacteria: *Macromonas*, *Thiospira* and, later, the large rapidly swimming *Thiovulum* which forms characteristic 0.1–0.2-mm thick white veils. Later, the unicellular forms are replaced by gliding *Beggiatoa* filaments of different diameters. This succession is driven at least in part by protozoan grazing since mass development of unicellular sulphur bacteria over a couple of days is followed by dense populations of bacterivorous ciliates that decimate the sulphur bacteria. The more slowly growing filamentous sulphur bacteria seem much less vulnerable to grazing by protozoa. *Beggiatoa* mats differ from cyanobacterial mats in that the former do not seem to produce large amounts of exopolymers and do not have a gelatinous matrix.[2] Eventually, such small mats disappear because the source of sulphide is depleted. When exposed to light (in shallow waters) the patches will eventually be colonised and overgrown by phototrophic sulphur bacteria and by oxygenic phototrophs.

Colourless sulphur bacteria are almost always a constituent of cyanobacterial mats. In the light the surface layers of such mats are aerobic, and sulphide oxidation is due to a large extent to phototrophic bacteria in deeper anoxic strata; the colourless sulphur bacteria are then found several millimetres beneath the surface at the oxic–anoxic interface, but they have limited access to sulphide. During night, however, oxygen penetration is less, the sediments are anoxic almost to the surface of the mat and the phototrophic bacteria cannot oxidise reduced sulphur. Colourless sulphur bacteria migrate to the surface which then appears white and they are responsible for sulphide oxidation in the absence of light (see also Section 6.2).

More permanent and extensive mats of filamentous colourless sulphur bacteria develop in deeper waters in darkness or in dim light where the water column, at some height above the sediment, is permanently or frequently hypoxic ($p_{O_2} < 5$–10% atm.sat.) thus preventing the colonisation of the sea floor by larger animals. Such conditions especially arise in stratified basins or fjords or where large amounts of organic debris tend to accumulate; the phenomenon is not rare although few detailed studies exist.[1,2,17,36] The most extensive mats are probably those found off the west coast of South America where the high water column productivity leads to hypoxic bottom waters. These thick mats consist mainly of *Thioploca* (a close relative of *Beggiatoa*, but in which several filaments are found together within a communal sheath).[22]

An example of the element cycling of a subtidal *Beggiatoa* mat is presented below.[17] The particular mat is situated at a depth of 6–7 m in a small basin in the Sound off the harbour of Helsingør (Denmark). The mat is about 0.6 mm thick and covers a sulphidic sediment rich in organic matter. *Beggiatoa* filaments constitute about 90% of the mat biovolume, the remainder being cyanobacteria, various other bacteria, nematodes and protozoa. Phototrophic sulphide oxidisers are absent due to the limited access to light at this depth. Sulphide oxidation takes place in two steps: deeper in the mat O_2 tensions are very low and sulphide oxidation is incomplete so that sulphur accumulates in the cells; closer to the surface sulphide is almost absent, O_2 tension is higher and the cells oxidise the stored elemental sulphur to sulphate. The individual *Beggiatoa* filaments continuously move up and down between the bottom and the surface of the mat over a distance of about 0.6 mm.

Corresponding pairs of the downward fluxes of O_2 and the upwards fluxes of HS^- (estimated from concentration gradients in the dark) are shown in Fig. 6.1 for different sites and times for this microbial mat. Since

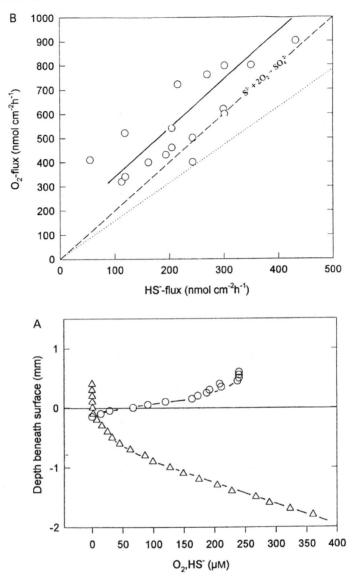

Figure 6.1   A. Gradients of oxygen and sulphide through a *Beggiatoa* mat (which occupies the upper 0.6 mm of the sediment). B. Corresponding values of oxygen and sulphide fluxes to the *Beggiatoa* mat calculated from concentration gradients. The dashed line represents the stoichiometric relation between sulphide and oxygen (assuming complete oxidation to sulphate). The dotted line represents sulphide that is oxidised directly by the bacteria (with O_2) assuming that the reduction equivalents of the remaining sulphide (represented by the distance between the dashed and the stippled lines) is used for the assimilatory reduction of CO_2. From Fenchel and Bernard[17]

it takes 2 mol of O_2 to oxidise 1 mol of HS^- completely to sulphate it is seen that about 85% of the O_2 consumption (the dashed line in Fig. 6.1) is directly or indirectly due to sulphide oxidation; this demonstrates the dominant role of the sulphur cycle in this type of community. The sulphur bacteria oxidise only about 80% of the sulphide directly (corresponding to about 67% of the total O_2 consumption); the remaining reducing power is used for the assimilatory reduction of CO_2 (the stippled line in Fig. 6.1; see Nelson *et al.*[42]). The O_2 consumption corresponding to the area between the dashed and the stippled lines, or about 17% of the total O_2 consumption, represents subsequent mineralisation of sulphur bacterial production (largely through grazing on the sulphur bacteria by protozoa and nematodes). Only about 15% of the total sediment O_2 uptake represents all other aerobic processes not related to the sulphur cycle. A more complete diagram of the mineral cycling of this *Beggiatoa* mat is shown in Fig. 6.2. The rates represent an estimate of an average situation and the (limited) role of oxygenic photosynthesis of cyanobacteria has also been taken into account. Sulphate is depleted at a depth of about 10 cm in the sediment and the terminal anaerobic mineralisation beneath this depth is due to methanogenesis which, however, accounts only for a few per cent of the total anaerobic mineralisation of organic carbon. The methane produced is largely oxidised anaerobically and this is assumed to happen via sulphate reduction. The system is principally driven by the allochtonous input of organic carbon in terms of macroalgal and seagrass debris, which tends to accumulate within this limited basin. The element cycling of such mats is therefore relatively open with respect to C and O_2 which are continuously supplied from the surroundings whereas the S cycle is largely self-contained. The diagram is incomplete and simplified in that it implies a steady-state situation and it does not take into account that some resilient organic matter and iron sulphides accumulate in the sediment. Another complication may be that some of the sulphide oxidation carried out by *Beggiatoa* may be due to denitrification rather than to O_2 respiration. Evidence that denitrification may be important in such mats has recently been found for *Thioploca* mats off the coast of Chile.[20]

Mats of lithotrophic sulphur bacteria which are supported by allochtonous organic carbon are by far the most widely distributed type. Such mats, however, may also be based on sulphide deriving from geological processes and reaching aquatic habitats via springs or seepage through the sea floor. Hydrothermal deep sea vents are discussed in Chapter 8. Seepage of sulphidic water also occurs in shallow sediments and may lead to mats of colourless sulphur bacteria, for example off the coast of the island Milos in the Aegean Sea.[9] Certain water-filled caves receive sulphidic waters through springs; depending on how this mixes with oxic water, it may result in suspensions of unicellular sulphur bacteria or to mats of filamentous sulphur bacteria covering the cave walls.[55]

$$\text{mmol}(C,O_2,S) \ m^{-2}day^{-1}$$

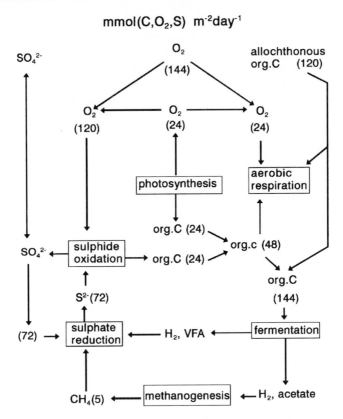

Figure 6.2   C, O and S cycling of a *Beggiatoa* mat; numbers in parentheses are average fluxes in mmol $m^{-2} d^{-1}$. From Fenchel and Bernard[17]

6.2   Cyanobacterial mats

The most studied mats are those dominated by filamentous cyanobacteria. They are widely distributed in protected intertidal sand flats where periodic desiccation discourages colonisation by marine invertebrates. The characteristic, sharply defined, coloured bands of such mats (green on the top, red or purple a few millimetres down and beneath this black) were recorded already before the middle of last century[45] and later given the name "Farbstreifen-Sandwatt".[52] Such shallow-water or intertidal mats have now been recorded from a variety of sites.[15,19,43,57,60] Subtidal, cyanobacterial mats which develop on coral debris and in seagrass beds are known from tropical waters.[63] These mats are all ephemeral or seasonal. Hyperhaline conditions may develop naturally in some tropical or subtropical lagoons and in hyperhaline lakes, or artificially in evaporation ponds used for salt production. Such habitats are more stable and may lead to laminated stromatolithic mats of considerable thickness;[5,30,35] see also Cohen and Rosenberg,[8] Schopf and Klein[51] and Stal and Caumette.[56] Concerning structure and function all these mats are very similar in most respects and a separate discussion of ordinary ephemeral and hyperhaline stromatolithic cyanobacterial mats is not necessary. The special cyanobacterial mats of hot springs and in Antarctic lakes are discussed in Chapter 8.

6.2.1 Structure of cyanobacterial mats

The characteristic and macroscopically visible colour banding of cyanobacterial mats crudely reflects the zonation of the most important phototrophic processes (Fig. 6.3). The uppermost layer typically has a yellow or brownish colour; the dominant phototrophs in this layer are

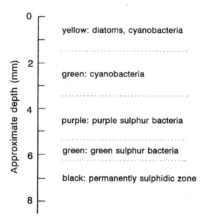

Figure 6.3　　The colour banding of typical cyanobacterial mats

diatoms (and other phototrophic protists) together with filamentous and unicellular cyanobacteria. The underlying dark green layer is totally dominated by filamentous blue-greens, including especially oscillatorians (*Microcoleus, Oscillatoria, Lyngbya, Spirulina*) and sometimes *Anabaena*, but also members of the Chroococcales (*Merismopedia, Chroococcus*). This layer is coherent due to intertangled filaments and the production of exopolymers which bind mineral particles together; such mats can be stripped off like a carpet. The surface layer may have a blistered appearance caused by O_2 bubbles which form under intense photosynthesis. Pigments extracted from these layers are first of all chlorophyll *a* and cyanobacterial phycobillins. The purple layer is dominated by purple sulphur bacteria (*Thiocapsa, Chromatium, Thiopedia*, etc.) and the principal photosynthetic pigment is bacteriochlorophyll *a*. The lower part of this zone is sometimes peach coloured and harbours the bacteriochlorophyll *b*-containing *Thiocapsa pfennigii*.[43,47] The lower green layer is dominated by green sulphur bacteria from which bacteriochlorophyll *b* can be extracted; this layer is not always macroscopically visible. Beneath the lower green layer the sediment is black due to ferrous sulphide. Other prokaryotes that are invariably present include (in addition to a variety of bacteria that cannot be identified microscopically) the filamentous *Chloroflexus* and *Beggiatoa*. Probably all mats also harbour various eukaryotic micro-organisms (amoebae, flagellates, ciliates) and meiofauna, especially nematodes.

The thickness of cyanobacterial mats depends primarily on light penetration. In principle, light intensity in a homogeneous medium decreases exponentially with depth and in a typical pure quartz sand values of 1% of surface irradiance are reached 3–4 mm beneath the surface. (One per cent of surface irradiance is traditionally considered to be the lower limit for a positive net photosynthesis; however, evidence suggests that cyanobacteria and anoxygenic phototrophs can grow at somewhat lower light intensities.) Owing to scattering, long-wave (infrared) light penetrates deeper than short-wave light, a fact that renders the infrared absorption of bacteriochlorophylls adaptive.[19] Close to the surface, scattering and collimation of the light cause deviations from a simple exponential attenuation with depth.[38] This also applies in the intact layered mat where selective absorption of light with different wavelengths takes place at different depth due to selective absorption spectra of different photosynthetic pigments. This has been demonstrated *in situ* using fibre-optic microprobes.[32] Light penetrates further into more coarse-grained sediments and so the thickness of the zonation pattern shown in Fig. 6.3 is somewhat variable. Gypsum crystals, forming on the bottom of evaporitic ponds, transport light well and in such substratum the purple layer of the mat may be situated beneath 2 cm depth.[44]

The above description of the vertical zonation of different functional groups of organisms is somewhat crude. For example, the fact that different purple sulphur bacteria vary with respect to their tolerance to O_2 and HS^- and that some are motile and others are not[58] suggests that the 5–10-mm thick active zone of cyanobacterial mats could be subdivided at a much finer scale. Also, as discussed below, opposing processes (sulphate reduction–sulphide oxidation, phototrophy–heterotrophy) occur at similar depths. Filamentous cyanobacteria are often covered by attached heterotrophs. The function of the mats depends to a large extent on syntrophic interactions between complementary types of metabolism, and spatial structures at a micrometre scale probably play an important role.

6.2.2 *Mineral cycling in cyanobacterial mats*

The major processes of cyanobacterial mats are shown in Figs 6.4 and 6.5. Cyanobacteria dominate oxygenic photosynthesis. Extended studies have

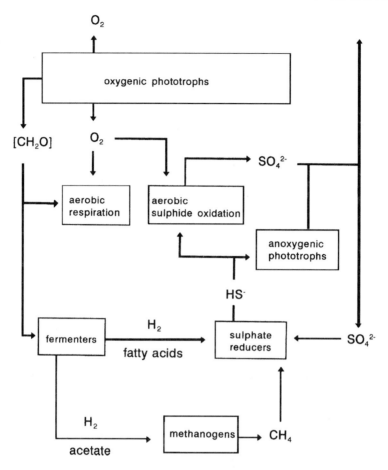

Figure 6.4   The principal cycling of C, O and S of cyanobacterial mats

shown that almost all the activity in terms of energy flow, and both photo-synthetic and heterotrophic element cycling, is confined to a surface layer < 1 cm thick; the slow anaerobic degradation of resilient organic matter in deeper layers of the sediment or in the deeper part of stromatolithic mats (producing sulphide and methane that diffuse upwards) plays a small role for mat metabolism which is primarily driven by the rapid turnover of cyanobacterial photosynthate. In the light, cyanobacterial mats are net producers of O_2 (thus accumulating reduced C and S) and during darkness the mat consumes O_2. Conversely, CO_2 is taken up during day and released during night. Most S, C and O_2, however, is recycled within the mats implying relatively closed systems.[5,31] The N cycle is probably also largely internal, but N fixation (by cyanobacteria and possibly by anoxygenic phototrophs and other anaerobic or microaerobic bacteria) compensates for

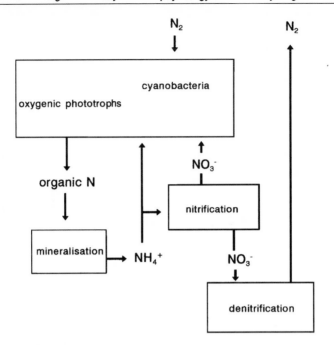

Figure 6.5   The N cycling of cyanobacterial mats

the losses of N due to denitrification. The N cycle varies diurnally due to the vertical migration of the oxic–anoxic interface and the fact that N fixation (carried to a large extent by oscillatorians without heterocysts) is largely confined to darkness when conditions close to the surface are micro-oxic or anaerobic.[46] Nitrogen fixation seems to be a particularly important process during the initial establishment of cyanobacterial mats; the first organisms to establish themselves on the sediment surface are species of *Oscillatoria*.[57] Established cyanobacterial mats approximate steady-state systems and the slow accretion of stromatolithic mats cannot be demonstrated directly by metabolic rate measurements, but only by following the mat over a longer time span (months to years).

During intense light exposure, photosynthesis leads to O_2 tensions exceeding atmospheric saturation by a factor of 2–3 (Fig. 6.6). The steepness of the O_2 concentration gradient (= the net production of O_2) has been interpreted as a direct measure of net production (the accumulation of reduced C or S during illumination). This is not quite true, however, because some of the produced O_2 is used for the reoxidation of immobilised, reduced end products of anaerobic mineralization (mainly as FeS), which have accumulated during the dark. Gross photosynthesis is not easily measured. However, the following considerations allow for a method that yields

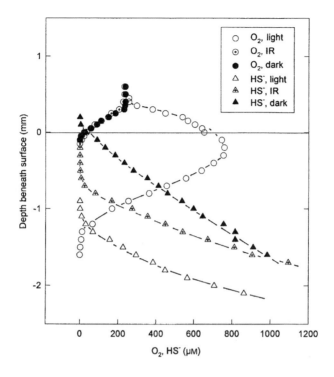

Figure 6.6   Oxygen and sulphide gradients in the upper millimetre of a cyanobacterial mat (from Nivå Bay, Denmark) exposed to darkness, to white incandescent light, and to infrared light, respectively (original data)

an estimate of total O_2 production, including that which is continuously consumed. A cyanobacterial mat exposed to constant illumination will exhibit a steady-state O_2 profile after some time ($\sim 1\,h$). If the mat is then suddenly exposed to darkness, the O_2 concentration at any point (as measured with an inserted O_2 microelectrode) will immediately decrease. Assuming steady state prior to the dark period, then the initial decrease in O_2 concentration measures the previous gross photosynthesis at the position of the electrode tip (Fig. 6.7). Integrating such measures from the surface to the bottom of the green layer will thus estimate gross photosynthesis. In Fig. 6.8 gross and net production of a cyanobacterial mat are shown as functions of surface illumination ("downwelling irradiance"). The difference between the two curves measures O_2 consumption (respiratory activity) which increases with increasing light intensity. This shows that most of the O_2 produced is recycled internally. Several mechanisms have been suggested to explain the increase in O_2 consumption with increasing light intensity. During photosynthesis the oxic zone expands considerably

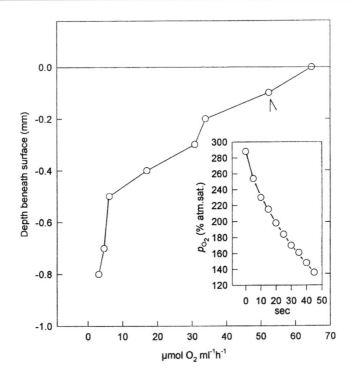

Figure 6.7   Gross production at different depths in a cyanobacterial mat (Nivå Bay, Denmark) with a surface illumination of $\sim 800\,\mu\text{Einstein m}^{-2}\,\text{s}^{-1}$. Insert: Decrease in p_{O_2} following a transition from light to dark 0.1 mm beneath the surface (original data)

so that there is a larger sediment volume in which O_2 uptake takes place.[14] Also, at a high light intensity and a high p_{O_2}, the cyanobacteria photorespire simultaneously with photosynthesis excreting glycolate, which is subsequently mineralised. Finally, under extreme CO_2 limitation the cyanobacteria may photoreduce water, leading to H_2 excretion; the H_2 is subsequently oxidised. It has recently been found that aerotolerant sulphate reducers are active within the cyanobacterial mat, presumably using glycolate or H_2;[21] it must be assumed that the sulphide formed is immediately oxidised with O_2. As mentioned above, however, much of the O_2 produced during light is actually used to cover an "oxygen debt" that has built up during the preceeding dark period and so the net O_2 production underestimates the real net production, which can be only be measured accurately as inorganic C uptake. The actual heterotrophic activity is still higher in the light than in the dark, but only by a factor of about 1.5 rather than by a factor of 3–4 as Fig. 6.8 would otherwise suggest (Fenchel, unpublished data).

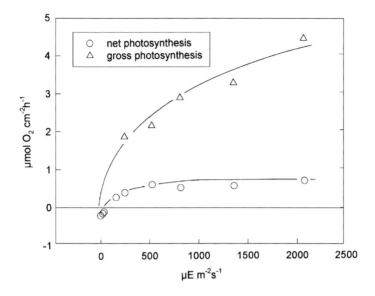

Figure 6.8   Net and gross production of O_2 in a cyanobacterial mat (Nivå Bay, Denmark) as a function of surface illumination; the difference between the two curves is the corresponding O_2 uptake of the mat (original data)

The oxidation of reduced sulphur can take place via either lithotrophic oxidation with O_2 (or NO_3^-) or via anoxygenic photosynthesis under anoxic conditions. The former process is carried out by colourless sulphur bacteria (in particular *Beggiatoa*) and the latter by purple or green sulphur bacteria and by some cyanobacteria under sulphidic conditions. Figure 6.6 shows O_2 and HS^- gradients in a cyanobacterial mat in the dark, and when illuminated by white incandescent light and by infrared light only. In the dark the oxic–anoxic interphase has migrated to the surface and all sulphide oxidation takes place at low O_2 concentrations (within the depth interval where the sulphide gradient is upwards concave) and this is due to lithotrophic sulphide oxidisers. With infrared light illumination O_2 penetrates only slightly further down (since the upwards sulphide flux is decreased, there is no oxygen production). Sulphide is almost completely oxidised under anoxic conditions by purple sulphur bacteria below 0.5 mm. During exposure to "white" light oxygenic photosynthesis is active and the oxic–anoxic boundary migrates down to a depth of about 1.3 mm. Now the sulphide gradient suggests the presence of aerobic sulphide oxidation at this depth as well as some phototrophic sulphide oxidation deeper down. In this mat both types of sulphide oxidation occur; as expected, only lithotrophic HS^- oxidation takes place in the dark while both phototrophic and lithotrophic HS^- oxidation may take place in the light.

The competition between phototrophic and lithotrophic sulphide oxidation in microbial mats has drawn some attention in the literature. Empirical evidence suggests that at permanently low light intensities (autumn to spring at higher latitudes, water depths below a couple of metres) purple sulphur bacteria are seemingly absent even if there is sufficient light for oxygenic photosynthesis and even if anoxygenic phototrophs can cope at very low light intensities. Under such circumstances all sulphide oxidation is probably due to lithotrophic oxidation (although cyanobacterial phototrophic sulphide oxidation cannot be excluded). This observation has been supported by theoretical considerations on the interaction between light limitation and competition for sulphide.[61] Another situation has been described[33] in which oxygenic photosynthesis in a cyanobacterial mat is so intense that the oxic zone extends to depths where light intensities are too low for anoxygenic phototrophs which consequently are absent. Some cyanobacteria are known to be capable of phototrophic sulphide oxidation under sulphidic and anoxic conditions.[7] Such cyanobacteria were found to be the most important phototrophic sulphide oxidisers in the stromatolithic mat in Solar Pond, Sinai, in which purple sulphur bacteria are quantitatively unimportant;[31] although not studied elsewhere in detail this seems to be an exception.

The "black band disease" of stony corals represents a special type of short-lived cyanobacterial mat. They occur as black bands that migrate over the infected coral (at velocities of several millimetres per day) leaving behind only the denuded calcareous skeleton. The dominant constituents of the bands are filamentous cyanobacteria (*Phormidium*) overlying a layer of *Beggiatoa* filaments. As the band progresses, newly covered polyps die (presumably due to the anoxic and sulphidic environment beneath the microbial community). The disintegrating polyps provide substrates for fermentative and sulphate-reducing bacteria and mineral N for the cyanobacteria. Infection of coral colonies requires an initial, localised injury and exposure to previously infected corals.[50]

6.3   Other types of mats

A type of mat which, from a qualitative point of view, largely harbours the same organisms as cyanobacterial mats, but in which purple sulphur bacteria play a dominant role, can be found during summer in protected shallow-water bays and lagoons where large amounts of organic debris accumulate. Sulphide is produced in copious amounts and the sediment and detrital particles become covered by dense masses of purple sulphur bacteria (especially *Thiocapsa roseopersicina* and *Chromatium*) so that everything looks as if covered by purple or pink paint. Such communities are referred to as "sulphureta". Closer examination reveals that other organisms including colourless sulphur bacteria, cyanobacteria and various eukaryotic micro-organisms are also present. First described by Warming in 1875[62] from inner Danish waters, the phenomenon has since been recorded in a variety of sites.[6,15,59]

The basic unit of trickling filters in aerobic sewage treatment plants (and of similar "slimy layers" found in organically polluted streams) is also a microbial mat which in most respects is similar to natural mats previously described. These mats maintain a net mineralisation of exogenous organic matter. Grazing by protozoa and meiofauna prevents the continuous accumulation of microbial biomass. Mats of several millimetres in thickness (often referred to as biofilms in the literature) develop on solid surfaces, and are continuously flushed with water containing a high load of organic and inorganic nutrients. At some depth (< 1 mm) in the mat conditions are anaerobic; a substantial part of the organic material is initially mineralised anaerobically and the resulting sulphide is subsequently oxidised at the surface of the mats. Where exposed to light the surface of the biofilm is colonised by microalgae and cyanobacteria.[37]

The last type of mats to be considered briefly here are those formed by iron bacteria, although in some respects they should be classified as biofilms rather than microbial mats in the terms of the definitions used here. Iron bacteria catalyse the oxidation of ferrous iron; the resulting ferric iron is deposited externally as ochre (ferric hydroxides). In ferruginous springs with slightly acid, mildly reducing water containing little dissolved organic matter, layers of the stalked bacterium *Gallionella* develop on solid surfaces and the bacteria may eventually be responsible for substantial ochre deposits. The oxidised iron is deposited in a long twisted band (stalk) excreted by each bacterial cell. Such microbial iron deposits have previously served as iron ore in some areas. They may also cause problems in terms of clogging drainage systems, wells and water pipes. It is still an open question whether *Gallionella* actually uses Fe^{2+} as a substrate in a respiratory process or whether the iron oxidation is coincidental. *Gallionella* has also been recorded from seawater. In more organic rich freshwater streams mats formed by the filamentous *Sphaerotilus* and *Leptothrix* occur. These bacteria excrete sheaths incrusted by ferric iron (or manganese); it is generally believed that they are organotrophs and that iron oxidation is catalysed by the exopolymers constituting the sheaths.[23,28]

More detailed studies of centimetre-thick mats of iron bacteria (including both *Gallionella* and *Leptothrix*) have recently been carried out.[12,13] A characteristic vertical zonation was demonstrated and it was shown that bacterial growth depended on the access to ferrous iron, thus providing evidence for the respiration of Fe^{2+} (at near neutral pH). Under anaerobic conditions the reduction of Fe^{3+} to Fe^{2+} took place in the presence of acetate or succinate; this suggests the presence of a closed iron cycle within the mat.

6.4   Stratified water columns

A vertical gradient of water density (a "pycnocline") stabilises the water column and effectively slows vertical transport of solutes. Sufficiently deep water bodies are at least transiently stratified during the warm season. The upper water layer is heated by insolation and the resulting "thermocline" resists mixing by wind energy. During autumn the water column becomes almost isothermic and wind-driven turbulent mixing of the entire water column takes place. Many coastal waters (and some lakes) have a deep layer of more saline water leading to a "halocline" which separates a mixed surface layer from a more stagnant deep part of the water column. Even a small difference in salinity causes a relatively large difference in water density and a vertical stratification based on a halocline is therefore more stable than one based on a thermocline alone. Temperate lakes may be "monomictic", meaning that the water column is mixed once a year during the period from autumn to spring. Some temperate lakes are "dimictic", meaning that the water column is mixed during spring and again during autumn and is stratified during summer and winter (in winter stratification is due to an inverse temperature gradient). Tropical lakes may show a more or less permanent thermal stratification or are mixed at irregular intervals. The deep water of "meromictic" lakes contains higher concentrations of dissolved salts and such lakes are permanently stratified.

The pycnocline represents a barrier for the vertical transport of solutes such as the downward flux of dissolved O_2 or the upward flux of mineral nutrients while the sedimentation of particulate matter is not affected. However, some turbulent diffusion above (and sometimes also below) the pycnocline secures that vertical transport across the pycnocline and vertical mixing is still several orders of magnitude higher than can be accomplished exclusively by molecular diffusion.

In productive waters, with a stratified water column, the deep layer may become anoxic. This happens because the aerobic mineralisation of particulate organic material which sinks down from the photic zone requires more oxygen than can be supplied by turbulent diffusion through the pycnocline. Consequently, biologically stratified systems develop in many respects resembling microbial mats. The major difference is that, while the characteristic scale of microbial mats is around a millimetre, corresponding zones in the water column are measured in metres. Also, the close spatial and temporal coupling between autotrophic and heterotrophic processes, which is characteristic of microbial mats, is less prevalent. This is because anaerobic mineralisation largely takes place in or on the sediment surface which may be separated by many metres from the zone inhabited by chemoautotrophic or phototrophic organisms and which is typically found at or immediately beneath the pycnocline.

Anoxic deep water is a common phenomenon in eutrophic lakes during the stratification period and in meromictic lakes as a permanent feature. It also occurs in many fjords with a halocline and a sill at the entrance and in some marine basins with a permanent deep halocline. The largest such marine anaerobic basins are represented by the Black Sea, the Cariaco Trench (Caribbean Sea) and the Gotland Basin in the Baltic Sea.[10,11,24,26,29,34,48,49,53,54]

A deep basin in the Mariager Fjord (east coast of Jutland, Denmark) provides a example of a stratified water body with anoxic deep water;[18] (see also Figs 6.9 and 6.10). The brackish fjord has a sill at its entrance;

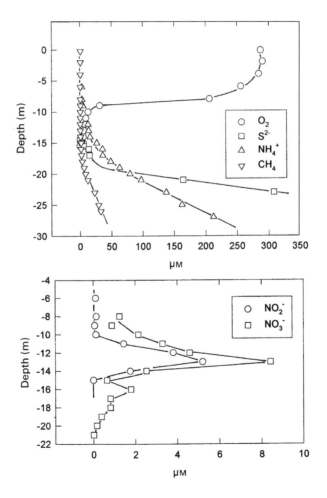

Figure 6.9   Concentration profiles of O_2, HS^-, NH_4^+, CH_4, NO_2^- and NO_3^- in the stratified Mariager Fjord in August 1994. From Fenchel *et al.*[18]

$$mmol\ (C,N,S)\ m^{-2}day^{-1}$$

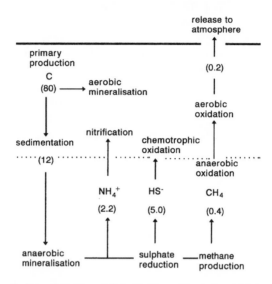

Figure 6.10   Cycling of C, N and S in Mariager Fjord; the thick horizontal line represents the water surface and the stippled line represents the oxic–anoxic interface. Data from Fenchel *et al.*[18]

the lower ∼ 15 m of a central basin contains water with a relatively high salinity and it is exchanged only under extreme weather conditions at intervals of several years. The water column is typically anoxic below depths of around 16 m. Vertical gradients of O_2, HS^-, NH_4^+, CH_4, NO_3^- and NO_2^- are shown in Fig. 6.9. It may be assumed that the steepness of the almost linear gradients of ammonia, sulphide and methane below 20 m reflects the relative vertical fluxes of these compounds. This assumption requires that the chemical species are neither produced nor consumed in the anaerobic water column and this is not quite true. Although sulphate reduction mainly takes place in the surface layers of the sediment, some sulphide is probably also produced in the water column. Also some anaerobic methane oxidation takes place in the water column and some methane escapes from the sediment via ebullition.

The ammonia is produced by deamination during fermentation of nitrogenous organic matter which sinks from the photic surface layers (dead algal cells, marine snow and other detrital particles) while the methane and sulphide represent the end products of terminal anaerobic mineralisation. Assuming a C:N ratio of 6 for the sedimenting material, it can be estimated that the amount of organic carbon, that is mineralised in the sediment, equals six times the production of ammonia (on a molar basis). The production of 1 unit of CH_4 or of 1 unit of HS^- each represents the terminal mineralisation of 2 units of $[CH_2O]$. It is therefore expected that $6 \times$ [flux of ammonia] $= 2 \times$ [fluxes of $CH_4 + HS^-$]. The data presented in Fig. 6.9 show that the relative slopes of the gradients of NH_4^+, HS^- and CH_4 are 1:2.3:0.2. The left-hand side of the above equation is therefore 6 and the right-hand side becomes $2 \times (2.3 + 0.2) = 5$. The relatively small discrepancy may be explained by anaerobic-mineralisation via denitrification or it may be due to underestimation of the CH_4 flux caused by ebullition. Direct measurements of some of the process rates allow for the calculation of absolute rates of C mineralisation via sulphate reduction and methanogenesis (Fig. 6.10).[18] A comparison with the *Beggiatoa* mat (Fig. 6.2) shows that the rate of anaerobic degradation is much smaller in the anoxic fjord basin on an areal basis. This is primarily due to the fact that in the latter case most of the primary production is mineralised under aerobic conditions in the upper mixed layer and only about 15% of the primary production sinks through the chemocline to the bottom. In the fjord a substantial amount of the produced methane escapes to the atmosphere rather than being oxidised in the water column (the flux of CH_4 across the water surface could be measured directly).[18]

The turbulent diffusion coefficient in the deep layer can be estimated as [flux/gradient] and is about $0.013 \, cm^2 \, s^{-1}$ or about 650 times higher than molecular diffusion. This is consistent with an average residence time for the reduced products in the anoxic zone (= depth-integrated concentration/flux) of 1–2 years.

Figure 6.9 shows the vertical distribution of nitrate and nitrite around the oxic–anoxic interface. Nitrification takes place at a depth around 13 m or about 1 m above the zone of maximum sulphide oxidation. The gradients suggest that about half of the produced nitrate moves upwards (to be used by primary producers in the photic zone) and the other half moves downwards and is denitrified. If so, then denitrification accounts for about $1 \, mM \, C \, m^{-2} \, day^{-1}$ or about 20% of the anaerobic mineralisation in this system. Sulphide oxidation in the water column is probably exclusively chemotrophic, since insufficient light penetrates through the turbid and chlorophyll-rich surface waters down to depths below 15 m. Profiles of sulphide, ammonia, nitrate and nitrite that are very similar to those of Mariager Fjord have been recorded from other fjords with a sill and from the Black Sea and Baltic basins.[11,48,53,54] Stratified lakes probably differ from these marine examples in that methanogenesis plays a larger role relative to sulphate reduction.

In systems where light penetrates to the anoxic layer, sulphide oxidation is largely phototrophic during the day. This has been studied especially in meromictic (sulphate rich) lakes.[3,4,26] In such systems a typically 0.5–1-m thick "bacterial plate", consisting mainly of phototrophic sulphur bacteria, is found immediately beneath the oxic–anoxic interface. The bacterial plate is vertically zonated: the upper part includes layers dominated by different types of purple sulphur bacteria, while green sulphur bacteria occur in the deeper part. Cyanobacteria frequently constitute the dominant chlorophyll *a*-containing phototrophs in the zone above the oxic–anoxic interface. These systems can therefore be considered a planktonic analogue of cyanobacterial mats.

In all these systems the chemocline harbours high densities of phagotrophic eukaryotes which depend on the production of phototrophic or chemoautotrophic bacteria for food. The consumers are mainly protozoa because many species tolerate low O_2 tensions or anoxia, but some zooplankters (rotifers, copepods) may also exploit the production of autotrophic bacteria in the water column.

References

1. Ankar S, Jansson BO (1973) Effect of an unusual natural temperature increase on a Baltic soft-bottom community. *Mar Biol* **18**: 9–18.
2. Bernard C, Fenchel T (1995) Mats of colourless sulphur bacteria. II. Structure, composition of biota and successional patterns. *Mar Ecol Prog Ser* **128**: 171–179.
3. Caldwell DE, Tiedje JM (1975) A morphological study of anaerobic bacteria from the hypolimnia of two Michigan lakes. *Can J Microbiol* **21**: 362–376.
4. Caldwell DE, Tiedje JM (1975) The structure of anaerobic bacterial communities in the hypolimnia of several Michigan lakes. *Can J Microbiol* **21**: 377–385.
5. Canfield DE, DesMarais DJ (1993) Biogeochemical cycles of carbon, sulfur, and free oxygen in a microbial mat. *Geochim Cosmochim Acta* **57**: 3971–3984.
6. Caumette P (1986) Phototrophic sulfur bacteria and sulfate–reducing bacteria causing red waters in a shallow brackish coastal lagoon (Prévost Lagoon, France). *FEMS Microbiol Ecol* **38**: 113–124.
7. Cohen Y, Jørgensen BB, Revsbech NP, Poblawski R (1986) Adaptation to hydrogen sulfide of oxygenic and anoxygenic photosynthesis among cyanobacteria. *Appl Environ Microbiol* **51**: 398–407.
8. Cohen Y, Rosenberg E (eds) (1989) *Microbial Mats*. American Society for Microbiology, Washington, DC.
9. Dando PR, Hughes JA, Thiermann F (1995) Preliminary observations on shallow hydrothermal vents in the Aegean Sea. In: Parson LM, Walker CL, Dixon D-R (eds) *Hydrothermal Vents and Processes. Special Publication of the Geological Society, London* **87**: 303–317.
10. Deuser WG (1975) Reducing environments. In: Riley JP, Skirrow G (eds) *Chemical Oceanography*, 2nd edn, Vol. 3. Academic Press, London, pp. 1–37.
11. Dyrssen D (1986) Chemical processes in benthic flux chambers and anoxic basin water. *Netherl J Sea Res* **20**: 225–228.
12. Emerson D, Revsbech NP (1994) Investigation of an iron-oxidizing microbial community located near Aarhus, Denmark: field studies. *Appl Environ Microbiol* **60**: 4022–4031.

13. Emerson D, Revsbech NP (1994) Investigation of an iron-oxidizing microbial mat community located near Aarhus, Denmark: laboratory studies. *Appl Environ Microbiol* **60**: 4032–4038.

14. Epping EH, Jørgensen BB (1996) Light-enhanced oxygen respiration in benthic phototrophic communities. *Mar Ecol Prog Ser* **139**: 193–203.

15. Fenchel T (1969) The ecology of marine microbenthos IV. Structure and function of the benthic ecosystem, its chemical and physical factors and the microfauna communities with special reference to the ciliated protozoa. *Ophelia* **6**: 1–182.

16. Fenchel T (1994) Motility and chemosensory behaviour of the sulphur bacterium *Thiovulum majus*. *Microbiology* **140**: 3109–3116.

17. Fenchel T, Bernard C (1995) Mats of colourless sulphur bacteria. I. Major microbial processes. *Mar Ecol Prog Ser* **128**: 161–170.

18. Fenchel T, Bernard C, Esteban G, Finlay BJ, Hansen PJ, Iversen N (1995) Microbial diversity and activity in a Danish fjord with anoxic deep water. *Ophelia* **43**: 45–100.

19. Fenchel T, Straarup BJ (1971) Vertical distribution of photosynthetic pigments and the penetration of light in marine sediments. *Oikos* **22**: 172–182.

20. Fossing H, Gallardo VA, Jørgensen BB, Hüttel M, Nielsen LP, Schulz H, Canfield DE, Forster S, Glud RN, Gundersen JK, Küver J, Ramsing NB, Teske A, Thamdrup B, Ulloa O (1995) Concentration and transport of nitrate by the mat-forming sulphur bacterium *Thioploca*. *Nature* **374**: 713–715.

21. Fründ C, Cohen Y (1992) Diurnal cycles of sulfate reduction under oxic conditions in cyanobacterial mats. *Appl Environ Microbiol* **58**: 70–77.

22. Gallardo VA (1977) Large benthic microbial communities in sulphide biota under Peru–Chile subsurface counter current. *Nature* **268**: 331–332.

23. Ghiorse WC (1986) Biology of iron- and manganese-depositing bacteria. *Ann Rev Microbiol* **38**: 515–550.

24. Grasshoff K (1975) The hydrochemistry of landlocked basins and fjords. In: Riley JP, Skirrow G (eds) *Chemical Oceanography*, 2nd edn, Vol. 2. Academic Press, London, pp. 455–597.

25. Grotzinger JP, Rothman DH (1996) An abiotic model for stromatolite morphogenesis. *Nature* **383**: 423–425.

26. Guerro R, Montesinos E, Pedrós-Alió C, Esteve I, Mas J, van Gemerden H, Hofman PAG, Bakker JF (1985) Phototrophic sulfur bacteria in two Spanish lakes: vertical distribution and limiting factors. *Limnol Oceanogr* **30**: 919–931.

27. Hamilton WA (1987) Biofilms: microbial interactions and metabolic activities. In: Fletcher M, Gray TRG, Jones JG (eds) *Ecology of Microbial Communities*. Cambridge University Press, Cambridge, pp. 361–385.

28. Hanert HH (1981) The genus *Gallionella*. In: Starr MP, Stolp H, Trüper H, Balows A, Schlegel HG (eds) *The Prokaryotes*, Vol. 1. Springer, New York, pp. 509–515.

29. Indrebø G, Pengerud B, Dundas I (1979) Microbial activities in a permanently stratified estuary. I. Primary production and sulfate reduction. *Mar Biol* **51**: 295–304.

30. Javor BJ, Castenholz RW (1981) Laminated microbial mats, Laguna Guerrero, Mexico. *Geomicrobiol J* **2**: 237–273.
31. Jørgensen BB, Cohen Y (1977) Solar Lake (Sinai). 5. The sulfur cycle of the benthic cyanobacterial mats. *Limnol Oceanogr* **22**: 657–666.
32. Jørgensen BB, DesMarais DJ (1986) A simple fiber-optic microprobe for high resolution light measurements: applications in marine sediments. *Limnol Oceanogr* **31**: 1376–1386.
33. Jørgensen BB, Des Marais DJ (1986) Competition for sulfide among colorless and purple sulfur bacteria in cyanobacterial mats. *FEMS Microbiol Ecol* **38**: 179–186.
34. Jørgensen BB, Kuenen JG, Cohen Y (1979) Microbial transformation of sulfur compounds in a stratified lake (Solar Lake, Sinai). *Limnol Oceanogr* **24**: 799–822.
35. Jørgensen BB, Revsbech NP, Cohen Y (1983) Photosynthesis and structure of benthic microbial mats: microelectrode and SEM studies of four cyanobacterial mats. *Limnol Oceanogr* **28**: 1075–1093.
36. Juniper SK, Brinkhurst RO (1986) Water-column dark CO_2 fixation and bacterial mat growth in intermittently anoxic Saanich Inlet, British Columbia. *Mar Ecol Prog Ser* **33**: 41–50.
37. Kühl M, Jørgensen BB (1992) Microsensor measurements of sulfate reduction and sulfide oxidation in compact microbial communities of aerobic biofilms. *Appl Environ Microbiol* **58**: 1164–1174.
38. Kühl M, Lassen C, Jørgensen BB (1994) Light penetration and light intensity in sandy marine sediments measured with irradiance and scalar irradiance fiber-optic microprobes. *Mar Ecol Prog Ser* **105**: 139–148.
39. Møller MM, Nielsen LP, Jørgensen BB (1985) Oxygen responses and mat formation by *Beggiatoa* spp. *Appl Environ Microbiol* **50**: 373–382.
40. Nelson DC (1989) Physiology and biochemistry of filamentous sulfur bacteria. In: Schlegel HG, Bowien B (eds) *Autotrophic Bacteria*. Springer, Berlin, pp. 219–238.
41. Nelson DC, Revsbech NP, Jørgensen BB (1986) Microoxic–anoxic niche of *Beggiatoa* spp: microelectrode survey of marine and freshwater strains. *Appl Environ Microbiol* **52**: 161–168.
42. Nelson DC, Jørgensen BB, Revsbech NP (1986) Growth pattern and yield of a chemoautotrophic *Beggiatoa* sp in oxygen–sulfide microgradients. *Appl Environ Microbiol* **52**: 225–233.
43. Nicholson JAM, Stolz JF, Pierson BK (1987) Structure of a microbial mat at Great Sippewisett Marsh, Cape Cod, Massachusetts. *FEMS Microbiol Ecol* **45**: 343–364.
44. Oren A, Kühl M, Karsten U (1995) An endoevaporitic microbial mat within a gypsum crust: zonation of phototrophs, photopigments, and light penetration. *Mar Ecol Prog Ser* **128**: 151–159.
45. Ørsted AS (1842) Beretning om en Excursion til Trindelen. *Naturhist Tidsskrift* **3**: 552–569. An even earlier description of cyanobacterial mats is Hofman NB

(1826) Om Confervernes Nytte i Naturens Husholdning. *Kgl Danske Vidensk Selsk* **4(2)**: 207–220.

46. Paerl HW, Pinkney JL (1996) A mini-review of microbial consortia: their roles in aquatic production and biogeochemical cycling. *Microb Ecol* **31**: 225–247.

47. Pierson B, Oesterle A, Murphy GL (1987) Pigments, light penetration, and photosynthetic activity in the multi-layered microbial mats of Great Sippewissett Salt Marsh, Massachusetts. *FEMS Microbiol Ecol* **45**: 365–376.

48. Rheinheimer G, Gocke K, Hoppe H-G (1989) Vertical distribution of microbiological and hydrographic–chemical parameters in different areas of the Baltic Sea. *Mar Ecol Prog Ser* **52**: 55–70.

49. Richards FA (1975) The Cariaco Basin (Trench). *Oceanogr Mar Biol Ann Rev* **13**: 11–67.

50. Rützler K, Santavy DL, Antonius A (1983) The black band disease of Atlantic reef corals. *PSZNI: Mar Ecol* **4**: 329–358.

51. Schopf JW, Klein C (eds) (1992) *The Proterozoic Biosphere*. Cambridge University Press, Cambridge.

52. Schulz E (1937) Das Farbstreifen-Sandwatt und seine Fauna, eine ökologische biozönosiche Untersuchung an der Nordsee. *Kieler Meeresforsch* **1**: 359–378.

53. Skei JM (1983) Permanently anoxic marine basins — exchange of substances across boundaries. *Ecol Bull (Stockholm)* **35**: 419–429.

54. Sorokin YI (1972) The bacterial population and the process of hydrogen sulphide oxidation in the Black Sea. *J Cons Int Explor Mer* **34**: 423–455.

55. Southward AS, Kennicutt MC, Alcalà-Herrera J, Abbiati M, Arnoldi L, Cinelli F, Bianchi CN, Morni C, Southward EC (1996) On the biology of submarine caves with sulphur springs: appraisal of 13C/12C ratios as a guide to trophic relations. *J Mar Biol Assoc UK* **76**: 265–285.

56. Stal LJ, Caumette P (eds) (1994) *Microbial Mats*. Spinger, Berlin.

57. Stal LJ, van Gemerden H, Krumbein WE (1985) Structure and development of a benthic marine microbial mat. *FEMS Microbiol Ecol* **31**: 111–125.

58. van Gemerden H (1993) Microbial mats: a joint venture. *Mar Geol* **113**: 3–25.

59. van Gemerden, de Wit R, Tughan CS, Herbert RA (1989) Development of mass blooms of *Thiocapsa roseopersecina* on sheltered beaches on the Orkney Islands. *FEMS Microbiol Ecol* **62**: 111–118.

60. van Gemerden H, Tughan CS, de Wit R, Herbert RA (1989) Laminated microbial ecosystems on sheltered beaches in Scapa Flow, Orkney Islands. *FEMS Microbiol Ecol* **62**: 87–102.

61. Visscher PT, van den Ende FP, Schaub BEM, van Gemerden H (1992) Competition between anoxygenic phototrophic bacteria and colorless sulfur bacteria in a microbial mat. *FEMS Microbiol Ecol* **101**: 51–58.

62. Warming E (1875) Om nogle ved Danmarks Kyster levende bakterier. *Vidensk Meddr dansk naturh Foren* **20**: 307–420.

63. Westphalen D (1993) Stromatolitoid microbial nodules from Bermuda — a special microhabitat for meiofauna. *Mar Biol* **117**: 145–157.

7

Symbiotic systems

We define "symbiosis" in accordance with De Bary as a specific association between two kinds of organisms involving physical contact. Generally, there is a size difference between the partners so that individuals of the smaller species (the *symbiont*) live within or are attached to the surface of another, larger organism (the *host*). This definition does not imply any particular functional significance. It is useful if only because the functional significance of many symbiotic associations (e.g. many examples of endosymbiotic bacteria within eukaryotic cells) is still not understood.

Symbiosis does not necessarily imply "mutualism", in which both partners are supposed to gain fitness from the association, and mutualism need not necessarily imply that the partners are physically attached to each other. Syntrophic associations between bacteria with complementary types of metabolism are a kind of mutualism, but this does not imply physical *attachment* (although physical *proximity* between syntrophic pairs is required; cf. Sections 1.2.1 and 1.3). Syntrophy is not necessarily specific: for example, in pairs involved in syntrophic hydrogen transfer, the substitution of one species of hydrogen scavenger for another may not impair function. Therefore, we do not include syntrophic interactions between bacteria in this chapter. However, the functional significance of symbiotic associations is often a kind of syntrophy (i.e. the association implies complementary types of metabolism, for example phototrophic symbionts within phagotrophic hosts or hydrogen scavengers within hosts with a fermentative metabolism).

Whether a symbiotic association benefits both constituents is often, in fact, a meaningless question, even when the association is fully understood. Typically, once such an association has been established the symbionts are evolutionary and physically separated from their free-living ancestors and comparisons in terms of Darwinian fitness make no sense. A more accurate description of many symbiotic associations is that the host parasitises the symbiont. However, clearcut definitions and generalisations are somewhat difficult due to the diversity of the phenomena referred to as symbiosis.

The present definition of symbiosis implies that the hosts are almost always eukaryotes since only they can take up and maintain intracellular symbionts or (in the case of multicellular organisms) provide an intestinal tract, body cavity or organ that can host symbionts. An exception that could qualify as a symbiotic relationship between two kinds of bacteria is the syntrophic association between a motile sulphur reducer, the surface of which is covered by cells of a green phototrophic sulphide oxidiser (*Chlorochromatium*; see Section 1.3).

The ultimate evolutionary consequence of intracellular symbionts in unicellular eukaryotes may be that they evolve into organelles: many structural and functional properties of the symbiont are lost during evolution and parts of the symbiont genome are transferred to the nuclear genome of the host. This is not likely to happen for intracellular symbionts of multicellular organisms because the symbionts will rarely reside within the cells of the germ line. Chloroplasts and mitochondria are the most important (and documented) examples of organelles evolving from intracellular bacterial symbionts; it is now generally accepted that these organelles are derived from bacteria which were members of the cyanobacteria and the α-group of proteobacteria, respectively. Thus, intracellular bacterial symbiosis has had a profound significance for the evolution of eukaryotes (and for flow of energy and materials in the extant biosphere). Examples that represent an intermediate stage between a symbiont and an organelle are known. For example, the flagellated protist *Cyanophora paradoxa* was previously considered as an organism containing "cyanella" (endosymbiotic cyanobacteria); more recently it has been found that in this case the cyanella (although structurally resembling a cyanobacterium with reduced cell walls) have largely evolved into organelles; the length of its genome is only about 1/10 that of free-living cyanobacteria, but it is still much longer and includes more functional genes than are found in typical chloroplasts.

Further discussion is limited to prokaryotic endosymbionts, although some examples involving eukaryote symbionts (e.g. mycorrhizal fungi, algal symbionts of lichens, and cellulose degrading protists in some herbivorous animals) are important in some ecological systems. There are numerous examples of bacterial symbionts in protists and animals. In many cases the functional significance is not understood. Other examples are more or less peripheral to the theme of this book (e.g. bacteria that synthesise essential amino acids or vitamins in insects feeding on blood or plant sap, or symbiotic bacteria that confer bioluminescence to certain species of fish). Therefore, we concentrate on examples in which prokaryotic symbionts play an essential role in ecosystems (polymer degradation in herbivorous animals, symbiotic nitrogen fixation) or which illustrate important aspects of mineral cycling and syntrophy (various examples of photo- and chemoautotrophic symbionts). Excellent recent reviews on symbiosis can be found in Douglas[19] and Smith and Douglas.[55]

7.1 Symbiotic polymer degradation

Animals are (with few exceptions) incapable of hydrolytic degradation of quantitatively important structural plant polymers (cellulose, xylan, pectin, alginates); that is, animals cannot produce the necessary enzymes. Also, plant and macroalgal tissues (in particular, tissue of vascular plants and especially wood) have a critically low content of nitrogen when considered as a food source. Finally, many plants produce toxic "secondary metabolites", a property that has evolved as a protection against herbivory. In consequence, some herbivores (e.g. many insects) must eat copious amounts of plant material to compensate for its low digestibility and low nitrogen content. Such herbivores have also often evolved detoxification mechanisms against specific secondary plant metabolites (alkaloids, cruciferin, etc.); this again means that the animals must specialise on a limited number of (usually related) plant species.

However, these obstacles to herbivory are overcome by many mammals and some animals belonging to other taxa by maintaining consortia of fermenting micro-organisms in a specialised part of their digestive system. The primary function of these microbial biota is to degrade structural carbohydrates; in some cases this type of symbiosis also allows for a more effective utilisation of nitrogen and for the microbial degradation of plant toxins. The most important and best established examples are shown in Tables 7.1 and 7.2.

Symbioses with carbohydrate-degrading bacteria have evolved independently in different animal groups. This type of digestion requires a relatively long retention time of the food in the digestive system, and many herbivores therefore have an unusually long intestinal tract, often with one or more caecae or an extension of the oesophagus (a rumen) which can hold a large volume of ingested food during the relatively slow process of microbial degradation and fermentation. Since the time needed for microbial degradation is more or less fixed, symbiotic digestion (especially "pregastric fermentation"; see below) occurs mainly in relatively large animals. The need for a voluminous and heavy fermentation chamber also explains why symbiotic polymer degradation is rare among flying birds. Thus, most herbivorous birds and insects rely on a high rate of food intake and an efficient mechanical degradation of the food, but accept a low digestion efficiency.

Table 7.1   Established distribution of symbiotic polymer digestion

Mammals	Almost all herbivorous groups; have evolved independently in at least 12 taxa
Birds	Ostrich, rhea, grouse, ptarmigan, hoatzin
Reptiles	Iguanas, green turtle
Fish	Surgeon fish
Insects	Termites, cockroaches
Molluscs	Shipworms
Echinoderms	Various echinids

Table 7.2   Symbiotic polymer digestion in mammals

Taxon	Pregastric fermentation	Postgastric fermentation
Artiodactyls:		
Ruminants	+	
Camels	+	
Hippopotami	+	
Primates:		
Colubine monkeys	+	+
Lemurs		+
Howler monkeys		+
Edentates:		
Sloths	+	
Marsupials:		
Kangaroos	+	+
Koalas		+
Wombats		+
Perissodactyls:		
Horses, rhinos, tapirs		+
Elephants		+
Lagomorphs (rabbits, hares)		+
Rodents		+

While there is some variation with respect to detail, one property is common to all these systems: they are based on anaerobic (fermentative) degradation. An obvious reason for this is that the maintenance of an oxic environment in, for example, a rumen or caecum with degrading organic material is not possible. However, anaerobic degradation can also be considered adaptive, since the growth efficiency of fermentative microbes is low. Thus, they convert the undigestible carbohydrates into metabolic end products (mainly volatile fatty acids) in which most of the potential chemical energy of the food is retained; these end products are then utilised by the aerobic host for energy metabolism and growth.

A characteristic of this type of symbiosis is that it includes many species of symbionts with different functional roles and complex interactions. Furthermore, in almost all cases the microbial biota include not only prokaryotes, but also a variety of protists, especially ciliates, chytrids and flagellates (e.g. trichomonads and, in insects, the related hypermastigines and oxymonads). The phagotrophic protozoa are often bacterivorous, but some species (and the fungi-like chytrids) also play a role in the primary degradation of plant material (cellulolytic activity) and subsequent fermentation. Indeed, in most termite groups the flagellate biota in the hind gut constitute the principal cellulose degraders.

The best understood (and economically most important) of such systems is the rumen of sheep and cattle. The rumen holds a special position in microbial ecology and physiology because our current understanding of fermentative microbial systems and the interactions among different types of fermenters and between hydrogen-evolving and hydrogen-consuming bacteria (see Section 1.2.3) originally derived from the classical studies on the rumen system.[18,36,37]

7.1.1 Symbiotic digestion in mammals

Symbiotic digestion is believed to have evolved independently within several mammal groups (Table 7.2). There are two principal types: *pregastric* and *postgastric* fermentation. In the former the fermentation chamber is placed before the proper stomach in an extension of the oesophagus. Thus, food constituents (carbohydrates, proteins) are mostly converted into fatty acids plus microbial cells, and acid digestion takes place afterwards in the true stomach. The volatile fatty acids are absorbed and constitute the main carbon source for host metabolism and growth (acetate is used primarily for energy, propionate for biomass via gluconeogenesis). The necessary amino acids derive from the digestion of microbial cells; vitamin supply is often dependent on microbial synthesis. In postgastric fermentation the food first undergoes acid digestion in the stomach and products of hydrolysis and soluble sugars are absorbed. A prolonged colon or a caecum (or both) then functions as a fermentation chamber in which non-digested plant polymers are converted into volatile fatty acids which are subsequently absorbed by the host animal. Postgastric fermentation is more widespread than pregastric fermentation. This, perhaps, is due to the fact that it was simpler to evolve: almost all animals harbour microbial biota in their hind guts and even humans probably benefit slightly from bacterial metabolites in this way. Kangaroos and colubine monkeys have pre- as well as postgastric fermentation.

The ruminant system (cows, sheep, etc.) is the most studied type of pregastric fermentation, but it is probably representative of the other examples. The rumen is an extension of the oesophagus and it constitutes up to 15% of the volume of the animal. It is followed by a smaller part, the omasum, which mainly serves for the absorption of solutes (especially volatile fatty acids). After the omasum the material passes into the abomasum (the true stomach) in which acid digestion takes place. The remaining intestinal tract is no different from that found in other animals.

The rumen content is almost neutral (due to the buffering capacity of the bicarbonate-containing saliva) and strongly reducing ($E_h \sim -350\,mV$). The bulk of the material is anaerobic, but some oxygen diffuses through the rumen wall and, at times, traces of oxygen can be detected. The rumen contains a complex mixture of rumen liquid and particulate material derived from plant tissue and microbes, including up to 10^{11} bacteria and 10^6 protists per millilitre. Ruminants may sometimes regurgitate part of the particulate material for further mechanical maceration of plant fibres. Carbohydrates (including cellulose and hemicelluloses) are effectively hydrolysed and fermented principally into acetate, propionate and butyrate ($+ CH_4$ and CO_2). Lipids are hydrolysed; glycerol is fermented and the fatty acids are hydrogenated mainly to stearic and palmitic acids. Proteins are hydrolysed, deaminated and fermented. Lignin, however, is not degraded and can be used as a conservative constituent for estimation of digestion efficiency (e.g. of cellulose).

These processes result from the concerted activities of a number of physiological types of bacteria and protists (for general treatments of rumen bacteriology see references[12,18,35-37]). More than 100 species of bacteria have been isolated from the rumen, but they are not necessarily all specific for this habitat. Cellulolytic bacteria constitute a relatively small fraction of the bacteria (5-15%) in spite of the fact that cellulose may constitute 30-40% of the food. The reason may be that cellulolytic bacteria excrete cellulases so that some of the resulting soluble sugars are used by other bacteria; cellulolytic bacteria may also depend on syntrophic interactions with other bacteria. Important cellulolytic bacteria include species of *Ruminococcus*, *Bacteroides* and *Butyrovibrio* which may also be responsible for the deamination and fermentation of proteins. *Selenomonas* and *Streptococcus bovis* exemplify bacteria that ferment the easily degradable carbohydrates (soluble sugars, starch); *Anaerovibrio lipolytica* ferments glycerol. Many rumen bacteria that use the fermentative pathway normally leading to propionate (*Bacteroides succinogenes, B. amylophila*) tend to produce succinate as a terminal metabolite rather than decarboxylating it to propionate. Succinate is an important intermediate which is decarboxylated to propionate by *Veillonella gazogeneous*.[5]

Many micro-organisms, for example the different kinds of cellulose degraders (which even include some of the protozoa and the chytrids), seem functionally redundant. However, there is evidence for niche diversification among them. Thus, it has been found that different cellulolytic bacteria have differential attachment mechanisms and tend to attach to different sites on degrading plant material;[3] also differential substrate affinities and growth yields may confer differential fitness according to substrate availability.[51,52]

Hydrogen is removed through H_2 / CO_2 methanogenesis in the rumen by *Methanobacterium ruminantium*. This process maintains a low p_{H_2} and so fermentation leads to relatively oxidised end products. Methane production in the rumen amounts to about 10% of the fermented carbon; this represents (a necessary) loss to the ruminant which rids itself of the gas through belching.

The metabolic pathways in the rumen include a number of intermediate compounds which are released in the rumen by some microbes and utilised by others; in particular H_2 is maintained at a low concentration with a very rapid turnover (Table 7.3). Intermediates like lactate and ethanol normally play only a small role and they are then rapidly fermented on to propionate, butyrate or acetate. However, if the fodder of ruminants

Table 7.3   Turnover of intermediates in rumen fermentation[37]

Intermediate	Concentration ($nmol\ ml^{-1}$)	Turnover (min^{-1})	Product	Production accounted for (%)
Lactate	12	0.03	Propionate, acetate	? 1
Ethanol	Trace	0.003	Acetate	?
Succinate	4	10	Propionate	33
H_2	1	710	Methane	100
Formate	12	10	H_2	18

is abruptly replaced by starch, the high level of lactate produced (e.g. by *Streptococcus bovis*) can inhibit further metabolism by rumen bacteria. The resulting acidity may eventually lead to lethal acidosis. A more gradual change to a starch-rich diet does not have this detrimental effect. The reason is that the microbial community is normally substrate limited, energy-efficient types of fermentation are favoured, and small amounts of lactate are rapidly fermented. But if large amounts of easily degradable carbohydrates are suddenly available, the rapidly dividing lactic acid bacteria are competitively superior.

A generalised scheme of rumen fermentative pathways is shown in Fig. 7.1 together with an approximate stoichiometric account of the

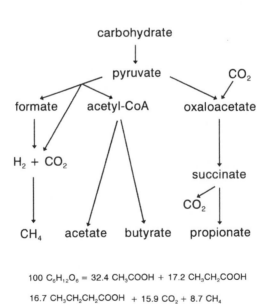

$$100\ C_6H_{12}O_6 = 32.4\ CH_3COOH + 17.2\ CH_3CH_2COOH$$

$$16.7\ CH_3CH_2CH_2COOH + 15.9\ CO_2 + 8.7\ CH_4$$

$$+\ 9.1\ \text{microbial cells}$$

Figure 7.1   The principal pathways of carbohydrate fermentation in the rumen

processes. Comparison with the previous discussion on communities of fermenting bacteria (Section 1.2.3 and Fig. 1.2) shows that anaerobic degradation in the rumen is incomplete. Complete anaerobic degradation of carbohydrates would (in the absence of other electron acceptors such as sulphate or nitrate) lead to a stoichiometric conversion of carbohydrates into CH_4 and CO_2. It would clearly be unfortunate for ruminants if this were to happen: the animals benefit from a maximum yield of volatile fatty acids.

It has not yet been possible to make a model that precisely predicts the actual stoichiometry of rumen fermentation processes. Two aspects are important: that the overall turnover time of the rumen contents is relatively short and different rumen constituents have a differential turnover time. Thus, the relatively rapid turnover (which in the case of the rumen liquid is about 15 h) prevents the complete fermentation of the substrate to acetate $+ H_2$ as well as acetoclastic methanogenesis since the growth rates of obligatory acetogens and acetoclastic methanogens are too low.

The rumen is often compared to a continuous culture such as a chemostat. This is inaccurate because in a chemostat all constituents have the same turnover time. In a chemostat, high dilution rate means a low efficiency of substrate conversion (due to low steady-state densities of bacteria) whereas a low dilution rate means that substrates are converted efficiently, but at a very slow rate. The ruminant must maximise effective substrate conversion (but not all the way to $CH_4 + CO_2$) as well as the rate of substrate processing. This is accomplished by differential retention times for different constituents of the rumen. The rumen is compartmentalised into lobes and material is moved between them by muscular action. In the process, the material is strained and larger particles are retained longer in the rumen whereas the liquid part has a shorter residence time. The microbial biota are also maintained longer in the rumen (relative to the liquid) because they are to a large extent attached to particulate material or to the rumen wall and this allows for larger steady-state population densities.[13] A complete understanding of the rumen metabolism must therefore include the hydraulics and mechanics of the rumen.

Rumen microbes preferentially use ammonia as a nitrogen source.[4] Ammonia derives in part from amino acid deamination and in part from urea, which is supplied with saliva and degraded to NH_4^+ and CO_2 by certain rumen bacteria. The nitrogen source of the ruminant is constituted by the rumen microbes which are digested in the abomasum and postduodenum. Mammals excrete excess nitrogen in the form of urea. In ruminants some (up to 50%) of the urea thus formed is excreted into the saliva. It is thus recycled in order to supply rumen microbes with nitrogen; this constitutes a mechanism for preserving nitrogen in animals that live on a nitrogen-depleted diet. Bacteria that have the potential to fix N_2 are present in the rumen, but nitrogen fixation is not believed to be important in the nitrogen budget.

Newborn ruminants do not have rumen microbes; they are acquired when the animals start to feed on plant material and thus ingest saliva (containing live rumen microbes) which has been deposited by adults. Ruminants without microbes can thus be reared experimentally. Such animals cannot survive on their normal plant diet. In the normal animal, eukaryotic micro-organisms play a role as cellulose decomposers and as a protein source. In ruminants without protozoa (but with a normal bacterial community) bacterial densities are increased and the animals are not adversely affected and, according to some reports, may show enhanced growth.

Ruminants originated relatively recently (during Miocene) and this is usually considered as an evolutionary response to the contemporary origin of grasses. These events have had a profound effect on terrestrial ecosystems.

Postgastric fermentation is often considered to be more primitive and less efficient than pregastric fermentation. Mammals with hindgut fermentation include perissodactyls, lagomorphs, rodents, manatees, elephants and some primates and marsupials. They are less effective on a roughage diet than ruminants. On the other hand, they do not require a very voluminous fermentation chamber and they can utilise some constituents in the food (proteins, vitamins) directly without depending on microbial synthesis, while cellulose and hemicellulose are still utilised relatively efficiently. The microbiology of carbohydrate degradation and metabolism in the colon or a caecum is quite similar to that of the rumen. Postgastric fermentation in mammals is discussed in Janis,[39] Moir[43] and Murray *et al*.[44]

7.1.2 Symbiotic fermentation in other animals

Fermentative polymer degradation is not as widespread among non-mammals (Table 7.1) although it is conceivable that more examples will be discovered. Grouse, ptarmigan, ostrich and rheas are birds known to have postgastric fermentation in paired caecae while the leaf-eating hoatzin is reported to have pregastric fermentation. In all cases, these birds are flightless or not very dependent on flight. Postgastric fermentation has been documented in iguanas and in the seagrass-eating green turtle. Herbivorous fish are generally not dependent on microbial symbiosis, but there is evidence of its presence in surgeon fish.[26,28,30,38]

The hindgut of wood-eating termites and cockroaches represents the classical and well-studied example in insects. In these systems the hydrolysis and fermentation of cellulose and other structural carbohydrates is primarily due to a remarkable assemblage of flagellates. However, bacteria (among which spirochaetes are conspicuous) also occur at high densities in the termite hindgut and they undoubtedly play an important role. The flagellates have numerous bacterial endosymbionts, some of which are methanogens (see Section 7.3.2). Free-living methanogens and CO_2/H_2

acetogens also play a role in maintaining a low p_{H_2}. Microbial nitrogen fixation appears to be of significance in some termites. Bacterial recycling of uric acid (the principal nitrogenous excretion product in insects) has also been demonstrated.[6,50]

Cellulolytic activity and nitrogen fixation have been demonstrated for bacteria isolated from intestinal "glands" in shipworms (*Teredo*, wood-boring bivalves) and so it is likely that microbial fermentation plays a role for these animals.[61] Regular sea urchins, which feed on macroalgae, depend on microbial degradation of structural carbohydrates. Among irregular sea urchins, the burrowing *Echinocardium* has an intestinal tract with two caecae which (like other parts of the intestine) are filled with various microbes; in one of the caecae there are sulphide-oxidising bacteria.[17,60] It is unclear how this peculiar system works and it warrants further study. General considerations on the role of bacteria in invertebrate guts can be found in Penry and Jumars[47] and Plante *et al*.[48]

7.2 Symbiotic nitrogen fixation

The fundamental properties of N_2 fixation were discussed briefly in Section 1.1.2. The ability to fix atmospheric N_2 is widespread among free-living bacteria. However, although free-living N_2-fixing bacteria (e.g. *Azotobacter*) are common in soils, it has been estimated that in terrestrial habitats > 90% of all biological N_2 fixation is due to symbiotic relationships between plants and specific types of bacteria. N_2 fixation requires a large energy expenditure and anaerobic or microaerobic conditions; both these conditions can be provided by plants and this explains the prevalence of symbiotic N_2 fixation. It is also estimated that about half of the biological N_2 fixation is due to legume cultivation.

Symbiotic N_2 fixation involves a limited number of plant and bacterial taxa. In the most important types of symbiotic N_2 fixation the plant host is infected by special free-living soil bacteria, maintains them as intracellular symbionts, and provides them with suitable conditions for N_2 fixation. At the same time the hosts inhibit ammonia assimilation in the symbionts; consequently, the symbionts excrete NH_4^+ which is then assimilated in host cells. The associations between particular plant and bacterial species are more or less specific. Established examples of symbiotic N_2 fixation are listed in Table 7.4.

The practical importance of symbiotic N_2 fixation is immense; it was already known in antiquity that legumes improve soil fertility and the phenomenon has drawn considerable attention since Beijerinck in 1888 first demonstrated that legume root nodules contain bacteria that fix atmospheric N_2. There is a large amount of literature on the subject; some important reviews are listed in the references.[20,49,57,58]

Table 7.4   Established examples of symbiotic nitrogen fixation

Hosts	N_2-fixing symbionts	Symbiont-containing organs
Legumes	*Rhizobium, Bradyrhizobium*	Root nodules
Various woody angiosperms (*Alnus, Hippophae, Myrica, Dryas, Casuarina,* and others)	*Frankia*	Root nodules
Cycads	*Nostoc, Anabaena*	Coralloid roots
Azolla (fern)	*Anabaena*	Cavities in leaves
Gunnera (angiosperm)	*Nostoc*	Glands in stems
Planktonic diatoms (*Rhizosolenia* and others)	*Richelia*	Intracellular
Codium (green alga)	*Azotobacter*	On surface of thallus
Some ascomycete and basidiomycete lichens	*Nostoc, Calothrix,* and others	Surrounded by hyphae
Some marine sponges (*Theonella, Siphonochalina*)	*Phormidium*	Intracellular

In addition to the classic examples of symbiotic N_2 fixation (which are discussed in more detail below) some cases of less integrated syntrophic relations between plants and N_2-fixing bacteria are known. Thus, otherwise free-living N_2-fixing soil bacteria may accumulate on roots of plants, apparently receiving organic root exudates in return for combined nitrogen. *Azospirillum lipoferum* is an N_2-fixing bacterium that is found in association with roots of certain tropical grasses, and this has been shown to provide the hosts with nitrogen. Enhanced N_2 fixation has been found in the rhizosphere of seagrasses (*Zostera, Thalassia*) and marsh grass (*Spartina*). Nitrogenase activity has been detected on the surfaces of macroalgae; in the case of the green alga *Codium* this is due to *Azotobacter*.[8]

7.2.1   Symbiotic N_2 fixation in legumes

This is the most important and best known type of symbiotic nitrogen fixation. Most (about 17 500) species belonging to the Leguminosae form root (or in few cases stem) nodules containing N_2-fixing rhizobias. These are now classified into two genera, *Rhizobium* and *Bradyrhizobium* (or "fast" and "slow" rhizobia, respectively). They belong to the α-group of the proteobacteria; *Rhizobium* is closely related to *Agrobacterium*, an organism that can invade plant tissue and form tumerous growth. This, perhaps, illuminates how the rhizobia–legume relationship originally evolved. Among non-legumes only a tree belonging to the elm family (*Parasponia*) is known to form rhizobia nodules.

Rhizobia occur as free-living bacteria in soils. However, they are relatively rare in soils in which legumes have not grown over a period of many years and they are especially numerous in the rhizosphere (i.e. the soil surrounding roots) of legumes; presumably they are stimulated by root exudates. In microaerobic cultures rhizobia can be induced to fix N_2 to a variable degree. The extent to which they fix N_2 in soils is not known, but it would seem likely that this property is adaptive under some circumstances.

Rhizobia multiply around roots of germinating legumes. Infection and subsequent nodule formation requires adhesion to root hairs and whether this takes place depends on the species of legume and the rhizobium strain. Some strains ("cross-inoculation groups") can infect several species of legumes and some legumes can form nodules with different rhizobia strains. Adhesion depends on specific lectins (produced by the host plant) and on specific polysaccharide cell coatings produced by the rhizobia.[65] Following adhesion, the root hair forms an "infection thread" through which the bacteria enter the roots. The bacteria then invade root cells and transform into "bacteroids": they swell and become deformed in various ways and they lose the ability to divide. These events also induce the root to form nodules which host the infected cells. Nodulation is inhibited by high ambient concentrations of combined nitrogen, acidic conditions and low availability of phosphate.

A special feature of the root nodules is *leghaemoglobin*, a haemoglobin in which the haem part is synthesised by the bacterium and the protein part is synthesised by the host plant. Leghaemoglobin is responsible for the pink colour that is seen when the nodules are sectioned. The functional significance of this respiratory pigment, which has a high affinity for O_2, is to maintain a low p_{O_2} (thus protecting the nitrogenase) while at the same time supplying the bacteria with oxygen so that they can maintain a high rate of aerobic metabolism.

A substantial part (13–28%) of the photosynthate of legumes is supplied to the nodules.[42] This is partly as carbohydrates serving as an energy source for N_2 fixation in the rhizobia and also as carbon skeletons for ammonia assimilation in the surrounding root cells. The plant assimilates ammonia as glutamine, other amino acids, or urea derivates ("ureides").

The density of free-living rhizobia increases in soils where legumes grow. While the bacteroids are incapable of growth, some untransformed rhizobia are always present in the nodules and in the infection thread; it has been suggested that these cells are liberated when the nodules senesce and decay, thus maintaining a high local population density. However, it is likely that the high density of rhizobia in the rhizosphere is primarily due to the stimulation of rhizobial growth by root secretion.

7.2.2 Actinorhizal N_2-fixing symbionts

Nitrogen-fixing nodules occur in > 100 species of somewhat unrelated woody angiosperms; many have been known for a long time, but the responsible bacterium (*Frankia*) has been isolated only relatively recently.[7] This is an actinomycete with filamentous growth. In contrast to rhizobias it will fix N_2 at atmospheric p_{O_2}; this is because the cells are compartmentalised and N_2 fixation takes place in vesicles that are protected from oxygen exposure.

The nodules can be several centimetres thick and contain host cells filled with the symbiont filaments. The nodules do not contain leghaemoglobin. Infection takes place through root hairs as in rhizobial infections. The *Frankia* filaments then multiply in the root hair and eventually invade host cells.

The best-known examples of woody plants with actinorhizas include *Alnus* (alder), *Myrica* (sweet gale), *Hippophae rhamnoides* (sea-buckthorn) and the tropical tree *Casuarina*. Many are "pioneer plants" which initially colonise barren and nutrient-poor soils.

7.2.3 Symbiosis with cyanobacteria

Both phototrophic and heterotrophic eukaryotes may harbour cyanobacterial symbionts. In the former case the functional significance is probably always symbiotic N_2 fixation; the symbionts typically have a large number of heterocysts and they fix N_2 at high rates. In the latter case, N_2 fixation may or may not be an important aspect of the association: in some cases the cyanobacterial symbionts are incapable of this process and in others the relative importance for the host of phototrophic CO_2 fixation and N_2 fixation, respectively, is not known. In this section we only discuss cases in which N_2 fixation has been established; other types of cyanobacterial symbionts are discussed briefly in Section 7.3.1.

Certain mosses (*Sphagnum* spp.) are known to harbour the cyanobacterium *Nostoc* in non-photosynthetic cells or on surfaces and this genus is also known to be associated with some species of liverworts; although not studied in detail the functional significance is probably N_2 fixation. A better-known example of some economic importance is the fern *Azolla*, which occurs floating on water in warm climates. The leaves form pockets which are colonised by the cyanobacterium *Anabaena azollae* from a colony at the apex of the stem; eventually the pockets close and form cavities containing the symbiont. In some places *Azolla* plays a role as green manure in rice cultivation. It grows on flooded rice paddies until it is outgrown by the rice plants; when the ferns die and decompose they release combined N_2.

Cycads also have cyanobacterial symbionts (*Anabaena, Nostoc*). These occur in cavities in branched "coralloid roots" which are situated laterally.

In the angiosperm genus *Gunnera,* glands in stems and the base of leaves harbour *Nostoc punctiforme.*

Some planktonic diatoms (and possibly other phytoplankters) are known to harbour intracellular cyanobacteria. This association seems to be common in oligotrophic seawater and in some cases N_2 fixation has been demonstrated.[29,40,59]

There are two cases of cyanobacterial symbionts of heterotrophic organisms in which N_2 fixation has been demonstrated, although here the functional significance of the symbiosis is also one of phototrophic C fixation. Lichens represent symbiotic consortia consisting of a fungal host (either an ascomycete or a basidiomycete) and phototrophic symbionts. In most lichens the symbionts are eukaryotic algae, but some lichens harbour only cyanobacteria or cyanobacteria as well as eukaryotic phototrophs. Symbiotic N_2 fixation has been convincingly demonstrated in some species.[41] Many marine tropical shallow-water sponges contain cyanobacterial symbionts; while the primary advantage for the host is phototrophic C fixation, N_2 fixation has been demonstrated in some cases.[63]

7.3   Autotrophic bacteria as symbionts

This section discusses some examples of mainly autotrophic bacterial symbionts which are associated with animals or protists. These associations generally have a limited impact on their ecosystems, but they are interesting from an evolutionary point of view and illustrate principles (notably syntrophy) discussed elsewhere in the book. Some of these examples (hydrogen scavengers in anaerobic protists, chemotrophic bacteria of certain invertebrates and protists) have been discovered only recently.

7.3.1   Phototrophic symbionts

A very large and diverse assemblage of protists and invertebrates harbour phototrophic endosymbionts. In most cases these are eukaryotic algae or flagellates, usually green algae in limnic hosts (for example in *Chlorohydra* and *Paramecium bursaria*) and dinoflagellates in marine hosts such as stony corals, giant clams, etc., but several other groups of eukaryotic phototrophs also occur as endosymbionts. In comparison, cyanella seem to be a rarer and more exotic phenomenon. Symbiotic cyanobacteria have recently been reviewed.[53]

Among protists, endosymbiotic cyanobacteria have been observed in several taxa, for example in the thecate amoeba *Paulinella,* some water moulds and marine planktonic dinoflagellates (in addition to the diatom–*Richelia* association discussed in the context of symbiotic N_2 fixation). Cases of cyanella, that have obtained organelle status ("cyanoplasts") through evolution, have already been discussed in the introduction to this chapter.

Among invertebrates, the endosymbiotic cyanobacteria of marine sponges were mentioned in Section 7.2.3; undoubtedly phototrophic C fixation is an important aspect of these associations.[62,64] Endosymbiotic cyanobacteria are also known from some echiurid worms and have been observed in a few other invertebrate groups. Ectosymbiotic prochlorophytes are known from the tunic of ascidia.

7.3.2   Hydrogen scavengers in anaerobic protists

A few protist taxa seem to be primarily amitochondrate and thus obligate anaerobes (diplomonads, rhizomastigines). Several other protist taxa include species that have reverted to become obligatory anaerobes. They have a fermentative metabolism and (in the case of secondary anaerobes) the mitochondria have often been transformed into an organelle referred to as a "hydrogenosome". Hydrogenosomes cannot perform oxidative phosphorylation (cytochrome c oxidase is lost); instead they are capable of fermenting pyruvate and malate into H_2 and acetate.[21,25] Organisms with hydrogenosomes include trichomonad, hypermastigid and oxymonad flagellates, a flagellate of obscure taxonomic affinity (*Psalteriomonas*), several ciliate taxa and anaerobic chytrids. Almost all anaerobic ciliates and many anaerobic flagellates also harbour hydrogen-scavenging bacterial symbionts (Table 7.5). The basic significance of this type of symbiosis is that it maintains a low intracellular p_{H_2} which allows the hosts to ferment their food almost completely into acetate and H_2. The topic is reviewed in detail in Fenchel and Finlay.[25]

Hydrogen-scavenging bacteria are often H_2/CO_2 methanogens. They are easy to detect microscopically because they contain the coenzyme F_{420} which fluoresces in violet light; it is also easy to measure CH_4 production of the symbiotic consortium. Methanogens occur in (at least some) hypermastigid symbionts of the termite hindgut, in *Psalteriomonas*, in almost all studied free-living, anaerobic freshwater ciliates, and in about 50% of their marine counterparts. The symbiosis has evolved independently

Table 7.5   Hydrogen-scavenging bacteria in anaerobic protists

Type of symbiont	Hosts
Methanogens	Many anaerobic ciliates, some trichomonad and hypermastigid flagellates, some rhizomastigid amoebae (e.g. *Pelomyxa*); the symbionts are intracellular
Sulphate reducers	Several marine anaerobic ciliates; the bacteria are ectosymbionts
Purple non-sulphur bacteria	Only in the ciliate *Strombidium purpureum*; the symbiont is intracellular and the significance of the symbiosis may also be that of a phototroph and a phagotrophic heterotroph

many times, even among the ciliates, and several groups of methanogens are represented. The methanogens are often greatly modified (frequently with a reduced or no cell wall, synchronisation of symbiont and host division). In most cases they lie adjacent to the hydrogenosomes or form complexes with hydrogenosomes so that hydrogen transfer is facilitated. Symbionts are always present *in situ*. In the laboratory, aposymbiotic hosts can be made artificially; these grow slower, and have lower cell yields and lower rates of hydrogen production than normal cells.[23,24] Some rumen ciliates also harbour endosymbiotic methanogens. A looser association is sometimes found in rumen ciliates on which methanogens attach temporarily. Anaerobic chytrids do not have endosymbiotic methanogens, but their growth and yield is enhanced when they are maintained in co-cultures with methanogens.

Some rhizomastigid amoebae also harbour intracellular methanogens. This association is not quite understood because these protists do not have hydrogenosomes and presumably do not generate hydrogen. It has been suggested that other endosymbiotic bacteria provide the H_2 for methanogenesis, but this has not been confirmed.

In limnic systems, symbiotic methanogenesis plays a very small role relative to CH_4 generation by free-living methanogens. In marine systems with a high ambient SO_4^{2-} concentration, methane production by free-living methanogens is relatively small, but under some circumstances anaerobic ciliates with methanogenic symbionts are responsible for a substantial part of the total methanogenesis.

Some marine anaerobic ciliates do not harbour endosymbiotic methanogens; instead they have a dense cover of ectosymbiotic bacteria. These are probably in all cases sulphate reducers, but this has been demonstrated only twice. Although experimental evidence is still not available it is likely that the role of these ectosymbionts is similar to that of the endosymbiotic methanogens in other species.

There is finally one example of an anaerobic marine ciliate (*Strombidium purpureum*) which harbours a purple non-sulphur bacterium, probably a *Pseudorhodomonas* sp. The interactions between the host and the symbionts are not known in detail: the symbiotic consortium has an absolute requirement for light and it shows a photosensory behaviour, being attracted to light with an action spectrum corresponding to that of bacteriochlorophyll *a*. The symbiont may play a role for the host in terms of phototrophic CO_2 fixation and simultaneously function as a sink for metabolically produced hydrogen.

7.3.3 Symbiotic oxidisers of reduced sulphur compounds and methane

Symbiosis between marine benthic invertebrates and chemolithotrophic bacteria was first discovered in animals collected from deep sea hydrothermal vents;[9,10] later it was recognised that this type of symbiosis is

Table 7.6   Chemotrophic symbionts on or in invertebrates and protists

Type of symbiont	Hosts
Oxidisers of reduced sulphur compounds	Ciliates: *Kentrophoros* spp., *Zoothamnium niveus* Turbellaria: catenulids Nematodes: members of Stilbonematinae, *Astomonema* Oligochaetes: members of Phallodrillinae Polychaetes: *Alvinella* Bivalves: *Solemya*, members of Lucinidae and Thyasiridae (*Lucina, Lucinoma, Thyasira*), some mytilids (*Bathymodiolus*) and *Calyptogena* Pogonophorans and vestimentiferans
Methanotrophs	Bivalves: some mytilids, Pogonophorans: *Siboglinum poseidoni*

widespread among marine invertebrates from all depths (the associations with bacteria had in some cases been described earlier, but their significance was not understood). The best-known examples are listed in Table 7.6; comprehensive reviews are found in Felbeck and Distel,[27] Ott and Novak,[45] and Southward.[56]

Most symbioses involve the oxidation of reduced sulphur compounds (sulphide, thiosulphate, sulphur), but a few have methanotrophic symbionts. The significance of the association is basically that the hosts exploit bacterial production. The bacteria may excrete organic matter or may be phagocytosed; in some cases ectosymbiotic bacteria are simply ingested by the host. The hosts are positioned in sediment gradients so that the symbionts receive an appropriate mixture of O_2 and reduced sulphur compounds (or CH_4). Sulphide detoxification may also confer a significant benefit. In some cases, the production of the symbionts is a supplement to host nourishment; in the most advanced cases the hosts are gutless and depend entirely on chemotrophic CO_2 fixation by the symbionts. Many other invertebrates (certain bivalves, polychaetes, etc.) living close to the sulphidic zone of sediments are more or less covered by sulphide-oxidising bacteria (such as *Thiothrix*). The significance and specificity of these associations are not known. It has been speculated that they protect their host against sulphide toxicity; these loose associations suggest how more specific and complex interactions originally arose.

Among protists there are only two examples of this sort of symbiosis. Species belonging to the ciliate *Kentrophoros* live at the chemocline in marine sandy sediments. One side of these extremely flattened cells is covered by a dense layer of sulphur bacteria. The ciliate is mouthless and lives exclusively from phagocytosing the attached bacteria through the cell surface.[22] More recently, it has been shown that the peritrich ciliate *Zoothamnium niveum* (which lives attached to the surface of mangrove peat) is covered by colourless sulphur bacteria; these are somehow detached and filtered by the host.[2]

In contrast, many invertebrates belonging to several taxa harbour symbiotic sulphur bacteria. Nematodes belonging to the Stilbonematinae are widely distributed in marine sediments. They are covered by colourless sulphur bacteria; depending on species these are filamentous or short rods. The worms scrape the bacteria from their own body surface and ingest them. The nematode *Astomonema* is mouthless, but a rudimentary gut contains symbiotic sulphur bacteria and a similar situation is found in the turbellarian worm *Paracatenula*.[45,46] Gutless oligochaetes (*Phallodrilus* and relatives) with symbiotic sulphur bacteria are also known.[31,33,34]

The pogonophorans (which occur in sediments in deeper water) and the related vestimentiferans (found at hydrothermal vents) are all gutless and depend completely on chemoautotrophic bacteria for nutrition; they occur intracellularly in a special organ, the "trophosome". Haemoglobin-containing blood carries sulphide and oxygen from extensive gills to the trophosome. Only the pogonophoran *Siboglinum poseidoni* harbours methanotrophs instead of sulphide oxidisers.[54] The polychaete *Alvinella*, which lives at deep sea hydrothermal vents, is covered by sulphur bacteria; how these are used by the host is not known.

Finally, many bivalves depend more or less completely on symbiosis with sulphur bacteria and this trait has evolved independently within several taxa. The symbionts are found intracellularly in expanded fleshy gills. Members of the lucinids and thyasirids are common in relatively shallow water sediments. They are also filter-feeders like "normal" bivalves. They draw oxygenated water from the sediment surface with their siphon and obtain reduced sulphur compounds through a system of channels within the sediment that are produced by the foot.[14-16,32] *Solemya* is a gutless bivalve found in sulphidic shallow-water sediments. Sulphide taken up by the clam is first partly oxidised in the blood, a process that is catalysed by an oxidase; the bacteria then metabolise the thus produced thiosulphate.[1] Some mytilid bivalves also harbour chemotrophic sulphur bacteria (e.g. *Bathymodiolus* from deep-sea vents); a relative is known to harbour methanotrophic bacteria.[11]

References

1. Anderson AE, Childress JJ, Favuzzi JA (1987) Net uptake of CO_2 driven by sulphide and thiosulphate oxidation in the bacterial symbiont-containing clam *Solemya reidi. J Exp Biol* **133**: 1–31.
2. Bauer-Nebelsick M, Bardele CF, Ott JA (1996) Redescription of *Zoothamnium niveum* (Hemprich & Ehrenberg, 1831), Eherenberg 1938 (Oligohymenophora, Peritrichida), a ciliate with ectosymbiotic, chemoautotrophic bacteria. *Eur J Protostol* **32**: 18–30.

3. Bhat S, Wallace RJ, Ørskov ER (1990) Adhesion of cellulolytic ruminal bacteria to barley straw. *Appl Environ Microbiol* **56**: 2698–2703.
4. Blackburn TH (1965) Nitrogen metabolism in the rumen. In: Dougherty RW (ed.) *Physiology of Digestion in the Ruminant.* Butterworths, Washington, pp. 322–334.
5. Blackburn TH, Hungate RE (1963) Succinic acid turnover and propionic production in the bovine rumen. *Appl Microbiol* **11**: 132–135.
6. Breznak JA (1984) Biochemical aspects of symbiosis between termites and their intestinal microbiota. In: Anderson JM, Rayner ADM, Walton DWH (eds) *Invertebrate–Microbial Interactions.* Cambridge University Press, Cambridge, pp. 173–204.
7. Callahan D, Del Tredici P, Torrey JG (1978) Isolation and cultivation *in vitro* of the actinomycete causing root nodulation in *Comptonia*. *Science* **199**: 899–902.
8. Capone DG (1988) Benthic nitrogen fixation. In: Blackburn TH, Sørensen J (eds) *Nitrogen Cycling in Coastal Marine Environments.* Wiley, Chichester, pp. 85–123.
9. Cavanaugh CM (1983) Symbiotic chemotrophic bacteria in marine invertebrates from sulfide rich habitats. *Nature* **302**: 58–61.
10. Cavanaugh CM, Gardiner SL, Jones ML, Jannasch HW, Waterbury JB (1981) Prokaryotic cells in the hydrothermal vent tube worm. *Science* **213**: 340–342.
11. Cavanaugh CM, Levering PR, Maki JS, Mitchell R, Lidstrom ME (1987) Symbiosis of methylotrophic bacteria and deep sea mussels. *Nature* **325**: 346–348.
12. Clarke RTJ, Bauchop T (eds) *Microbial Ecology of the Gut.* Academic Press, London.
13. Czerkawski JW, Cheng K-J (1989) Compartmentation in the rumen. In: Hobson PN (ed.) *The Rumen Microbial Ecosystem.* Elsevier, London, pp. 361–385.
14. Dando PR, Southward AJ, Southward EC, Terwilliger NB, Terwilliger RC (1985) Sulphur-oxidizing bacteria and haemoglobin in gills of the bivalve mollusc *Myrtea spinifera*. *Mar Ecol Prog Ser* **23**: 85–98.
15. Dando PR, Southward AJ, Southward EC, Barret RL (1986) Possible energy sources for chemoautotrophic prokaryotes symbiotic with invertebrates from a Norwegian fjord. *Ophelia* **26**: 135–150.
16. Dando PR, Ridgeway SA, Spiro B (1994) Sulphide "mining" by lucinid bivalve molluscs: demonstrated by stable sulphur isotope measurements and experimental models. *Mar Ecol Prog Ser* **107**: 169–175.
17. de Ridder C, Jangoux M, De Vos L (1985) Description and significance of a peculiar intradigestive symbiosis between bacteria and a deposit-feeding echinid. *J Exp Mar Biol Ecol* **96**: 65–75.
18. Dougherty RW (ed.) (1965) *Physiology of Digestion in the Ruminant.* Butterworths, Washington.
19. Douglas AE (1994) *Symbiotic Interactions.* Oxford University Press, Oxford.
20. Eady RR (1991) The dinitrogen-fixing bacteria. In: Balows A, Trüper HG, Dworkin M, Harder W, Scheifer K-H (eds) *The Prokaryotes*, 2nd edn, Vol. 1. Springer, New York, pp. 534–553.

21. Fenchel T (1996) Eukaryotic life: anaerobic physiology. In: Roberts DMcL, Sharp P, Alderson G, Collins M (eds) *Evolution of Microbial Life*. Cambridge University Press, Cambridge, pp. 185-203.

22. Fenchel T, Finlay BJ (1989) *Kentrophoros*: a mouthless ciliate with a symbiotic kitchen garden. *Ophelia* **30**: 75-93.

23. Fenchel T, Finlay BJ (1991) Endosymbiont methanogenic bacteria in anaerobic ciliates: significance for the host. *J Protozool* **38**: 18-22.

24. Fenchel T, Finlay BJ (1992) Production of methane and hydrogen by anaerobic ciliates containing symbiotic methanogens. *Arch Microbiol* **157**: 475-480.

25. Fenchel T, Finlay BJ (1995) *Ecology and Evolution in Anoxic Worlds*. Oxford University Press, Oxford.

26. Fenchel T, McRoy CP, Ogden JC, Parker P, Rainey WE (1979) Symbiotic cellulose degradation in green turtles, *Chelonia mydas* L. *Appl Environ Microbiol* **37**: 348-350.

27. Felbeck H, Distel L (1991) Prokaryotic symbionts of marine invertebrates. In: Balows A, Trüper HG, Dworkin M, Harder W, Schleifer K-H (eds) *The Prokaryotes*, 2nd edn, Vol. 4. Springer, New York, pp. 3891-3906.

28. Fischelson L, Montgomery L, Myrberg A (1985) A unique symbiosis in the gut of tropical herbivorous surgeon fish (Acanthuridae: Teleostei) from the Red Sea. *Science* **229**: 49-51.

29. Floener L, Bothe H (1980) Nitrogen fixation in *Rhopalodia gibba*, a diatom containing blue-greenish inclusions symbiotically. In: Shenk HEA, Schwemmler W (eds) *Endocytobiology, Endosymbiosis and Cell Research*, Vol. 1. Walter de Gruyter, Berlin, pp. 541-552.

30. Gasaway WC (1976) Seasonal variation in diet, volatile fatty acid production and size of the cecum of rosk ptarmigan. *Comp Biochem Physiol* **A53**: 109-144.

31. Giere O (1981) The gutless marine oligochaete *Phallodrilus leukodermatus*. Structural studies on an aberrant tubificid associated with bacteria. *Mar Ecol Prog Ser* **5**: 353-357.

32. Giere O (1985) Structure and position of bacterial endosymbionts in the gill filaments of lucinidae from Bermuda (Mollusca, Bivalvia). *Zoomorphology* **105**: 296-301.

33. Giere O, Langheld C (1987) Structural organisation, transfer and biological fate of endosymbiotic bacteria in gutless oligochaetes. *Mar Biol* **93**: 641-650.

34. Giere O, Wirsen CO, Schmidt C, Jannasch HW (1988) Contrasting effects of sulfide and thiosulfate on symbiotic CO_2-assimilation of *Phallodrilus leukodermatus* (Annelida). *Mar Biol* **97**: 413-419.

35. Hobson PN (ed.) (1989) *The Rumen Microbial Ecosystem*. Elsevier, London.

36. Hungate RE (1966) *The Rumen and its Microbes*. Academic Press, New York.

37. Hungate RE (1975) The rumen microbial system. *Ann Rev Ecol Syst* **6**: 39-66.

38. Iverson JB (1980) Colic modifications in iguanine lizards. *J Morphol* **163**: 79-93.

39. Janis C (1976) The evolutionary strategy of Equidae and the origins of rumen and cecal digestion. *Evolution* **30**: 757-774.

40. Mague TH, Weare NM, Holm-Hansen O (1971) Nitrogen fixation in the North Pacific Ocean. *Mar Biol* **24**: 109–119.

41. Millbank JW (1984) Nitrogen fixation by lichens. In: Subba Ras NS (ed.) *Current Development in Nitrogen Fixation*. Edward Arnold, London, pp. 197–218.

42. Minchin FR, Summerfield DJ, Hadley P, Roberts EH, Rawsthorne S (1981) Carbon and nitrogen nutrition of nodulated roots of grain legumes. *Plant Cell Environ* **4**: 5–26.

43. Moir RJ (1965) The comparative physiology of ruminant-like animals. In: Dougherty RW (ed.) *Physiology of Digestion in the Ruminant*. Butterworths, Washington, pp. 1–23.

44. Murray RM, Marsh H, Heinsohn GE, Spain AV (1977) The role of the midgut caecum and large instestine in the digestion of sea grasses by the dugong (Mammalia: Sirenia). *Comp Biochem Physiol* **A56**: 7–10.

45. Ott JA, Novak R (1989) Living at an interface: meiofauna at the oxygen/sulfide boundary of marine sediments. In: Ryland JS, Tyler PA (eds) *23rd Europ Mar Biol Symp*, Olsen & Olsen, Fredensborg, Denmark, pp. 415–422.

46. Ott JA, Rieger G, Rieger R, Enders F (1982) New mouthless interstitial worms from the sulfide system: symbiosis with prokaryotes. *PSZNI-Mar Ecol* **38**: 313–333.

47. Penry DL, Jumars PA (1987) Modelling animal guts as chemical reactors. *Am Nat* **129**: 69–96.

48. Plante CJ, Jumars PA, Baross JA (1989) Rapid bacterial growth in the hindgut of a marine deposit feeder. *Microb Ecol* **18**: 29–44.

49. Postgate J (1982) *The Fundamentals of Nitrogen Fixation*. Cambridge University Press, Cambridge.

50. Potrikos CJ, Breznak JA (1981) Gut bacteria recycle uric acid nitrogen in termites: a strategy of nutrient conservation. *Proc Natl Acad Sci* **78**: 4601–4605.

51. Russel J, Baldwin RL (1979) Comparison of maintenance energy expenditures and growth yields among several rumen bacteria grown in continuous culture. *Appl Environ Microbiol* **37**: 537–543.

52. Russel J, Baldwin RL (1979) Comparison of substrate affinities among several rumen bacteria: a possible determinant of rumen bacterial competition. *Appl Environ Microbiol* **37**: 531–536.

53. Schenk HEA (1991) Cyanobacterial symbioses. In: Balows A, Trüper HG, Dworkin M, Harder W, Schleifer K-H (eds) *The Prokaryotes*, 2nd edn, Vol. 4. Springer, New York, pp. 3819–3854.

54. Schmaljohann R, Flügel HG (1987) Methane-oxidizing bacteria in pogonophora. *Sarsia* **72**: 91–98.

55. Smith DC, Douglas AE (1987) *The Biology of Symbiosis*. Edward Arnold, London.

56. Southward EC (1987) Contribution of symbiotic chemoautotrophs to the nutrition of benthic invertebrates. In: Sleigh MA (ed.) *Microbes in the Sea*. Wiley, New York, pp. 83–118.

57. Sprent JI, Sprent P (1990) *Nitrogen-fixing Organisms: Pure and Applied Research*. Chapman and Hall, London.

58. Stewart WDP, Gallon JR (eds) (1980) *Nitrogen Fixation.* Academic Press, London.
59. Taylor FJR (1982) Symbiosis in marine microplankton. *Ann Inst Océanogr Paris* **58**(s): 61–90.
60. Temara A, de Ridder C, Kuenen JG, Robertson LA (1993) Sulfide-oxidizing bacteria in the burrowing echinoid *Echinocardium cordatum* (Echinodermata). *Mar Biol* **115**: 179–185.
61. Waterbury JB, Calloway CB, Turner RD (1983) A cellulolytic nitrogen-fixing bacterium cultured from the gland of Deshayes in shipworms (Bivalvia: Teredinidae). *Science* **221**: 1401–1403.
62. Wilkinson CR (1983) Net primary productivity in coral reef sponges. *Science* **219**: 410–412.
63. Wilkinson CR, Fay P (1979) Nitrogen fixation in coral reef sponges with symbiotic cyanobacteria. *Nature* **279**: 527–529.
64. Wilkinson CR, Vacelet J (1979) Transplantation of marine sponges to different conditions of light and current. *J Exp Mar Biol Ecol* **37**: 91–104.
65. Young JPW, Johnston AWP (1989) The evolution of specificity in the legume–rhizobium symbiosis. *Trends Ecol Evol* **4**: 341–349.

8

Biogeochemistry and extreme environments

Extreme environments are a focus of considerable attention for a number of reasons:

1. increased awareness of the widespread distribution of extreme environments;
2. increased accessibility of extreme environments, especially those of the subsurface and deep sea;
3. the availability of suitable methods for experimental manipulation of organisms under extreme conditions;
4. increased recognition of the commercial value of organisms from extreme environments or products of such organisms;
5. general recognition that Earth's earliest organisms may have originated in extreme environments.

In addition, the possibility that life might exist elsewhere (e.g. Mars) under extreme conditions (by terrestrial standards) has prompted renewed interest in potentially analogous habitats on Earth.

Microbiologists and biochemists have been particularly intrigued by the phylogenetic, physiological and biochemical novelty of the biota in extreme environments,[1,6,7,17,22−25,27,29,33,39,43,44,46,47,59,60] as are applied microbiologists using high-temperature fermentation for waste digestion.[49,53] However, extreme environments have also attracted the attention of microbial ecologists and biogeochemists, some of whom were among the first to explore many of these systems. For ecologists, extreme environments were initially attractive as presumably simple model systems for addressing basic questions about ecosystem dynamics. Not surprisingly, close examination of several different extreme environments has indicated that they are often rather complex, and as challenging to understand as less extreme systems.[61,66,67]

What then can be learned about biogeochemical principles from an analysis of extreme environments? Does phylogenetic novelty in extreme environments imply novel biogeochemical processes? Do such systems serve as models of processes or dynamics occurring in all systems, or do extreme environments represent exceptions, the heuristic value of which derives primarily from that fact? Before addressing these and other questions specifically, it is perhaps useful to define the term "extreme environment".

As used traditionally, extreme environments are characterised by one or more physical or chemical parameters that routinely exceed growth limits for most bacteria and eukaryotes. Examples include environments with pH < 2 or > 10; environments with temperatures > 40 °C; environments with water potential less than about −2 MPa (see Chapter 4): aquatic systems with salt concentrations > 1 M. Obviously, extremes of oxidation–reduction potential, substrate availability, pressure, toxic organics or inorganics and UV radiation among others can be used to define extreme environments. Of course, these definitions are somewhat arbitrary, and must be considered provisional to the extent that they are based on responses of known microbes. This is because the diversity of extreme environments is only now beginning to be addressed quantitatively. Conceivably, prokaryotes in some extreme environments could prove more diverse than those of some "moderate" environments; this would leave the tolerance limits of eukaryotes as a metric for defining extremes. In this case, the term "extreme" might be inapt, reflecting only the relative tolerances of eukaryotes and prokaryotes. None the less, we adopt here the conventional view of extreme environments and direct our attention to a few points that transcend the specific attributes of any given extreme environment while commenting on select "type" examples.

8.1   General overview and summary of hypersaline systems

Prior to the advent of techniques for routinely isolating and sequencing 16S ribosomal RNA, relatively little was known about the diversity of microbes inhabiting extreme environments. While microbial diversity in extreme environments is clearly much greater than originally imagined, the decrease in microbial diversity with increasingly extreme conditions remains a more or less accurate general principle. For instance, eukaryotes do not tolerate high temperatures (> 60 °C for microeukaryotes; > 50 °C for metazoans) or high ionic strengths; accordingly, prokaryotes dominate the biota of environments with these characteristics. Fungi and the relatively few halophilic microalgae that tolerate high salt concentrations (e.g. > 4 M NaCl by *Dunaliella tertiolecta*) are notable eukaryotic exceptions, occurring in xeric environments such as the Antarctic dry valleys and hypersaline lagoons, ponds and salterns.[15,30,63,64] At the upper limits of tolerable temperature and ionic strength ranges, diversity is reduced even further,[42] with Archaea (archaebacteria) more numerous than Bacteria (eubacteria); however, greater diversity need not mean greater biomass.[47] Diversity is also reduced at the lower limits of temperature and ionic strength, but in these cases numerous eukaryotes are important components of the microbiota. Patterns similar to those for temperature have been reported for pH, but extremes in other parameters (e.g. high pressure, low substrate availability) do not appear to evoke as dramatic a change in diversity. At present it is not clear whether archaeal dominance at the highest temperature and lowest pH regimes can be accounted for by intrinsic biochemical properties (e.g. cell wall or membrane composition), or whether ecological and evolutionary constraints dictate dominance.

The impacts of observed changes in diversity on the biogeochemistry of extreme environments are also unclear. This is due to the lack of absolute estimates of microbial diversity as well as to limited research on biogeo-chemical cycling. Biogeochemical assays are especially difficult in certain extreme environments, such as sea ice, Antarctic endolithic communities and hyperthermal systems. Samples are not only a challenge and expensive to obtain, but manipulation and incubation of material while maintaining *in situ* conditions often pose substantial problems.

Of the many extreme environments, hypersaline systems are perhaps the best studied due to their relative accessibility and ease of manipulation. Shallow, mat-forming systems have attracted the greatest attention, with numerous physical, geochemical and microbiological studies establishing several general trends[9,11,61,65,67] (see also Chapter 6 for a more detailed summary of mat systems). Distinct vertical differentiation of microbial communities, strong chemical gradients and high rates of metabolic activity characterise many hypersaline systems, the majority of which are shallow and dominated by populations of cyanobacteria and microeukaryotic algae. Typical primary producer–microbial heterotroph interactions control much of the biogeochemistry of such systems. For example, in hypersaline systems with an abundance of sulphate, sulphur transformations play a major role in organic matter dynamics;[16,50,54] rates of sulphate reduction in such systems are among the highest reported, and account for a large fraction of net primary production. Sulphide is also effectively oxidised in shallow hypersaline systems by both photosynthetic and chemolithotrophic sulphur oxidisers.

Although they are less well documented, submarine brines are more exotic, and perhaps more common, than shallow hypersaline systems.[5] Submarine brines occur commonly in the Gulf of Mexico and around the globe. These brines are often formed by seepage of high-salinity waters in contact with salt deposits; they are typically associated with hydrocarbon seeps and sulphide-rich sediments. The brines are stabilised by density differences due to salt concentrations more than five-fold that of seawater. Brine pools, such as the East Flower Garden and those of the Florida Escarpment, support dense animal assemblages reminiscent of hydrothermal vent fauna, including mytilid mussels with methanotrophic symbionts.[8,40] However, these assemblages appear restricted to the periphery of brine pools, since few macrofauna tolerate high salinities.

Microbial activity within the brines depends on organic inputs from hydrocarbon seeps or settling detritus, and chemolithotrophic metabolism of reduced sulphur species.[41] While substrates appear abundant, microbial activity in submarine brines is relatively low compared to shallow photic hypersaline systems.[38] Visual observations indicate that algal detritus and animal carcasses are preserved within brines; in addition, an apparent stability of dissolved ATP in brine pools has been attributed to inhibitory effects of high salinity on microbes.[62] Differences in activity between submarine and surface brines may also be a function of lower temperatures in the former (about 1–4 °C, depending on water depth). However, much remains unknown about the biogeochemistry and microbiology of both submarine brines and surface hypersaline systems. The nitrogen cycle is largely unexplored, the role of metals (e.g. iron and manganese) is uncertain, and the dynamics of methane have been rarely addressed.

The lack of comprehensive biogeochemical analyses in extreme environments leaves unanswered a variety of basic questions. Is the inventory of processes similar to that in less extreme systems, or are certain processes absent? What limits the activity of specific processes in a given type of environment? Patterns of methane oxidation in hypersaline environments illustrate limitations of current understanding. Methane oxidation cannot be readily detected in hypersaline Solar Pond (Sinai, Egypt), even though methane is produced within the system.[10,19,34] Since methanotrophic bacteria are able to use very low methane concentrations and tolerate at least moderate water stresses (see Chapter 9), there is no apparent physiological constraint that mitigates against activity in hypersaline environments. To what, then, can the absence of methanotrophic activity be attributed? Are there other extreme environments that mitigate against methanotrophs? Does the distribution of ammonia-oxidising bacteria, which are similar in many respects to methanotrophs, exhibit a similar pattern? Answers to these and related questions require a focused research effort combining conventional and molecular approaches. The former are needed to establish rates of activity and to obtain isolates for specific transformations; the latter are necessary to assess more completely the diversity of functional groups not amenable to cultivation.

More detailed biogeochemical analyses of extreme environments are not only important for understanding systems that have played an important role in mineral and hydrocarbon deposits over geological time (e.g. hypersaline and hydrothermal systems),[35] but also for answering basic questions about the structure and function of microbial communities. Interpretation of the geological record in a number of instances has depended on biogeochemical evidence derived from contemporary analogues of Proterozoic and later mat systems. Many conclusions appear sound, but hypotheses about the antiquity of methane oxidation provide an example of the need for additional research. Conjecture about the origins of methanotrophy is based on isotopic evidence derived from fossil stromatolithic systems,[21] yet methanotrophic activity is low or absent in contemporary stromatolithic and hypersaline systems.[10,55] This discrepancy indicates that an alternative interpretation of the isotopic evidence is needed, that early methanotrophs were halophilic to a greater extent than today, or that methane oxidation does occur in contemporary hypersaline systems, but just not commonly. Analyses of microbial communities in extreme environments may also help reveal the minimum constraints for biogeochemical stability, and the ability of relatively simple systems to respond to and recover from disturbance. In addition, assays of extreme environments may help frame hypotheses about exobiogeochemical systems. The dry valleys of the Antarctic have long been considered analogues of Martian soils. Althought the Martian surface seems lifeless based on evidence from the Viking landers, additional research on the dry valleys as well as the terrestrial subsurface may help resolve controversy about putative microfossils from the meteorite, ALH-84001, and guide future efforts to explore the martian subsurface.

The paucity of metazoans in extreme environments raises additional questions. For example, is the efficiency of carbon mineralisation reduced in the absence of a significant faunal community? Evidently, contemporary hypersaline environments are analogous to similar systems believed to account for many of the Tertiary petroleum deposits;[35] the latter also lacked a significant fauna (as indicated by the fossil record). Does this imply that a faunal community is essential for efficient organic matter degradation (or that the absence of fauna enhances organic burial)? Does the absence of bioturbation in particular promote organic matter preservation in subsurface sediments by limiting oxygen availability?

Though limited in number, metazoans are not entirely excluded from extreme environments.[18] Certain hypersaline environments occasionally support large grazer populations that can consume substantial amounts of biomass. In these cases, grazing undoubtedly contributes to nutrient recycling that would otherwise depend on microbial hydrolytic enzyme systems, and perhaps protozoan grazers (e.g. Chapter 2). However, most of the grazers in extreme environments (typically insects [e.g. brine flies] or crustaceans, but also flamingoes in some African hypersaline lakes) do not mix sediments, and therefore have little impact on subsurface organic matter.

Metazoan grazers are essentially absent from highly acidic (pH < 2) and hyperthermal (> 60 °C) environments; metazoans also appear to be absent in submarine hypersaline systems. In these cases, mineralisation of organic matter originating from primary or chemolithotrophic production depends largely on microbial processes. Inefficient mineralisation results in organic matter accumulation within the system, or export and utilisation elsewhere. Hydrothermal vents are classical examples of the latter.

Although considerable bacterial production occurs within vent plumes under moderate conditions, some biomass is produced directly within the cooler regions of vent chimneys;[28] some of this material can be entrained in the seawater outflow, and undoubtedly accounted for some of the early (now discounted) claims about life at 250 °C.[4,56,58,68]

Organic exports and burial both require some compensatory input of nutrients, such as nitrogen and phosphorus, that are not recycled. In systems characterised by fluid flow (e.g. hydrothermal vents or springs), nutrient limitation may seldom prove a problem. However, in many other systems, for example dry valley soils, ice floes and hypersaline lakes, nutrient limitations may constrain organic matter production. In at least some of these systems, nitrogen fixation can alleviate nitrogen losses. Although nitrogen fixers are present in virtually all extreme environments due to the ubiquity of nitrogenase in the Archaea, the significance of nitrogen fixation is unclear due to the lack of synoptic input–output budgets. Likewise, the physiology of nitrogen fixation under extreme conditions has not been adequately explored, and represents an area for additional research effort.

8.2 Subsurface environments

For a number of reasons, terrestrial subsurface environments and deeply buried marine sediments have drawn a great deal of attention. While once considered largely devoid of life, it is now apparent that the subsurface harbours an impressive diversity of microbes that appear to be taxonomically and phylogenetically unique.[2,13,14,20] Although the term "subsurface" is often used to describe widely varying types of systems, when used as a descriptor for strata effectively isolated from the surface for time scales of many years, it encompasses environments that can be considered extreme in at least several respects. First, the thermal gradient with depth (about $15 °C \, km^{-1}$) results in temperatures suitable for thermophiles and hyperthermophiles at moderate crustal depths. Second, organic matter availability in isolated formations or strata is typically very low, resulting in profound long-term starvation regimes relative to conditions at the surface. Whether such starvation regimes select for novel physiologies or simply produce largely moribund populations is uncertain. Third, subsurface systems that are not water-saturated may experience low water potentials leading to additional stresses. The combination of moderate to high temperature, starvation substrate regimes and water stress clearly justifies designation of at least some subsurface systems as extreme.

A number of remarkable examples of subsurface systems exist. Krumholz et al.[36] have obtained rocks from Cretaceous period shales and sandstones at depths of up to 250 m. The rocks had apparently been isolated from contact with the surface for at least 10^4 years based on groundwater ages. Since the parent materials were deposited in a marine environment,

Krumholz *et al.*[36] conducted an assay for sulphate reduction. Activity was readily detected in crushed material incubated in slurries and, perhaps more significantly, in relatively undisturbed freshly exposed rock faces. They also documented *in vitro* stimulation of sulphate reduction in sandstones by organic matter contained in added shales; in these analyses, the response of sulphidogens was relatively rapid, which indicates that cells *in situ* exist in an active state. *In situ* activity appears to be controlled in part by the diameter of pore sizes in the sandstones and adjacent shales, and by changes in pore-size distribution over time.[12] Pore-size distribution controls both microbial movement and organic fluxes in groundwater. In addition, Krumholz *et al.*[36] have provided evidence for a potentially complex community of anaerobes, including fermentors and acetogens, as well as sulphidogens. Other ancient formations presumably isolated from the surface for $\gg 10^6$ years have also yielded viable microbes, though the nature of any substrates used *in situ* is uncertain. Conceivably, cells in such systems may remain in a moribund state indefinitely, responding only when (or if) conditions permit. However, the mechanisms by which cells or resting stages (e.g. cysts or spores) remain viable for millions of years is unknown. Further analysis of "suspended animation" is essential since the limits on microbial longevity have implications for exobiological research programmes. If viable cells can persist for $\gg 10^6$ years, then remnant populations may still exist in formerly habitable surfaces or subsurfaces of planets such as Mars.

A relatively young formation (Miocene age) within the Columbia River Basalt Group has yielded additional, remarkable insights about processes in the subsurface.[57] In this case, anoxic groundwaters apparently react chemically with basalts containing ferrous silicate resulting in hydrogen production. Hydrogen can accumulate to high concentrations (up to 60 μM), and consequently support communities of lithotrophic bacteria, including methanogens and acetogens, present in contiguous groundwaters.[57] In addition, experimental results suggest that basalt alone appears to be sufficient to promote hydrogen-based microbial growth. Continued validation of these results may reveal widespread subsurface ecosystems essentially independent of photosynthesis. Such systems could provide additional models for the origin of life, especially in high-temperature formations.

Comprehensive surveys of subsurface microbial diversity are few in number. A variety of fermentative, acetogenic, sulphidogenic and methanogenic bacteria representing the Archaea and Bacteria as well as heterotrophic and lithotrophic metabolic modes have been reported from several sites. Some of these populations are novel. Balkwill *et al.*[2] have conducted the most extensive physiological, taxonomic and phylogenetic characterisations of subsurface microbes, mostly from deep coastal plain sediments. Their isolates include aerobes and anaerobes, the former exhibit substantial physiological diversity in terms of substrate utilization. One particularly intriguing and unusual isolate has been identified as a *Sphingomonas* sp. that degrades a number of aromatics, including toluene and naphthalene.[13] While much remains unknown about subsurface microbiology and *in situ* activity, the physiological diversity observed thus far suggests that the array of available substrates is complex if scarce, and that subsurface bacteria may be adapted to use many substrates simultaneously, a trait that has obvious competitive advantages when the total substrate supply is extremely limited.

A variety of microbes and microbial processes also occurs deep in marine sediments. Core samples collected from > 500 m sediment depth contain viable populations of sulphate-reducing bacteria that actively (though very slowly) reduce sulphate.[48] Whether or not such populations are distinct from those near the sediment surface is unclear. However, it appears that they represent populations isolated by burial for millions of years, much like populations from deep terrestrial sites. Several analyses of microbes associated with seamounts and oil-producing formations in deep sediments of the North Sea and North Slope, Alaska support assessments from Pacific Ocean cores.[3,26,37,45,51] Results from oil fields strongly suggest that both thermophilic archaeal (e.g. *Archaeoglobus* and *Thermodesulforhabdus* spp.) and bacterial (e.g. *Desulfotomaculum* spp.) sulphate reducers grow within oil reservoirs, presumably on hydrocarbons or various organic acids and other substrates in the reservoirs. Considerable diversity in microbial populations among oil reservoirs appears to exist, but in virtually all cases sulphate reducers, important at depth, are not common or important in surface sediments. Given the elevated temperatures characteristic of deeply buried sediments as well as the nature of available substrates, such differences are not surprising. However, they may reflect more generally the existence of unique subsurface microbial communities throughout the deep sediments of the oceans.

8.3 Microbial diversity and high-temperature environments

Comprehensive surveys of microbial diversity in extreme environments are lacking not only for subsurface environments, but for extreme environments more generally. However, traditional enrichment and isolation methods as well as molecular approaches have been used to characterise specific physiological or phylogenetic groups in certain systems, especially high-temperature environments.[22–26,31,32,43,51,59,60,67] Results from such analyses provide insights into biogeochemical transformations observed within a given system, or an indication of processes that might occur when local conditions permit. For instance, isolation and characterisation of methylamine-degrading methanogens from hypersaline sites supports and extends sediment slurry analyses of hypersaline methanogenesis showing strong stimulation by trimethylamine.[19,33] Routine isolation of lithotrophs from acidic and hyperthermal systems indicates that reduced inorganic oxidation dominates metabolism, even though the relevant transformation rates have seldom been assessed.[52]

High-temperature environments (e.g. various hot springs and hydrothermal vents) have been subject to the most intensive analyses of microbial diversity. These systems are characterised by bacterial and archaeal thermophiles and hyperthermophiles, most of which are chemolithotrophs; relatively few hyperthermophilic chemoorganotrophs have been isolated. The paucity of chemoorganotrophs may reflect ecological attributes of high-temperature systems rather than inherent limitations of temperature on organotrophs *per se*. High-temperature systems are usually based on flows of heated water with low concentrations of organics, but high concentrations of reduced inorganic species (e.g. H_2, H_2S, Fe^{2+}) arising from geochemical reactions. The latter conditions favour the evolution and proliferation of lithotrophic metabolism; in contrast, water flow regimes may limit organic accumulation, and therefore chemoorganotroph proliferation. Greater abundance of chemoorganotrophs at lower temperatures, especially temperatures conducive to photolithotrophic metabolism (less than about 75 °C), may be attributed to cyanobacterial and microeukaryotic algal excretion of a variety of simple organics, and formation of biofilms that provide diverse microhabitats, as well as a matrix that retards diffusive losses of organics sufficiently to promote microbial uptake.

Isolates from high-temperature systems include genera forming distinct lineages within the domain Bacteria, as well as numerous examples of the Archaea (Fig. 8.1). The Bacteria are represented by chemoorganotrophic aerobes (e.g. *Thermoleophium, Thermomicromicrobium*) and anaerobes (*Thermosipho, Thermotoga*) of the Thermotogales, and

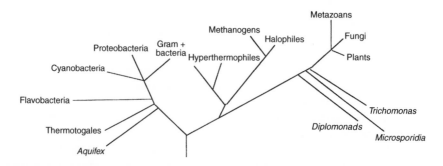

Figure 8.1   Phylogenetic tree based on comparative sequences of 16S and 18S rRNA sequences indicating relationships of major domains and selected taxonomic groups. Note the thermophilic and hyperthermophilic taxa at the base of the Archaea and Bacteria domains. Data from Woese[69] and other sources

chemolithotrophic aerobes (e.g. *Aquifex*) of the Aquificales. The aerobic character of *Aquifex* as well as its ability to denitrify raise a number of important questions, since it is the most deeply divergent lineage of the Bacteria according to 16S rRNA molecular phylogenetic analysis. Presumably, this lineage predates the anaerobic photolithotrophs (e.g. purple sulphur bacteria), which in turn predate oxygenic photosynthesis. Some have argued that respiratory pathways followed the advent of photosynthesis, and that aerobic respiration in particular postdated oxygenic photosynthesis; likewise, denitrification presumably evolved subsequent to the advent of an ocean and atmosphere sufficiently oxic to facilitate nitrate production (see Chapters 9 and 10). To the extent that these arguments are valid, the metabolic characteristics of *Aquifex* appear enigmatic. The fact that it is microaerophilic and inhibited by contemporary levels of oxygen does not provide a solution to the enigma, even though some have postulated that very low oxygen concentrations existed from water photolysis in Earth's early atmosphere. While this may have been the case, it seems implausible that molecular oxygen in biologically meaningful concentrations could have existed at the high-temperature sites necessary for evolution of *Aquifex's* hyperthermophilic trait.

Photolytic and chemical production of oxidants other than molecular oxygen offer a possible resolution for the discrepancy between the physiological traits and phylogenetic position of *Aquifex*. As described in more detail subsequently (see Chapter 10), UV photolysis of ferrous iron can produce ferric iron readily; ferric iron can in turn serve as an oxidant for ammonium, resulting in nitrite or nitrate formation. Though speculative, such processes could have provided an early source of oxidant in high-temperature systems at concentrations sufficient to promote evolution of respiratory systems predating (and perhaps leading to) the electron transport systems used in photosynthesis and oxygen respiration. Alternatively, horizontal gene transfer could have resulted in the acquisition of physiological properties that are not concordant with 16S rRNA phylogenies. At present, the limited number of bacterial hyperthermophiles precludes a more definitive answer.

Among the Archaea, the most deeply divergent lineages, for example the Desulfurococcaceae and Thermoproteales, have physiological traits more or less consistent with their phylogenetic position. The genera of these orders are primarily anaerobic chemoorganotrophs or chemolithotrophs that use elemental sulphur as an electron acceptor. These modes of metabolism are compatible with current understanding of conditions that might have prevailed at high-temperature systems early in Earth's history (e.g. possible availability of abiologically synthesised organics and an abundance of reduced inorganic species). The phylogenetic positions of other Archaea found in extreme environments, including the methanogens and halophiles, also pose no particular problems regarding their origin and likely biogeochemical conditions that existed at the time. None the less, with only a small fraction (< 1%) of total microbial diversity documented and the microbiology of extreme environments and molecular phylogeny both still in their infancy, it is premature to assume that existing phylogenetic patterns and the associated distribution of physiological traits necessarily reflect a linear evolutionary progression. Perhaps the key exception to this caveat is the apparently central role of thermophily in the origins and early evolution of life (see Chapter 10). The possibility that Earth's first organisms inhabited analogues of contemporary hydrothermal vents and hot springs will ensure continued research on hyperthermophiles and the microbiota of other extreme environments for the foreseeable future. Increased attention to the biogeochemistry of these systems may also reveal novel processes central to the origins of life.

References

1. Achenbach-Richter L, Gupta R, Stetter KO, Woese CR (1987) Were the original eubacteria thermophiles? *Syst Appl Microbiol* **9**: 34–39.
2. Balkwill DL, Fredrickson JK, Thomas JM (1989) Vertical and horizontal variations in the physiological diversity of the aerobic chemoheterotrophic bacterial microflora in deep southeast plain subsurface sediments. *Appl Environ Microbiol* **55**: 1058–1065.
3. Beeder J, Nilsen RK, Thorstenson T, Torsvik T (1996) Penetration of sulphate reducers through a porous North Sea oil reservoir. *Appl Environ Microbiol* **62**: 3551–3553.
4. Bernhardt G, Lüdemann H-D, Jaenicke R, König H, Stetter KO (1984) Bio-molecules are unstable under "black smoker" conditions. *Naturwissenschaft* **71**: 583–585.
5. Brooks JM, Bright TJ, Bernard BB, Shwab CR (1979) Chemical aspects of a brine pool at the East Flower Garden Bank, northwestern Gulf of Mexico. *Limnol Oceanogr* **24**: 735–745.
6. Burggraf S, Jannasch HW, Nicolaus B, Stetter KO (1990) *Archaeoglobus profundus* sp. nov., represents a new species within the sulfate-reducing archaebacteria. *Syst Appl Microbiol* **13**: 24–28.
7. Burggraf S, Stetter KO, Rouviere P, Woese CR (1991) *Methanopyrus kandleri*: an archael methanogen unrelated to all other known methanogens. *Syst Appl Microbiol* **14**: 346–351.
8. Cary C, Fry B, Felbeck H, Vetter RD (1989) Multiple trophic resources for a chemoautotrophic community at a cold water brine seep at the base of the Florida Escarpment. *Mar Biol* **100**: 411–418.
9. Cohen Y (1989) Photosynthesis in cyanobacterial mats and its relation to the sulfur cycle: a model for microbial sulfur interactions. In: Cohen Y, Rosenberg E (eds) *Microbial Mats: Physiological Ecology of Benthic Microbial Communities*. American Society for Microbiology, Washington, DC, pp. 3–15.

10. Conrad R, Frenzel P, Cohen Y (1995) Methane emission from hypersaline microbial mats: lack of aerobic methane oxidation activity. *FEMS Microbiol Ecol* **16**: 295-305.

11. D'Amelio ED, Cohen Y, Des Marais DJ (1989) Comparative functional ultrastructure of two hypersaline submerged cyanobacterial mats: Guerrero Negro, Baja California Sur, Mexico, and Solar Lake, Sinai, Egypt. In: Cohen Y, Rosenberg E (eds) *Microbial Mats: Physiological Ecology of Benthic Microbial Communities*. American Society for Microbiology, Washington, DC, pp. 97-113.

12. Fredrickson JJ, McKinnley JP, Bjornstad BN, Long PE, Ringelberg DB, White DC, Krumholz LR, Suflita JM, Colwell FS, Lehman RM, Phelps TJ (1996) Pore-size constraints on the activity and survival of subsurface bacteria in a late Cretaceous shale-sandstone sequence, northwestern New Mexico. *Geomicrobiol J* **14**: 183-202.

13. Fredrickson JK, Balkwill DL, Drake GR, Romine MF, Ringleberg DB, White DC (1995) Aromatic-degrading *Sphingomonas* isolates from the deep subsurface. *Appl Environ Microbiol* **61**: 1917-1922.

14. Fredrickson JK, Balkwill DL, Zachara JM, Li SW, Brockman FJ, Simmons MA (1991) Physiological diversity and distributions of heterotrophic bacteria in deep Cretaceous sediments of the Atlantic coastal plain. *Appl Environ Microbiol* **57**: 402-411.

15. Friedmann EI (1980) Endolithic microbial life in hot and cold deserts. *Orig Life* **10**: 223-235.

16. Fründ C, Cohen Y (1992) Diurnal cycles of sulfate reduction under oxic conditions in cyanobacterial mats. *Appl Environ Microbiol* **58**: 70-77.

17. Gerhardt M, Svetlichny V, Sokolova TG, Zavarzin GA, Ringpfeil M (1991) Bacterial CO utilization with H_2 production by the strictly anaerobic lithoautotrophic thermophilic bacterium *Carboxydothermus hydrogenus* DSM-6008 isolated from a hot swamp. *FEMS Microbiol Lett* **83**: 267-272.

18. Gerdes G, Spira Y, Dimentman C (1985) The fauna of the Gavish Sabkha and the Solar Lake — a comparative study. In: Friedman GM, Krumbein WE (eds) *Hypersaline Ecosystems: The Gavish Sabkha*. Springer, Berlin, pp. 322-345.

19. Giani D, Giani L, Cohen Y, Krumbein WE (1984) Methanogenesis in the hypersaline Solar Lake (Sinai). *FEMS Microbiol Lett* **25**: 219-224.

20. Ghiorse WC (1997) Subterranean life. *Science* **275**: 789-791.

21. Hayes JM (1983) Geochemical evidence bearing on the origin of aerobiosis, a speculative hypothesis. In: Schopf JW (ed.) *Earth's Earliest Biosphere: Its Origins and Evolution*. Princeton University Press, Princeton, pp. 291-301.

22. Henry EA, Devereux R, Maki JS, Gilmour CC, Woese CR, Mandelco L, Schauder R, Remsen CC, Mitchell R (1994) Characterization of a new thermophilic sulfate-reducing bacterium, *Thermodesulfovibrio yellowstonii*, gen. nov. and sp. nov.: its phylogenetic relationship to *Thermodesulfobacterium commune* and their origins deep within the bacterial domain. *Arch Microbiol* **161**: 62-69.

23. Huber G, Huber R, Jones BE, Lauerer G, Neuner A, Segerer A, Stetter KO, Degens ET (1991) Hyperthermophilic archaea and bacteria occurring within Indonesian hydrothermal areas. *Syst Appl Microbiol* **14**: 397-404.

24. Huber G, Stetter KO (1991) *Sulfolobus metallicus*, sp nov, a novel strictly chemolithoautotrophic thermophilic archaeal species of metal-mobilizers. *Syst Appl Microbiol* **14**: 372-378.

25. Huber R, Burggraf S, Mayer T, Barns SM, Rossnagel P, Stetter KO (1995) Isolation of a hyperthermophilic archaeum predicted by in situ RNA analysis. *Nature (Lond)* **376**: 57-58.

26. Huber R, Stoffers P, Cheminee JL, Richnow HH, Stetter KO (1990) Hyperthermophilic archaebacteria within the crater and open-sea plume of erupting MacDonald Seamount. *Nature (Lond)* **345**: 179-182.

27. Isaksen MF, Jørgensen BB (1996) Adaption of psychrophilic and psychrotropic sulfate-reducing bacteria to permanently cold marine environments. *Appl Environ Microbiol* **62**: 408-415.

28. Jannasch HW, Mottl MJ (1985) Geomicrobiology of deep-sea hydrothermal vents. *Science* **229**: 717-725.

29. Javor BJ (1984) Growth potential of halophilic bacteria isolated from solar salt environments: carbon sources and salt requirements. *Appl Environ Microbiol* **48**: 352-360.

30. Johnston CG, Vestal JR (1991) Photosynthetic carbon incorporation and turnover in Antarctic cryptoendolithic microbial communities: are they the slowest-growing communities on Earth? *Appl Environ Microbiol* **57**: 2308-2311.

31. Jørgensen BB, Isaksen MF, Jannasch HW (1992) Bacterial sulfate reduction above 100 °C in deep-sea hydrothermal vent sediments. *Science* **258**: 1756-1757.

32. Jørgensen BB, Zawacki LZ, Jannasch HW (1990) Thermophilic bacterial sulfate reduction in deep-sea sediments at the Guaymas Basin hydrothermal vent site (Gulf of California). *Deep-Sea Res* **37**: 695-710.

33. Kimura T, Horikoshi K (1988) Isolation of bacteria which can grow at both high pH and low temperature. *Appl Environ Microbiol* **54**: 1066-1067.

34. King GM (1988) Methanogenesis from methylated amines in a hypersaline algal mat. *Appl Environ Microbiol* **54**: 130-136.

35. Krumbein WE (1985) Applied and economic aspects of sabkha systems — genesis of salt, ore and hydrocarbon deposits and biotechnology. In: Friedman GM, Krumbein WE (eds) *Hypersaline Ecosystems: The Gavish Sabkha*. Springer, Berlin, pp. 426-436.

36. Krumholz LR, McKinley JP, Ulrich GA, Suflita JM (1997) Confined subsurface microbial communities in Cretaceous rock. *Nature (Lond)* **386**: 64-66.

37. L'Haridon S, Reysenbach A-L, Glénat P, Prieur D, Jeanthon C (1995) Hot subterranean biosphere in a continental oil reservoir. *Nature (Lond)* **377**: 223-224.

38. LaRock PA, Lauer RD, Schwarz JR, Watanabe KK, Wiesenburg DA (1979) Microbial biomass and activity distribution in an anoxic hypersaline basin. *Appl Environ Microbiol* **37**: 466-470.

39. Lazcano A, Fox GE, Oró JF (1992) Life before DNA: the origin and evolution of early archaean cells. In: Mortlock RP (ed.) *The Evolution of Metabolic Function*. CRC Press, Boca Raton, pp. 237–296.

40. Macdonald IR, Reilly JF, Guinasso NLJ, Brooks JM, Carney RS, Bryant WA, Bright TJ (1990) Chemosynthetic mussels at a brine-filled pockmark in the northern Gulf of Mexico. *Science* **248**: 1096–1099.

41. Martens CS, Chanton JP, Paull CK (1991) Biogenic methane from abyssal brine seeps at the base of the Florida escarpment. *Geology* **19**: 851–854.

42. Martinez-Murcia AJ, Acinas SG, Rodriguez-Valera F (1995) Evaluation of prokaryotic diversity by restrictase digestion of 16S rDNA directly amplified from hypersaline environments. *FEMS Microbiol Ecol* **17**: 247–256.

43. Miroshnichenko ML, Bonch-Osmolovskaya EA, Neuner A, Kostrikina NA, Chernych NA, Alekseev VA (1989) *Thermococcus stetteri* sp. nov., a new extremely thermophilic marine sulfur-metabolizing archaebacterium. *Syst Appl Microbiol* **12**: 257–262.

44. Mooers AØ, Redfield RJ (1996) Digging up the roots of life. *Nature (Lond)* **379**: 587–588.

45. Nilsen RK, Beeder J, Thorstenson T, Torsvik T (1996) Distribution of thermophilic marine sulfate reducers in North Sea oil field waters and oil field waters and oil reservoirs. *Appl Environ Microbiol* **62**: 1793–1798.

46. Nisbet EG, Fowler CMR (1996) Some like it hot. *Nature (Lond)* **382**: 404–405.

47. Pace NR (1997) A molecular view of microbial diversity and the biosphere. *Science* **276**: 734–740.

48. Parkes RJ, Cragg BA, Bale SJ, Getliff JM, Goodman K, Rochelle PA, Fry JC, Weightman AJ, Harvey SM (1994) Deep bacterial biosphere in Pacific Ocean sediments. *Nature (Lond)* **371**: 410–413.

49. Petersen SP, Ahring BK (1991) Acetate oxidation in a thermophilic anaerobic sewage-sludge digestor: the importance of non-aceticlastic methanogenesis from acetate. *FEMS Microbiol Ecol* **86**: 149–157.

50. Ramsing NB, Kühl M, Jørgensen BB (1993) Distribution of sulfate-reducing bacteria, O_2 and H_2S in photosynthetic biofilms determined by oligonucleotide probes and microelectrodes. *Appl Environ Microbiol* **59**: 3840–3849.

51. Rosnes JT, Torsvik T, Lien T (1991) Spore-forming thermophilic sulfate-reducing bacteria isolated from North Sea oil field waters. *Appl Environ Microbiol* **57**: 2302–2307.

52. Sandbeck KA, Ward DM (1982) Temperature adaptations in the terminal processes of anaerobic decomposition of Yellowstone National Park and Icelandic hot spring microbial mats. *Appl Environ Microbiol* **44**: 844–851.

53. Schmidt JE, Ahring BK (1991) Acetate and hydrogen metabolism in intact and disintegrated granules from an acetate-fed, 55 °C UASB reactor. *Appl Microbiol Biotechnol* **35**: 681–685.

54. Skyring GW, Lynch RM, Smith GD (1989) Quantitative relationships between carbon, hydrogen, and sulfur metabolism in cyanobacterial mats. In: Cohen Y, Rosenberg E (eds) *Microbial Mats: Physiological Ecology of Benthic*

Microbial Communities. American Society for Microbiology, Washington, DC, pp. 170–177.

55. Sokolov AP, Trotsenko YA (1995) Methane consumption in (hyper)saline habitats of Crimea (Ukraine). *FEMS Microbiol Ecol* **18**: 299–304.

56. Stetter KO, Fiala G, Huber R, Huber G, Segerer A (1986) Life above the boiling point of water? *Experientia* **42**: 1187–1191.

57. Stevens TO, McKinley JP (1995) Lithoautotrophic microbial ecosystems in deep basalt aquifers. *Science* **270**: 452–454.

58. Straube WL, Deming JW, Somerville CC, Colwell RR, Baross JA (1990) Particulate DNA in smoker fluids: evidence for existence of microbial populations in hot hydrothermal systems. *Appl Environ Microbiol* **56**: 1440–1447.

59. Svetlichny VA, Sokolova TG, Gerhardt M, Kostrinkina NA, Zavarzin GA (1991) Anaerobic extremely thermophilic carboxydotrophic bacteria in hydrotherms of Kuril Islands. *Microb Ecol* **21**: 1–10.

60. Svetlichny VA, Sokolova TG, Gerhardt M, Ringpfeil M, Kostrinkina NA, Zavarzin GA (1991) *Carboxydothermus hydrogenoformans* gen. nov., sp. nov., a CO-utilizing thermophilic anaerobic bacterium from hydrothermal environments of Kunashir Island. *Syst Appl Microbiol* **14**: 254–260.

61. Turner S, DeLong EF, Giovannoni SJ, Olsen GJ, Pace NH (1989) Phylogenetic analysis of micro-organisms and natural populations by using rRNA sequences. In: Cohen Y, Rosenberg E (eds) *Microbial Mats: Physiological Ecology of Benthic Microbial Communities*. American Society for Microbiology, Washington, DC, pp. 390–401.

62. Tuoliva BJ, Dobbs FC, Larrock PA, Siegel BZ (1987) Preservation of ATP in hypersaline environments. *Appl Environ Microbiol* **51**: 2749–2753.

63. Vestal JR (1988) Biomass of the cryptoendolithic microbiota from the Antarctic desert. *Appl Environ Microbiol* **54**: 957–959.

64. Vestal JR (1988) Carbon metabolism of the cryptoendolithic microbiota from the Antarctic desert. *Appl Environ Microbiol* **54**: 960–965.

65. Ward DM, Weller R, Shiea J, Castenholz RW, Cohen Y (1989) Hot spring microbial mats: anoxygenic and oxygenic mats of possible evolutionary significance. In: Cohen Y, Rosenberg E (eds) *Microbial Mats: Physiological Ecology of Benthic Microbial Communities*. American Society for Microbiology, Washington, DC, pp. 3–15.

66. Ward DM, Bateson MM, Weller R, Ruff-Roberts AL (1992) Ribosomal RNA analysis of micro-organisms as they occur in nature. *Adv Microb Ecol* **12**: 219–286.

67. Ward DM, Ferris MJ, Nold SC, Bateson MM, Kopczynski ED, Ruff-Roberts AL (1994) Species diversity in hot spring microbial mats as revealed by both molecular and enrichment culture approaches — relationship between biodiversity and community structure. In: Stahl LJ, Caumett P (eds) *Microbial Mats: Structure, Development and Environmental Significance*. Springer, Berlin, pp. 33–44.

68. White RH (1984) Hydrolytic stability of biomolecules at high temperatures and its implications for life at 250 °C. *Nature (Lond)* **310**: 430–431.

69. Woese CR (1987) Bacterial evolution. *Microbial Rev* **51**: 221–271.

9

Microbial biogeochemical cycling and the atmosphere

9.1   The atmosphere as an elemental reservoir

The Earth's elements are distributed among four major reservoirs: atmosphere, biosphere, hydrosphere and lithosphere (Fig. 9.1); the cryosphere, or polar ice caps, represents a fifth reservoir important for Earth's climate, but largely unimportant as a repository for elements. Definitions of these reservoirs are generally familiar, but for our purposes the biosphere includes living organisms only, and excludes their abiological surroundings and extra-organismal products. Using such a definition, the biosphere is clearly dispersed within (or on) the hydrosphere and lithosphere, and to a minor extent within the atmosphere. In addition, the hydrosphere is defined as inclusive of all surface waters and contiguous groundwaters; physically isolated subsurface water is considered a component of the lithosphere. By analogy, soil gas phases physically isolated from the atmosphere are also a component of the lithosphere.

   The distribution of elements within and transfer rates among reservoirs are a function of the different forms in which elements exist. These forms include solid, dissolved and gas phases. Obviously, specific phases are characteristic of specific reservoirs, although each of these phases occurs within each of the major elemental reservoirs. Elements can also be categorised according to the relative importance of solid, dissolved and gas phases for biospheric exchanges. In particular, one can distinguish elements for which at least one volatile species is an essential component of biospheric exchange. With the exception of phosphorus, the major elements in organic matter occur in the atmosphere in one or more forms that can contribute significantly to biosynthesis (Table 9.1). CO_2, carbon monoxide (CO) and methane dominate carbon flows into and out of the biosphere; CO_2, O_2 and water dominate oxygen flows. Exchanges of nitrogen and sulphur involve gases (primarily N_2 and dimethyl sulphide, respectively), but for both of these elements a large fraction of biospheric exchange is based on non-volatile species (nitrate and ammonium; sulphate and bisulphide). However, it is important to note that many non-volatile species

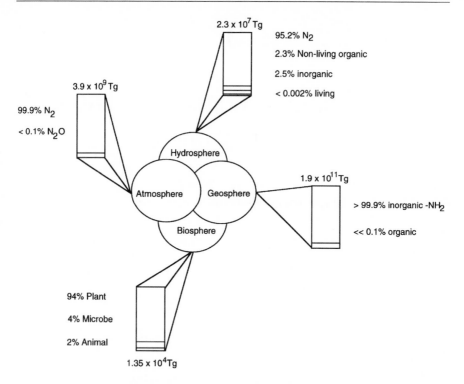

Figure 9.1   Conceptual illustration of overlapping elemental reservoirs, and distribution of nitrogen within and among reservoirs; data for nitrogen[133]

Table 9.1   Approximate composition of the atmosphere and fluxes of selected elements*

Constituent	Volume (%)	Total mass (g)	Element	Atmospheric flux	Industrial flux
Total atmosphere		5.136×10^{21}	H (as H_2O)	6×10^4	Large
Water vapour	Variable	0.017×10^{21}	C	200	8
Dinitrogen	78.084	3.866×10^{21}	N	0.25	0.1
Oxygen	20.94	1.185×10^{21}	O (as O_2)	300	0.1
Argon	0.934	6.59×10^{19}	O (as H_2O)	5×10^5	Large
Carbon dioxide	0.0315	2.45×10^{18}	Na	0	0.001
Neon	1.818×10^{-3}	6.48×10^{16}	Mg	0	3×10^{-4}
Helium	5.24×10^{-4}	3.71×10^{15}	Si	0	0.1
Methane	1.7×10^{-4}	4.8×10^{15}	P	0	0.15
Nitrous oxide	3×10^{-5}	2.3×10^{15}	S	0.1	0.15
Carbon monoxide	1.2×10^{-5}	5.9×10^{14}	As	2×10^{-5}	5×10^{-5}
Ammonia	1×10^{-6}	3×10^{13}	Hg	5×10^{-5}	8×10^{-6}
Ozone	Variable	3.3×10^{15}	Pb	0	0.004

*Fluxes are in units of 10^{12} kg year^{-1}.
Values of "0" indicate very small but finite[35,128]

are transported through the atmosphere, and ultimately enter the biosphere via wet and dry deposition. For example, inputs of nitrogen to coastal oceans and temperate forests by deposition have become increasingly important sources of eutrophication and pollution.[56,66,115,116] Inputs of sulphur by dry and wet deposition are likewise important in terrestrial systems both as a source of sulphur nutrition and, when in excess, as a source of acidity that decreases primary production.[19,94,128]

Biospheric exchanges of the minor elements in biomass (e.g. alkaline metals and earths, transition metals, metalloids, etc.) seldom involve truly volatile species. Exceptions include the halides (chlorine, bromine, iodine), for which several gases (e.g. methyl chloride, bromoform, methyl iodide) are an important component of biospheric exports.[3,106,144] Like nitrogen and sulphur, some trace metals, especially iron, and a variety of other elements may be transported through the atmosphere as eolian dust.[47,58] Variations in iron deposition have been proposed to contribute to the development of oceanic regions with low primary production, and to account for long-term climate changes via a negative feedback system involving consumption of atmospheric CO_2 and deposition of organic matter in marine sediments.[102,103] Briefly, during relatively warm, dry interglacial periods with elevated atmospheric CO_2, increased eolian transport of soil, resulting from expanded desertification, has been suggested as a source of iron and perhaps other limiting trace metals. This enhances primary production (primarily by diatoms), and the deposition and burial of marine organic matter. The ensuing depletion of atmospheric CO_2 promotes cooler, wetter climates and less iron transport. While there is clear support for iron limitation in certain regions of the ocean,[102] and evidence for increased production resulting from increased input of volcanic dust,[143] the link between eolian dust and climate remains unproven and controversial.[111]

Although the mass of most of the elements in the atmosphere is small relative to that in other reservoirs (nitrogen representing the major exception), transport within the atmosphere is more rapid and global in scale. Complete mixing within the atmosphere is extremely rapid, occurring with time constants < 1 year in contrast to scales > 10^2 and > 10^7 years for the hydrosphere and lithosphere, respectively.[64,142] Rapid mixing within the atmosphere combined with elemental export by deposition provide mechanisms for linking the biosphere globally. For instance, methane emitted from tropical wetlands contributes significantly to global temperature regimes, and provides up to 10% of the carbon oxidised by methanotrophs in temperate forest soils;[81,118] nitrous oxide production in soils and wetlands also contributes to global warming and to the regulation of stratospheric ozone concentrations as well.[19]

The pace of atmospheric mixing also means that regional and even local events can quickly become manifest at a planetary level. For example, sulphur and aerosol emissions from the eruption of Mt Pinatubo in 1991 were distributed worldwide within a few months, resulting in measurable disturbances in atmospheric composition, global climate and Southern and Pacific Ocean productivity.[10,21,43,143] The discovery of elevated copper in Greenland ice cores dating to 2500 years ago also illustrates the ability of the atmosphere to amplify seemingly local, small-scale events.[65] In this case, anthropogenic metal processing and mining activities in central Europe introduced into the atmosphere Cu-enriched particulates that were subsequently deposited globally. A similar phenomenon is at least partially responsible for increases in mercury concentrations within the hydrosphere and biosphere. The potential for large-scale mercury pollution results from the facts that it has a high vapour pressure in its elemental state (Hg^0), and that it can be transformed to an organic gas (dimethyl mercury) by microbial metabolism.[48] Mercury mobilisation by anthropogenic activity, when coupled with microbial activity, has a global impact due in part to atmospheric transport. Other toxic elements mobilised by anthropogenic activity and transported globally through the atmosphere include arsenic, selenium, cadmium, lead and beryllium.[2,46]

The atmosphere then is an intermediary in the flow of elements among reservoirs. As a consequence of rapid and global-scale mixing, local- to regional-scale variabilities in gas or particulate production and consumption are dampened (but not eliminated). This can be viewed as a mechanism contributing to the stability of biogeochemical cycles. However, it is also evident that the atmosphere can amplify changes in regional biogeochemical cycling through a variety of positive and negative feedbacks.[91,114] Further, the physical and chemical dynamics of the atmosphere, which are extremely sensitive to atmospheric composition, represent a potent source of disturbance for the biosphere.

9.2   Atmospheric structure and evolution

9.2.1   Atmospheric structure

The Earth's atmosphere consists of well-defined regions with characteristic regimes of temperature, pressure and composition[64] (Fig. 9.2). The two regions adjacent to Earth's surface, the troposphere and stratosphere, are the most important with respect to climate and the biosphere. The troposphere extends to a height of about 10–15 km, and is separated by the tropopause from the stratosphere, which extends to a height of about 50 km. The troposphere can be further subdivided into northern and southern hemispheres, each of which mixes separately on a time scale of a few months, and which mix together on a scale of about a year.

The troposphere contains the bulk of the atmospheric mass, and is dominated in its composition by nitrogen, oxygen, water vapour and the noble gases (Table 9.1), the relative abundances of which are a function of

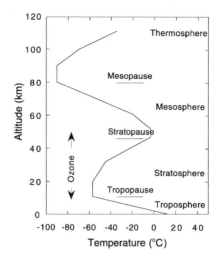

Figure 9.2 Diagrammatic representation of the structure of the atmosphere as defined by trends in temperature with elevation[64]

both biological and abiological processes. This remarkably thin elemental reservoir (barely reaching beyond the distance of a good foot race) is also the primary site for the development of the Earth's climate, which results in large part from the presence of infrared-absorbing tropospheric gases (e.g. water vapour, CO_2, methane, nitrous oxide). The chemistry of the troposphere includes photochemical reactions driven by wavelengths < 330 nm, but is dominated by a variety of free-radical oxidations, gas phase reactions, and reactions occurring on particles and in solutions.

In contrast to the troposphere, the stratosphere contains only a small fraction of the total atmospheric mass, and is characterised by an abundance of ozone, which strongly absorbs UV radiation between about 220 and 330 nm, thus permitting the biosphere to flourish. The presence of ozone also accounts for the increase in stratospheric temperature with elevation (e.g. −56 °C at about 10 km to −2 °C at 50 km). The chemistry of the stratosphere is dominated by homogeneous and mixed-phase chemical reactions, free-radical reactions and photochemical processes based on UV radiation.

The troposphere and stratosphere are both extremely sensitive to interactions involving the lithosphere (e.g. volcanic eruptions), hydrosphere (e.g. water vaporisation) and biosphere, (e.g. trace gas production). Stratospheric ozone depletion and enhanced UV penetration to the biosphere, resulting from the transport of chlorofluorocarbons and nitrous oxide across the tropopause, provide but one example of the sensitivity of the atmosphere to perturbations,[53,123] and the potentially devastating implications of such perturbations for the biosphere.[89,127,139].

9.2.2 Atmospheric evolution

Earth's atmosphere, like that of any other planet, exists as a function of several parameters. Earth's mass determines the nature of tectonic activity due to the loss of internally generated heat; mass also determines the gravitational field strength, and constrains gas loss from the atmosphere to space.[22] Differences between the atmospheres of Earth and Mars can be attributed in part to both of these size-related phenomena. The production of gases within the mantle and degassing by volcanic activity are also size-related phenomena since volcanic activity reflects the dynamics of internally generated heat and heat dissipation. Volcanic emissions of crustal and mantle gases are particularly important for regulation of CO_2.[14] At present, volcanic emissions are sufficient to replace the total atmospheric CO_2 mass on a scale of about 10^4 years.[128] Volcanic emissions are also important in the budgets of other gases,[35] including halogens (e.g. as HCl), sulphur (as SO_2) and nitrogen (as N_2 and NH_3). Non-volcanic interactions between the atmosphere and lithosphere also play an important role in regulating atmospheric composition. This is especially true for CO_2, which interacts with lithospheric carbonates and silicates;[14,110] however, similar analogous processes can affect other gases.

Subsequent to the differentiation of the lithosphere (i.e. formation of crust and mantle) and the cooling that promoted formation of the hydrosphere, tectonic cycles and abiological weathering are thought to have contributed to temperature regulation of the prebiotic Earth.[52,69,70,71] Physical models of solar evolution suggest that illumination has increased through time,[59] requiring proportional changes in the Earth's thermal radiation budget for a stable, biologically hospitable climate maintenance. Such changes appear to have resulted to a large degree from depletion of atmospheric CO_2 by weathering of lithospheric calcium silicates according to the following simplified reaction scheme:

$$CaSiO_3 + CO_2 \rightarrow CaCO_3 + SiO_2 \qquad (9.1)$$

This temperature-sensitive process likely reduced the degree of thermal trapping (or "greenhouse effect") associated with the high CO_2 concentrations postulated for the early atmosphere. Calcium silicate weathering represents a negative-feedback mechanism since increased "greenhouse" warming results in higher rates of CO_2 removal and a compensatory cooling as the greenhouse effect is diminished. This process still provides a major control for atmospheric CO_2 on time scales $> 10^7$ years (Fig. 9.3).

The evolution and proliferation of life have resulted in an additional set of controls for atmospheric composition on time scales from $10^0 - > 10^7$ years (Fig. 9.4). These controls are also temperature sensitive, and provide both positive and negative feedbacks on climate change. Biological contributions to climate homoeostasis involve a number of mechanisms.

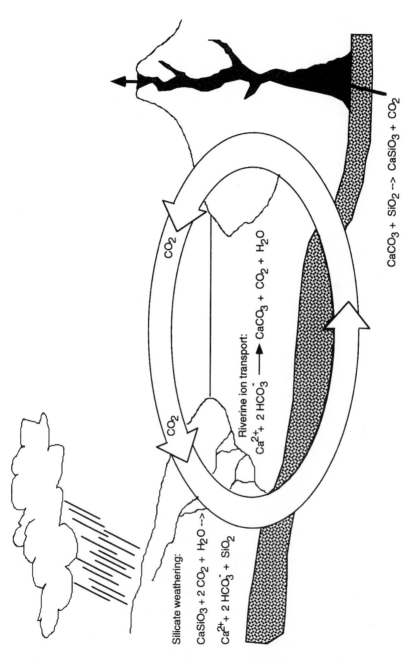

Figure 9.3 Diagrammatic representation of interactions among carbon dioxide, continental weathering, carbonate deposition and tectonic activity as controls of atmospheric CO_2[128]

Silicate weathering:

$CaSiO_3 + 2\ CO_2 + H_2O \longrightarrow$

$Ca^{2+} + 2\ HCO_3^- + SiO_2$

Riverine ion transport:

$Ca^{2+} + 2\ HCO_3^- \longrightarrow CaCO_3 + CO_2 + H_2O$

CO_2

CO_2

$CaCO_3 + SiO_2 \longrightarrow CaSiO_3 + CO_2$

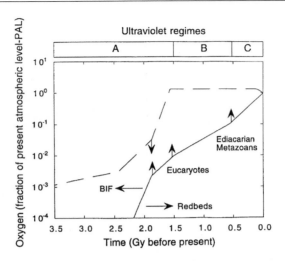

Figure 9.4   Historical reconstruction of trends in atmospheric oxygen and regimes of UV at the Earth's surface. Dashed and solid lines represent upper and lower limits for oxygen based on geological and biological constraints; arrows indicate probable directions for oxygen. UV regimes A indicate deep penetration of 210–285 nm wavelength; B, surface penetration only for 210–285 nm wavelengths; C, surface insulated from 230–290 nm wavelengths.[12,72] BIF, banded iron formation; PAL, present atmospheric level

The reflectance of the Earth's terrestrial surface (albedo) affects the radiation budget; terrestrial albedo is determined in part by the extent to which the surface is colonised by vegetation. Biologically enhanced weathering of silicate minerals may not only stimulate atmospheric CO_2 depletion, but prove essential for a level of depletion sufficient to maintain moderate global mean temperatures ($< 20 °C$) in the face of increases in solar illumination over the last 1 Gy.[120,130]

The Earth's biota has profoundly affected virtually all of the atmospheric gases, not just CO_2 (Table 9.1). The noble gases (e.g. Ar, Ne, He) represent the only major exceptions. Ignoring changes over the last two centuries due to anthropogenic activity,[97,124] micro-organisms have affected atmospheric composition more dramatically than any other biological group throughout the history of the biosphere. This undoubtedly reflects the extraordinary metabolic diversity of micro-organisms (bacteria in particular) relative to that of animals and plants. Though they are impressive in their phylogenetic diversity and their potential to modify rates of biogeochemical cycling, neither animals nor plants exhibit the array of gas production and consumption pathways known for bacteria.[5]

During the early Archaean (> 3.0 Gy), the composition of the atmosphere was mostly likely neutral to slightly reducing,[23] and dominated in its compositional dynamics by physical and chemical processes (e.g. volcanic emissions, photochemical oxidation). However, with the advent of methanogenic archaea, the atmosphere may have accumulated significant quantities of methane, which would have added to greenhouse warming by CO_2. This would possibly have marked a transition for the biosphere from a system regional in scale with little climatic impact to one global in scope with interactions involving the atmosphere.

The development of anaerobic photosynthesis and then oxygenic photosynthesis (by 3.5 Gy or earlier) marked another profound transition for the biosphere.[72] Oxygen evolution by photosynthetic micro-organisms radically changed the composition and chemistry of the atmosphere, resulting in the formation of hydroxyl radical, which is responsible for the oxidative capacity of the atmosphere, and UV-absorbing ozone, which promoted further proliferation of the biosphere (especially in terrestrial systems). Oxygenation of the oceans initially, and then later the atmosphere, resulted in a reorganization of many biogeochemical cycles by establishing pathways for complete oxidation of various reduced species (e.g. ferrous iron, sulphide). This is evinced in part by the episodic yet massive deposition during the middle Proteozoic of iron oxides in the so-called 'red beds'.[16,72] Further evidence exists in changes in the nature of organic matter preserved in early Proterozoic marine sediments; these changes have been interpreted as a shift in the locus of organic matter oxidation from an anoxic sea floor to an oxic water column (Fig. 9.5). Oxygenation of the atmosphere also produced conditions necessary for the evolution of processes such as aerobic methane oxidation, ammonium oxidation to nitrate, and subsequent nitrate reduction to nitrous oxide and dinitrogen.[25,54,62] Each of these latter processes is the exclusive domain of bacteria (see Chapters 1 and 2), and each influences both the composition of the atmosphere as well as the extent of thermal trapping by greenhouse gases (Table 9.1).

Microbial activity has also affected atmospheric concentrations of non-greenhouse gases (i.e. gases that absorb little or none of the Earth's infrared radiation). For example, CO is derived from natural and anthropogenic sources,[32] and its atmospheric concentration is determined in part by bacteria, some of which (the carboxydotrophs) are capable of using CO as a sole source of carbon and energy. Bacterial enzymes (hydrogenases) in soils appear partially to regulate concentrations of atmospheric hydrogen.[32,33] Soil bacteria also produce and consume NO as well as other reactive nitrogen oxides (NO_x). Emissions to the atmosphere by plants of methanol and various hydrocarbons[49,60] may be attenuated by bacterial populations (especially methylotrophs) living on plant surfaces.[34] In most of these instances, the relevant microbial populations likely evolved subsequent to oxygenation of the atmosphere.

Regardless of the specific sequence of events in microbial evolution, it is apparent that microbial processes in concert with geochemical processes

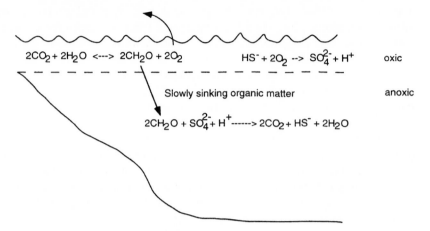

Proterozoic ocean: oxygen retained in surface waters

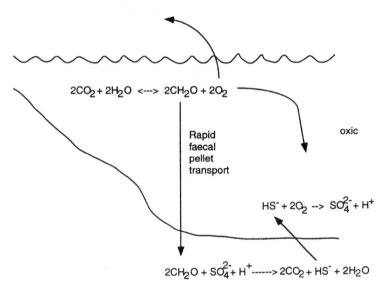

Phanerozoic ocean: oxygen release from surface waters

Figure 9.5   Proposed scheme for "reorganization of biogeochemical cycles" accompanying oxygen increases during the Proterozoic–Phanerozoic transition. Note that in the Proterozoic ocean (upper panel), slowly sinking organic matter that escapes fuels sulphate reduction in a largely anoxic water column. After increases in oxygen have led to evolution of grazing zooplankton (lower panel), rapid delivery of faecal pellets to the bottom promotes oxygenation of the water column due to decreased organic matter availability anaerobic processes retreat to sediments following the organic matter supply[98a]

have had substantial effects on atmospheric composition generally, as well as on the concentration of those gases that play a key role in global climate. Though changes in atmospheric composition have been dramatic (e.g. oxygen from $< 0.001\%$ to $> 20.9\%$), they have typically been gradual, occurring over periods $> 10^6$ years. In contrast, anthropogenic disturbances of microbial trace gas production and consumption have resulted in modest concentration changes, but at rates that are virtually instantaneous on a geological time scale. The full ramifications of these changes remain uncertain.

It is also evident that the extant microbial processes and pathways involved in trace gas metabolism are evolutionarily old. However, many taxa (i.e. oxygen-intolerant methanogens) with traits optimised for an early atmosphere have survived in spite of dramatic changes in atmospheric composition; further, some evolutionarily old taxa (e.g. sulphidogens) exhibit traits (i.e. oxygen tolerance) that reflect adaptation to a more modern atmosphere. Recent changes in surface UV fluxes may elicit additional adaptational responses in micro-organisms due to mutagenesis at a scale unparalleled since establishment of current levels of O_2.

9.3   Synopsis of trace gas biogeochemistry and linkages to climate change

The atmosphere exists in a state of gross chemical disequilibrium (Table 9.1). It is maintained in this state to a large extent by microbial and plant activity on short time scales (years to millennia), and in concert with geochemical and geophysical processes on longer time scales. One controversial view of atmospheric composition on any time scale is that it reflects more than the sum of the various gas-producing and consuming biogeochemical processes.

According to some, the atmosphere is an integral component of a regulatory system for global climate, and both climate regulation and control of atmospheric composition represent "emergent properties" manifest only at the level of the biosphere. Alternatively, atmospheric composition and climate regulation may be viewed as "epiphenomena", reflecting only the fact that biogeochemical dynamics involve closely coupled processes. Since the biosphere is an open system in thermodynamic terms, and driven by an energy flux, we expect cyclical phenomena (including cybernetic control systems) and disequilibrium as properties rather than characteristics that emerge only at a certain level of complexity (see Chapter 10). This view acknowledges the importance of the biosphere for atmospheric composition and climate control, but suggests that hypotheses based on "emergent properties" are unnecessary for understanding biosphere–atmosphere interactions or climate stability.

Regardless, it is evident that microbial activity affects the dynamics of gases that play key roles in regulating thermal trapping, the abundance of OH radical, and ozone concentrations. It is also evident that the various microbial processes involved in gas metabolism not only contribute to climate control, but are sensitive to climate perturbations arising from within the biosphere (e.g. fossil fuel combustion) or through external forces (e.g. oscillations in Earth's orbital parameters).[11] In some cases, these responses can serve to amplify or exacerbate climate change. For example, increases in atmospheric methane concentrations, due to the expansion of tropical wetlands during interglacial periods, likely contributed to warming trends associated with the Milankovitch cycles of the Earth's orbit.[18,27]

We now summarise briefly some of the microbial processes and interactions that determine the composition of the atmosphere, and the sensitivity of these processes to global scale disturbances. Basic aspects of the microbiology and ecology of these processes have been discussed in detail previously (see Chapters 1 and 2).

9.3.1 Oxygen–organic matter–CO_2

In the contemporary atmosphere the dynamics of O_2 and CO_2 are linked on short time scales through the balance between photosynthesis and the various respiratory processes summarised in Chapters 1 and 2. Since rates of photosynthesis appear to match estimates of respiration closely (both are approximately 2×10^6 Tg year^{-1} on a global basis),[128] oxygen concentrations exist in a steady state. Organic carbon burial, which accounts for only 10^2 Tg year^{-1}, has little short-term impact on either the carbon cycle or oxygen dynamics; for example, current burial rates yield depletion times of 7.2×10^3 and about 1.2×10^7 years for atmospheric carbon and oxygen, respectively.

Clearly, because of its relatively small pool size, atmospheric CO_2 is much more sensitive than oxygen to imbalances in photosynthesis and respiration and to changes in photosynthetic rates. For instance during the preceding 160 000 years, CO_2 has varied between about 180 and 280 ppm in association with glacial and interglacial fluctuations;[27] during this same period, variations in oxygen have been negligible; similarly, recent anthropogenically induced increases in CO_2 to levels of 350 ppm have not involved marked changes in oxygen.

In addition to biological processes, atmospheric CO_2 levels are determined by the balance between carbonate burial, as discussed previously (Fig. 9.3), and carbonate dissolution, which occurs during the weathering of terrigenous carbonate deposits, and the heating of deposits subducted into the crust.[14,72] Both biological and abiological processes are temperature dependent, with increasing temperature tending to increase the rate of carbonate burial by reaction 9.1 (see p. 208) in what constitutes a negative-feedback control system for temperature regulation. However, this system is clearly subject to perturbation by anthropogenic activity.

The dynamics of oxygen also depend on geochemical processes, especially the availability of inorganic reductants arising from abiological (i.e. geological) sources. Massive organic matter burial resulting from imbalances in photosynthesis and respiration have contributed to net oxygen accumulation during continental emergence and lithospheric weathering in the early Proterozoic, and organic matter burial in Carboniferous and Permian wetlands.[13,15,41,72] However, much of the oxygen produced historically (about 96%) has been sequestered in iron oxide deposits (banded iron formations, red beds) and in seawater sulphate (Fig. 9.6). However, oxygen concentrations have remained relatively stable over the past 100+ My in spite of dramatic changes in the biosphere and climate (e.g. large-scale extinctions, ascendancy of the Mammalia). This suggests that the oxygen–organic matter–CO_2 system is itself relatively stable. The stability of this system is likely due to the fact that organic matter metabolism is closely coupled to all of the major elemental cycles and is driven by the extraordinary diversity of biological processes, which include a substantial component of functional redundancy.

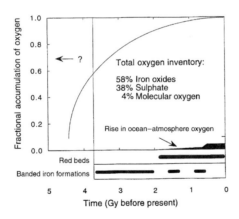

Figure 9.6   Oxygen accumulation (as a fraction of total production) as a function of time and distribution of oxygen in various reservoirs[128]

9.3.2 Oxygen–organic matter–methane

Atmospheric methane concentrations are determined by the balance between production and consumption in the biosphere, and by chemical destruction in the troposphere.[50,55,118] Geochemical processes, such as methane emission from hydrothermal vents and hydrocarbon seeps, appear to have a negligible impact on methane budgets at present, though this may not have been the case in the early Archaean. Chemical destruction of methane in the troposphere is primarily dependent on the availability of OH radical, the abundance of which is controlled by complex reactions involving a number of reduced gases (e.g. CO, terpenes, NO_x).[36,50]

The methane content of the atmosphere (about 1.7 ppm) has been increasing at a rate of about 1% year^{-1} mostly due to anthropogenic perturbations of methane production[118] (Fig. 9.7). Atmospheric methane concentrations have fluctuated historically from a low of about 0.3 ppm during recent glacial maxima to a high of about 0.7 ppm prior to the industrial period and during glacial minima.[27,28] The dynamics of methane in the biosphere are also linked to other gases. Methane production is based in part on consumption of hydrogen and to a much lesser extent, CO;[17,112] methane oxidation is related to oxygen, ammonia and CO, and can involve production of N_2O.[33,61,81,82,101]

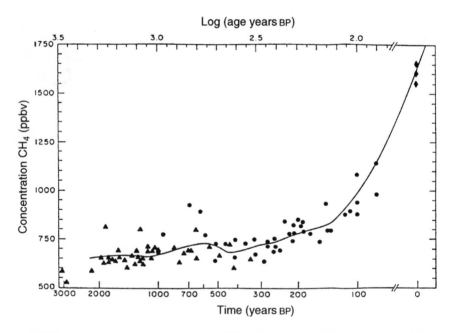

Figure 9.7   Recent changes (pre- and postindustrial) in atmospheric methane concentrations based on estimates from Greenland (circles) and Antarctic (triangles) ice cores[76]

At a global scale, the distribution and abundance of organic matter, oxygen and sulphate are the key determinants of methane production rates.[26,112] Competitive interactions with denitrifiers, iron reducers and sulphate-reducing bacteria limit the availability of substrate for methanogens.[99] The distribution of oxygen in freshwater sediments, and both oxygen and sulphate in marine sediments, also determines the relative activity of aquatic methane oxidation,[79,80,118] and therefore of exchanges of methane between the aquatic ecosystems and the atmosphere (see Chapters 1 and 2). Methane production and consumption both increase with increasing temperature, but the response depends on a number of edaphic factors. Elevated rates of methane production with rising temperature can accelerate climate change since methane is approximately 20-fold more potent a greenhouse gas than CO_2.[143] However, methane production and consumption also depend on a variety of other temperature-related parameters that act to increase or decrease atmospheric methane concentrations. Anthropogenic factors paramount in importance include biomass burning, natural gas production and distribution, ruminant husbandry, and expansion of rice culture, all of which increase the atmospheric methane burden.[118,143] Linkages between these parameters and climate change are uncertain, but in any case are likely less important than linkages of methane to human population growth.

9.3.3 N_2-ammonia-nitrous oxide

Like most gases in the atmosphere, N_2 exists in a state of gross thermodynamic disequilibrium.[128] The current and apparently stable steady state for N_2 concentrations is maintained by a balance between N_2 sources (denitrification and volcanic emissions) and N_2 sinks (microbial nitrogen fixation [see Chapters 1 and 2]; anthropogenic fixation from the Haber process and internal combustion engines; abiological fixation from atmospheric photochemical and lightening-induced oxidation). In general, biological fixation (about $170 \, \text{Tg N year}^{-1}$) substantially exceeds fixation by abiological processes ($< 20 \, \text{Tg N year}^{-1}$). Though microbes have dominated nitrogen fixation throughout the history of the biosphere, anthropogenic fixation has recently become important, reaching values comparable to estimates for microbes.[56]

Microbial nitrogen fixation is controlled by a number of biological and abiological parameters about which a great deal is known (see Chapter 1). In contrast, there is much less certainty about the response of nitrogen fixation to climate change. As with methane production and consumption, nitrogen fixation may respond positively to increased temperature. However, it is more likely that changes in nitrogen fixation will be driven by changes in parameters that are linked more or less strongly to temperature. For instance, symbiotic nitrogen fixation dominates activity in terrestrial systems.[136,137] Thus, total rates of fixation are more likely to respond to changes in plant community composition (the hosts for nitrogen-fixing symbionts) and host-symbiont interactions than to temperature *per se*. Since nitrogen-fixing symbionts are dependent on host photosynthate, increases in photosynthesis resulting from elevated p_{CO_2} and temperature could enhance rates of nitrogen fixation. Such an effect has been reported for marsh plants grown *in situ* under an atmosphere of elevated p_{CO_2}.[45]

Whether these results can be extrapolated to other systems or even to marshes experiencing gradual rather than experimental CO_2 fertilisation is an open question. Regardless, enhanced photosynthesis-linked nitrogen fixation represents a possible negative feedback control system on global temperature changes since increased plant production can lead to decreased atmospheric CO_2 levels.

Ammonia exists in the atmosphere mostly as a result of volatilisation from terrestrial systems with global rates estimated at about $50\,Tg\,N\,year^{-1}$.[93,128] Since net terrestrial primary production is estimated as about $60 \times 10^3\,Tg\,C\,year^{-1}$, it is apparent that ammonia losses are small relative to the demands for biosynthesis, even for nitrogen utilisation at a relatively high C:N ratio. Ammonia is not significant as a greenhouse gas, but plays an important role in controlling the pH of precipitation since it is the only source of alkalinity for the atmosphere.[56,128] Contemporary rates of ammonia volatilisation include a substantial perturbation due to anthropogenic activity, with fertilisation and animal feedlots contributing significantly to the total ammonia flux.[56,105] Microbial ureases are especially important for volatilisation from feedlots since much of the ammonia in this case originates in urine, resulting in locally large point sources. In contrast, fertiliser-linked losses are more direct, and to some extent abiological.

Ammonia volatilisation is likely sensitive to climate change for several reasons: warming decreases the solubility of ammonia in an aqueous phase; drier climates promote gas exchange; elevated temperature can increase microbial mineralisation of organic matter resulting in greater ammonium production. Microbial ammonia oxidation and immobilisation along with plant uptake from soils, water and the atmosphere represent sinks for ammonium. Each of these parameters is sensitive to climate change, but the overall impact of change on the balance of volatilisation and oxidation or uptake is far from clear. Increased rates of primary production in terrestrial systems due to elevated CO_2 and temperature, along with inputs to soils of high C:N organic matter, could result in greater ammonium immobilisation, potentially decreasing ammonia volatilisation. However, there is no indication that global-scale anthropogenic eutrophication by ammonium and nitrate is abating; on the contrary, eutrophication is increasing along with the demand for more intensive and productive agriculture.[56,105] Increased terrestrial plant production may ameliorate eutrophication since production in many systems is nitrogen limited.[66,140] However, other factors, for example phosphorus, may also play a role in determining nitrogen utilisation,[128] and ammonia may also be released directly from plants.[92]

Ammonium is indirectly related to the dynamics of two critical greenhouse gases, nitrous oxide and methane. Nitrous oxide is present in the atmosphere at about 300 ppb and is increasing at a rate of about 0.3% $year^{-1}$.[143] Although its mass is much smaller than that of CO_2, nitrous oxide is approximately 200-fold more potent in trapping thermal radiation. As a result, an increase of only 0.3 ppm is anticipated to contribute as much as 15% of the thermal trapping due to a 300 ppm change in CO_2. Nitrous oxide is also important in the dynamics of stratospheric ozone, since photochemical degradation of nitrous oxide is coupled to ozone destruction.[36,98] Although chlorofluorocarbons currently represent the greatest disturbance for stratospheric ozone,[53] nitrous oxide may become increasingly important as chlorofluorocarbon use continues to decline and anthropogenic disturbances leading to nitrous oxide production continue to increase.

The linkages between ammonium and nitrous oxide are two-fold. First, oxidation of ammonium to nitrite by ammonia-oxidising bacteria is accompanied by "leakage" of nitrous oxide which is formed as a metabolic byproduct.[39,148] Second, nitrous oxide is produced as a variable (minor to sole) product during the denitrification of nitrate produced from ammonium oxidation. Both of these processes are major sources of nitrous oxide in soils, sediments and the oceans. Emissions of nitrous oxide from these systems to the atmosphere are therefore dependent on a variety of parameters, including the distribution of oxygen; the availability of organic matter and its nitrogen content; interactions among wetland plant roots, ammonia-oxidising and denitrifying bacteria; soil water content and pH; and nitrate concentrations[20,39,104,117,121,125] (and see Chapters 2 and 4). Ammonium eutrophication and elevated temperatures are especially important factors in enhancing nitrous oxide emissions from the biosphere, and lead to a positive feedback for global warming.

Interactions between ammonium and methane also result in a positive feedback for global warming.[86] As discussed subsequently in more detail, ammonium inhibits atmospheric methane consumption by soils; the nature of the inhibition is such that increasing atmospheric methane concentrations increase the extent of inhibition, thus exacerbating the rise in atmospheric methane. Ammonium eutrophication can also contribute to increasing methane fluxes from wetlands, since rates of wetland methanogenesis and primary production are closely coupled.[38,67,146,147]

9.3.4   Hydrogen sulphide (H_2S)-sulphate-dimethyl sulphide (DMS)

H_2S production and consumption quantitatively dominate microbial sulphur transformations (see Chapters 1 and 2) in marine systems. However, in spite of relatively high rates of sulphate reduction in such systems, H_2S fluxes to the atmosphere are relatively small.[4] This is because H_2S is rapidly and extensively oxidised, often in conspicuous mats or biofilms, by aerobic and photosynthetic sulphide-oxidising bacteria (see Chapter 6); reactions of sulphide with metals (especially iron) play an important role in constraining emissions as well. Fluxes of H_2S are also relatively small in freshwater systems due to the lack of sulphate for dissimilatory sulphate reduction and to the relatively greater incorporation of the available sulphur into biomass. However, H_2S release is important for some wetlands and soils where sulphide is formed from the mineralisation of organic sulphur (putrefaction); in addition, uptake by plants of sulphur in excess of biosynthetic demands results in H_2S emission to the atmosphere from the plant canopy.[63,119] These latter processes may account for as much as 40% of total natural sulphur emissions.

Various methylated sulphur species (e.g. methanethiol, DMS and dimethyl disulphide), carbonyl sulphide (COS) and carbon disulphide (CS_2) are also important for exchanges between the biosphere and atmosphere.[4,8,94,95] DMS is perhaps the most important of these, as it dominates biogenic sulphur emissions from marine as well as some freshwater systems.[63,78] During the industrial period, SO_2 has become increasingly significant due to the combustion of sulphur-rich fossil fuels. Although none of these gases plays a major, direct role in thermal trapping, they are all oxidised within the troposphere to sulphuric acid, which contributes to acid precipitation. More significantly, aerosols arising from sulphuric acid salts are precursors of tropospheric cloud condensation nuclei (CCN). CCN affect the extent of cloud formation, and therefore Earth's albedo or reflectivity. In general, tropospheric clouds reflect incoming solar radiation due to their relatively high albedo. Thus, increased loading of sulphur gases has been proposed as a mechanism for inducing regional to global cooling.[29,30] If rates of sulphur gas production are directly correlated with temperature, then linkages among sulphur gases, CCN and albedo represent elements of a negative feedback system for controlling temperature. While some evidence supports such a system,[6] doubts have been raised about its significance.[7,9]

With the exception of DMS and direct emission of H_2S from sediments, soils and plants, the various sulphur gases originate primarily from microbial decomposition of sulphur-containing organics (Fig. 9.8). DMS arises from hydrolysis of dimethylsulphoniopropionate (DMSP) by plant and animal activity as well as microbial activity in marine environments (Fig. 9.9). Typical organic precursors for COS, CS_2 and the methylated sulphur gases include methionine and cysteine from proteins and isothiocyanates and thiocyanates from plant secondary metabolites.[75] Methanethiol and DMS are also formed in anoxic freshwater sediments from reactions based on H_2S and various methyl donors, for example methoxylated aromatic acids, such as syringic acid, that occur in lignins.[51] Although freshwater systems are generally considered to play only a minor role in sulphur gas fluxes, recent comparisons suggest that rates of DMS emission per unit area are similar for both the oceans and *Sphagnum*-dominated wetlands[78]. The relative importance of the latter then is limited by the relatively small area of wetlands.

Rates of DMS emission in freshwater wetlands appear to depend on sulphide methylation reactions that couple dissimilatory sulphate reduction

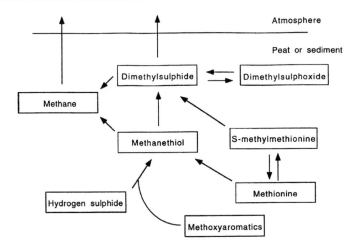

Figure 9.8   Scheme for organic sulphur cycling in a freshwater peatland; note the central role of methanethiol as opposed to dimethyl sulphide (DMS)[77]

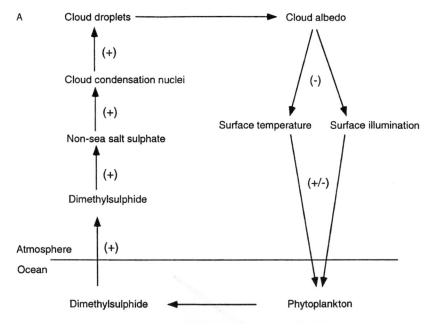

Figure 9.9   A. Relationships between dimethyl sulphide (DMS), cloud condensation nuclei and climate controls;[29] + and − signs indicate the effect of the preceding parameter on the downstream parameter. DMSO, dimethyl sulphoxide B. Interactions between DMS, inorganic sulphur gases and eolian iron in the marine atmosphere.[149,150] C. Illustration of the complex interactions affecting dissolved concentrations of DMS and its exchange with the atmosphere in marine systems.[78]

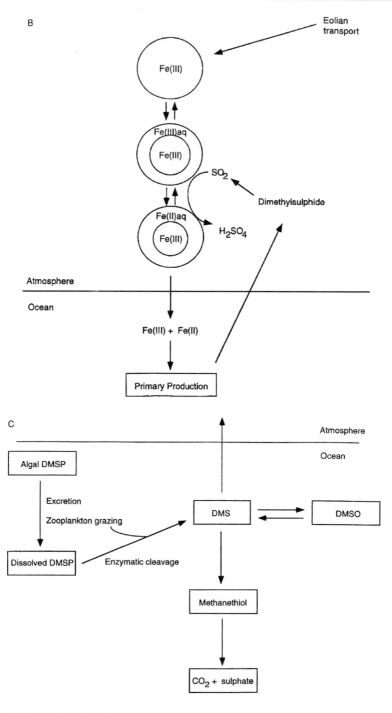

Figure 9.9   (*continued*)

to DMS production,[78] and involve acetogenic bacteria, some of which degrade aromatic acids to acetate. A wide diversity of other anaerobic and aerobic bacteria also contributes to sulphur gas production in soils and marine and freshwater systems. In addition, diverse aerobes (e.g. methylotrophs and sulphide oxidisers) and anaerobes (e.g. methanogens) consume sulphur gases, thereby regulating fluxes to the atmosphere.[74]

Controls of the balance between production and consumption and responses to climate change are uncertain. Increased sulphate deposition from acid rain might enhance sulphur gas production and emission resulting in a negative feedback on warming; alternatively, warmer, drier climates might enhance consumption, decreasing emissions and amplifying warming through decreased albedo (cloud cover). Similar uncertainties exist for terrestrial systems. However, at present it seems most likely that sulphur-linked climate changes are driven by anthropogenic disturbances in sulphur emission (e.g. via fossil fuel use) and marine DMS emissions.

9.3.5 Other gases

Microbes produce and consume several other gases that play an important role in the chemistry and radiative properties of the atmosphere. CO is present in the atmosphere at levels (< 100–200 ppb) considerably lower than those for methane (1.7 ppm). However, CO is a critical trace gas that plays a major role in determining the oxidative state of the atmosphere. This is because CO reacts very rapidly with OH radical, the primary tropospheric oxidant. Since OH is a major determinant of the atmospheric residence time for methane, the dynamics of CO are intimately related to those of methane. The relative importance of CO can be appreciated from the fact that global CO emissions to the atmosphere are estimated as 1000–3000 Tg year^{-1} (about 1–2% of net primary production), exceeding those for methane (400–500 Tg year^{-1}) by a factor of 2.5–6. CO is also important as a determinant of tropospheric ozone concentrations; tropospheric ozone not only poses problems for human health (and that of other organisms), but has recently been recognised as a potentially important greenhouse gas.

While CO is both produced and consumed by a wide diversity of microbes, fungi and plants, bacterial CO consumption in soils has the greatest impact on the atmosphere.[32] Soils are estimated to consume globally on the order of 300–400 Tg CO year^{-1} (10–40% of CO emission), well in excess of soil methane consumption (40–50 Tg year^{-1}; also about 10% of total emission). Soil CO consumption appears remarkably efficient, especially given the low concentrations of CO available. Although the microbiology and ecology of atmospheric CO consumption are not well understood,[32] it is important to note that enhanced CO consumption can help mitigate anthropogenic CO emissions and the rise in atmospheric methane. This is because decreased CO concentrations can increase the availability of OH radical for chemical methane oxidation; conversely, climate-related decreases in soil CO consumption could increase the residence time of atmospheric methane, thereby enhancing the contribution of methane to greenhouse warming.

Microbes may also contribute to the dynamics of natural (e.g. methyl chloride, bromoform) and anthropogenic (e.g. methyl bromide, freons [dichlorofluoromethane]) halomethanes. Several primarily marine sources of halomethanes are known.[106,144] Though not abundant in the atmosphere, natural halomethanes are greenhouse gases like their anthropogenic analogues,[90] and may contribute to global bromine and iodine redistribution and to natural "background" levels of stratospheric ozone depletion. Several methylotrophs, methanotrophs and methanogens degrade halomethanes, and have been suggested as possible controls of atmospheric halomethane emission.[113]

Microbial degradation may also play a role in the flux of anthropogenic halomethanes to the atmosphere. Methanotrophs and other soil bacteria appear substantially to degrade methyl bromide that has been applied to agricultural soils as a fungicide.[113] Rapid consumption in soils may also partially determine the residence time of atmospheric methyl bromide. Aerobic microbial consumption of atmospheric freons (CFCs) has not been documented; however, methanogenic degradation of CFCs appears to occur in landfills, termite guts and other anaerobic systems.[76,100] Thus, CFC degradation could limit to some extent fluxes to the atmosphere, although the magnitude of this process is uncertain. Given the potency of CFCs as greenhouse gases and agents of stratospheric ozone destruction, even slow rates of CFC degradation are significant.

Various soil and aquatic bacteria may also play an increasingly important role in the fate of CFC replacements. Since production and use of dichlorodifluoromethane and trichlorofluoromethane have been banned, a family of multicarbon, incompletely halogenated replacements has been introduced (e.g. the hydrochlorofluorocarbons [HCFCs] and hydrofluorocarbons [HFCs]). Though ozone "friendly", many of the replacements are greenhouse gases. Unlike the CFCs, some of these gases can be at least partially degraded by bacteria, including for example methanotrophs that dehalogenate both HFCs and HCFCs.[113]

Perhaps even more important is the fate of trifluoroacetate (TFA), a major product of the chemical decomposition of HFCs in the atmosphere. TFA is produced in the troposphere from several HFCs; it is then lost to soils and aquatic systems through precipitation.[113,138] TFA appears to be toxic to many plants and animals at relatively modest concentrations; it is also relatively stable, and thus can accumulate quickly to potentially toxic levels. The detoxification of TFA in anoxic systems appears to involve complete defluorination coupled with methanogenesis, but in oxic systems metabolism appears slower and more incomplete, possibly yielding trifluoromethane (TFM) and CO_2. However, under oxic conditions, TFM appears even more stable than

TFA.[85] Thus, any TFM formed from TFA would diffuse to the atmosphere where it could function as yet another potent greenhouse gas, the synthesis of which would be dependent in part on microbial detoxification of TFA.

9.4   Trace gas dynamics and climate change: an analysis of methane production and consumption

As discussed previously (Chapters 1 and 2), methane is the terminal product of anaerobic metabolism in a complex food web coupling the activity of fermentative bacteria with methanogens. Methane production is most active in anaerobic systems with relatively low concentrations (or availability) of oxygen, nitrate, ferric iron and sulphate. These conditions, along with relatively large supplies of degradable organic matter, occur in the sediments of freshwater lakes and wetlands (including rice paddies), the guts of many invertebrate and vertebrate herbivores, and various anaerobic waste digestors, including landfills. Not surprisingly, such systems are the primary sources of atmospheric methane (Table 9.2). Other important sources include methane losses during natural gas production and distribution,[96] and methane production during biomass burning. Historically, natural wetlands, ruminants and other herbivores, including termites, have accounted for the bulk of the atmospheric methane burden.[31] However, since the beginning of the industrial period, anthropogenically controlled sources have dominated the rise in concentrations. At present, methane emissions from rice culture, farmed ruminants, landfills, natural gas utilisation and biomass burning account for more than 50% of all sources.[118]

Table 9.2   Sources and sinks for atmospheric methane; all rates in Tg yr^{-1} Data from Reeburgh et al.[118]

Source or sink	Flux estimate
Wetlands	115
Rice paddies	100
Ruminants	80
Termites	20
Marine, lakes	10
Biomass burning	55
Coal and gas production	75
Methane hydrates	5
Total net emission	ca. 500
Uptake by soils	−40
Oxidation in atmosphere	−450
Total net oxidation	ca. −460
Net annual accumulation	ca. 40

The significance of anthropogenic sources notwithstanding, wetlands remain the single most important source of atmospheric methane, especially when emissions from farmed wetlands (rice culture) are added to emissions from natural wetlands.[118] Of the numerous natural wetlands, those in the tropics and at latitudes $> 40\,°N$ are most important.[55] The former are often active year-round, while the latter account for a large fraction ($> 20\%$) of total wetland area. Changes in the area of both tropical and northern wetlands in conjunction with glacial cycles have been implicated in preindustrial shifts in atmospheric methane.[27] Current climate model projections suggest that changes in northern wetlands could be particularly significant for future atmospheric methane concentrations, since these systems will likely experience greater warming and increased growing seasons relative to the tropics. However, the direction and magnitude of the changes in northern wetland methane dynamics are not clear.

Changes in the extent of rice culture are also expected to have an important impact on atmospheric methane. Asian rice culture is projected to double over a period of about 50 years, and methane emissions along with it.[143] The ability to grow rice with a high yield but low methane flux has thus become a priority research problem.[132] In addition to manipulation of plant characteristics, successful strategies to reduce methane emissions will likely require manipulation of the biogeochemical cycles of oxygen, nitrogen, iron and possibly sulphur. However, the global scale of rice culture is such that reductions in methane flux could be accompanied by changes in other gases, for example nitrous oxide, resulting in little or no amelioration of anthropogenic climate forcing.

Of course, predictions of future trends in methane emission from rice and natural wetlands are confounded by at least four major factors that introduce a number of uncertainties. For example, methanogenesis responds strongly to temperature, with Q_{10} values (relative changes in rate per $10\,°C$ change in temperature) usually > 2 (Fig. 9.10). Thus, rising ambient temperatures can stimulate methane production, resulting in a positive feedback on temperature change. However, in the absence of changes in net primary production in wetlands or organic inputs in rice paddies, such an effect would likely have only a short-term impact limited by the availability of readily degradable organic matter.

In this context, it is extremely important to note that primary production by wetland macrophytes and rice increases during growth in atmospheres with elevated $p - CO_2$ (Fig. 9.11). This results from the biochemistry and physiology of photosynthesis based on the "C_3" pathway which characterises most wetland plants and rice.[45] Since several studies have strongly linked plant production and methane emission (Fig. 9.12A, B), the combination of elevated temperature and atmospheric CO_2 could sustain increased methane emissions at a new steady-state level considerably higher than at present.

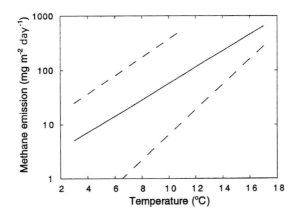

Figure 9.10 Relationship between methane emission and temperature for a variety of freshwater wetlands solid line indicates mean; dashed lines indicate upper and lower ranges[5a]

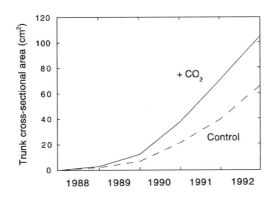

Figure 9.11 Enhanced net primary production (as increase in trunk cross-sectional area) by plants incubated *in situ* with elevated CO_2 versus controls with ambient CO_2 levels[69]; similar trends have been reported for marsh vegetation[45]

None the less, long-term trends in plant production are themselves uncertain since the availability of nitrogen or other nutrients can constrain photosynthetic responses to CO_2.[42] However, sustained increases in production over a period of 7 years have been observed in at least one field study,[45] and increased terrestrial production in northern forests[73] provides a strong explanation for the imbalance in the global carbon budget. Both phenomena may be attributed to increased microbial nitrogen fixation or anthropogenic eutrophication.[56,66,126] While inputs of ammonium from nitrogen fixation or anthropogenic sources could stimulate methanogenesis

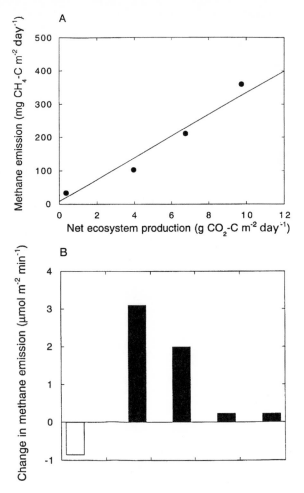

Figure 9.12   A. Relationship between methane emission and net ecosystem production (NEP) for diverse wetlands.[147] B. Differences in methane emission for five paired plots of a sedge incubated with elevated CO_2 during a continuous long-term study relative to controls with ambient CO_2 levels. A solid bar indicates that emission from a control plot exceeds emission from its paired elevated CO_2 plot; each of the five paired plots are represented by a single bar[38]

through impacts on plant production, inputs of nitrate might prove inhibitory due to the ability of nitrate-respiring bacteria to out-compete methanogens for substrate. Inhibition of methanogenesis due to competition between sulphate reducers and methanogens may also occur as a result of increased sulphur (sulphate) deposition. The effects of sulphate deposition suggest a possible mechanism for decreasing the otherwise positive effects of temperature, CO_2 and ammonium since sulphur is not typically considered an important limiting nutrient in wetland plant production.

Although the linkages between plant production, nitrogen and methane emission seem relatively clear, the strength of these linkages as well as the magnitude of future wetland methane emissions depends to a great extent on changes in local to regional hydrologic regimes. The degree and timing of water saturation in peats exerts a major control over the relative activity and spatial distribution of microbial processes (e.g. methanogenesis and methanotrophy) by determining the availability of inorganic electron acceptors, which in turn regulate fermentation and methanogenesis.[107,108,141] For example, decreased precipitation in boreal wetlands, as predicted by various climate models, is expected to lower water tables, increase peat aeration and oxic respiration, and decrease methane emission. These predictions are supported by results from a number of studies that document an inverse relationship between emission and water table both in undisturbed and drained peatlands (Fig. 9.13).

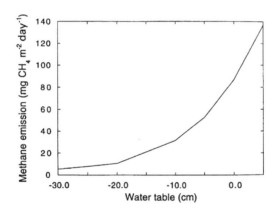

Figure 9.13 Relationship between methane emission and water table height (about the peat surface) for a boreal wetland[107]

In addition to hydrologic regimes, the linkages among wetland plant production, nitrogen and methane emission may be sensitive to plant species composition. Though little is known about the likely responses of wetland plant communities to future climate change, some species succession is inevitable. Since wetland methane emission is typically dominated by gas transport through plants,[24] and the capacity for transport varies significantly among species,[131] changes in the species mix could affect emission rates independent of other variables in a given system. Similarly, variations among plant species in the extent to which methanogenic and methanotrophic bacteria are associated with roots and rhizomes[40,57,83] can result in patterns of emission contrary to predictions based only on changes in abiological parameters (e.g. precipitation). The dynamics of root-associated methanotrophic bacteria are particularly important in this context as a variety of field and laboratory analyses indicate that these bacteria can attenuate methane emission by 25–90% (Fig. 9.14).

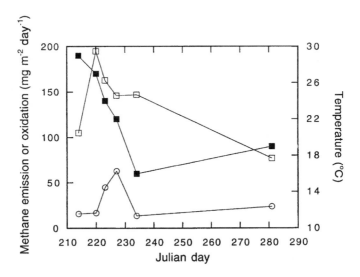

Figure 9.14   *In situ* rates of methane emission from the stems and leaves of the bur-reed, *Sparganium eurycarpum* (open squares); *in situ* rates of methane oxidation in the rhizosphere (open circles); ambient temperature (closed squares) for a Maine wetland[84]

In a more general sense, methanotrophic bacteria are an important component of methane dynamics at a global scale. They are essentially ubiquitous, occurring wherever both oxygen and methane coexist.[61,81] Thus, methanotrophs have been isolated from soils, oxic (and anoxic) sediments, various freshwater and marine systems, plant tissues, the gills of mussels growing at hydrocarbon seeps and the mouths of assorted ruminants. Owing to their ubiquity and significant attentuation of methane emissions from aquatic environments in particular, methanotrophs contribute to a biosphere–climate feedback system. The responses of methanotrophs to climate fluctuations can therefore amplify (positive feedback) or dampen (negative feedback) change.

Positive feedbacks on climate warming occur if methanogenesis increases to a greater extent than methane oxidation as a function of temperature. Though methane oxidation rates increase with temperature, the limited data available at present suggest that rates of methanogenesis may increase to a disproportionate extent (i.e. methanogenesis is characterised by a higher Q_{10}). Data showing increased methane emission as a consequence of CO_2 fertilisation (Fig. 9.12B) also indicate that the capacity for changes in methane oxidation may be less than that for methane production. Relative differences in response capacity might be attributed to the fact that methane oxidation at the surface of aquatic sediments and peats as well as in the rhizosphere is oxygen limited.[79,84,87,88] Thus, changes that further constrain oxygen availability (e.g. carbon loading from increased production) favour methane production. Obviously, changes that increase oxygen availability (e.g. benthic photosynthesis [Fig. 9.15]; lower water table [Fig. 9.13]) favour methanotrophs, resulting in a negative feedback on climate warming. Thus, predicting future trends in methane emission requires a much greater understanding of the relative impact of parameters that change simultaneously (e.g. temperature, hydrologic regime, CO_2) and that have contrasting effects.

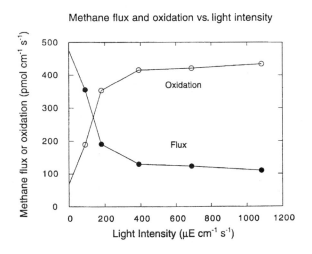

Figure 9.15   Methane oxidation and methane emission versus light intensity for illuminated cores from a Danish wetland[80]

The need to consider multiple interacting parameters is well illustrated by the responses of soil methane consumption to temperature, water content, ammonium and land use. Since soils currently represent the only net biological sink for atmospheric methane, consuming about 10% of the annual biospheric methane flux (Table 9.2), changes in consumption rate can amplify or dampen the rate of atmospheric methane accumulation.[33,81,85] Of the several parameters that control soil methane consumption, temperature is perhaps the least significant because consumption rates are diffusion limited. As a result, the direct effects of temperature change are manifest

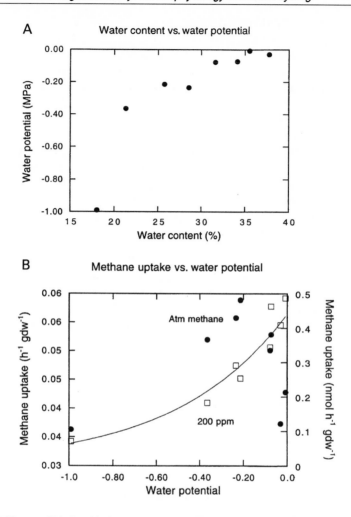

Figure 9.16   Relationship between water potential and water content for mineral soils from a temperate forest (A), and relationship between methane consumption by soils incubated with atmospheric methane (closed circles, left axis) or elevated methane (open squares, right axis) versus water potential (B). Note that activity increases initially with decreasing potential for soils with atmospheric methane, but decreases monotonically for soils with elevated methane[129]

through changes in gaseous diffusion coefficients.[1,37,145]. The diffusion coefficients for gases vary as a function of approximately T^2 (K), which means that over a range of 0–30 °C, changes of only 23% would be expected. Thus, responses of methane consumption to a temperature change of a few degrees may be difficult to detect. Correlations between temperature and methane uptake in field and laboratory studies generally agree with this prediction, although temperature responses can obviously be confounded by other parameters.

Because methane consumption in soils is diffusion limited, water content (and potential) have a profound effect on consumption rates.[129] Changes in water content could increase or decrease future atmospheric methane consumption rates. The direction of change depends on the extent to which current regimes reflect optimum conditions. Obviously, drying soils, with water contents at or less than the optimum, will decrease activity, while stimulation would occur for soils with water regimes above the optimum (e.g. Fig. 9.16).

Although increased gas transport in soils (see Chapter 4) increases methane consumption, the effect could be substantially limited in the future by continued nitrogen eutrophication or increased water stress as soils dry.[44,68,85,86,109,134,135] The effects of ammonium are especially important because ammonium inhibits methanotrophs by several mechanisms. Moreover, ammonium inhibition is enhanced at elevated methane levels. This suggests that the relative strength of the soil methane sink decreases with increases in atmospheric methane, resulting in a positive feedback on methane accumulation. On balance, data from a number of soils strongly indicate that the future capacity of soils to consume atmospheric methane will be lower than at present, resulting in a positive feedback on atmospheric methane accumulation and climate change.[85] Historical trends in land-use change, which invariably decreases methane uptake, suggest that the relative strength of the soil sink is lower at present than in the past, and that these changes have contributed to the rise in atmospheric methane concentrations.

9.5   Summary and conclusions

With the exception of phosphorus, virtually all of the major elements required for biomass occur in the atmosphere. Though the pool sizes of these elements are ususally small relative to those in other reservoirs (hydrosphere, lithosphere), transport through the atmosphere is rapid and global in scale. The dynamics of the atmosphere thus ensure continuity of elemental distribution within the biosphere, and promote extensive interactions between the biosphere, lithosphere and hydrosphere.

Biological and abiological processes regulate the specific composition of the atmosphere (Fig. 9.17). The latter are particularly important as controls of CO_2 at time scales $> 10^7$ years, while the former control the distribution of many elements on short time scales ($< 10^6$ years). Since the composition of the atmosphere determines Earth's climate in large part, the interactions between biological and abiological processes have a profound effect on climate stability, which is reflected in Earth's relatively constant

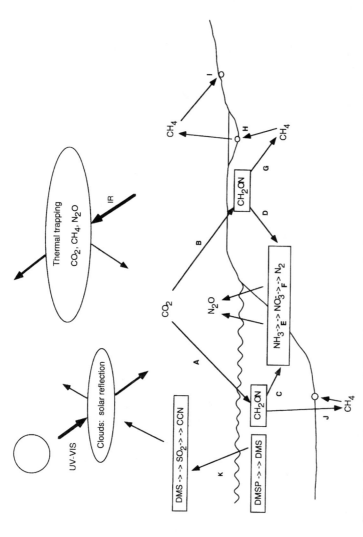

Figure 9.17 Simplified schematic representation of trace gas production, consumption and interactions with climate. Size of arrows for UV-VIS and IR indicate that a fraction of incoming solar radiation is reflected and a fraction of outgoing infrared radiation is absorbed within the atmosphere. Specific transformations include: A, oceanic primary production; B, terrestrial primary production; C, ocean nitrogen mineralization; D, terrestrial nitrogen mineralization; E, nitrification; F, denitrification; G, fermentation and methanogenesis; H, aerobic methane oxidation in aquatic sediments; I, atmospheric methane consumption by soils; J, fermentation and methanogenesis; K, dimethyl sulphide (DMS) production and atmospheric exchange

temperatures for about 3.5 Gy in spite of large changes in solar illumination, the structure of the lithosphere, and the composition and distribution of the biosphere.

Although plant and animal activities *per se* are an essential component of current biosphere–climate interactions, this has not always been the case. For much of Earth's history, micro-organisms dominated these interactions. Even at present this remains largely true since key trace gases such as methane and nitrous oxide are produced and consumed exclusively by bacteria; bacteria also contribute significantly to the biogeochemical cycles of other elements that affect climate (e.g. sulphur; Figs. 9.8, 9.9 and 9.17) or that involve an atmospheric phase (e.g. mercury). While many microbial processes (e.g. nitrogen fixation) are enhanced or even regulated by symbiotic associations or other interactions with plants or animals (see also Chapter 7), recent disturbances of atmospheric composition by anthropogenic activity reveal the fundamental role of micro-organisms in the dynamics of trace gases, and the sensitivity of balances between production and consumption to perturbations that rapidly become manifest at global scales. The only certainty about biospheric responses to perturbations in elemental cycles that affect climate is that micro-organisms will survive them, perhaps flourishing in a new steady-state temperature regime ill suited for extant macroorganisms.

References

1. Adamsen APS, King GM (1993) Methane consumption in temperate and subarctic forest soils: rates, vertical zonation and response to water and nitrogen. *Appl Environ Microbiol* **59**: 485–490.
2. Amouroux D, Donard FX (1996) Maritime emission of selenium to the atmosphere in eastern Mediterranean seas. *Geophys Res Lett* **23**: 1777–1780.
3. Anbar AD, Yung YL, Chavez FP (1996) Methyl bromide: ocean sources, ocean sinks and climate sensitivity. *Glob Biogeochem Cyc* **10**: 175–190.
4. Andreae MO (1990) Ocean–atmosphere interactions in the global biogeochemical sulfur cycle. *Mar Chem* **30**: 1–29.
5. Bachofen R (1991) Gas metabolism of micro-organisms. *Experientia* **47**: 508–514.
5a. Bartlett KB, Harriss RC (1993) Review and assessment of methane emissions from wetlands. *Chemosphere* **26**: 261–285.
6. Bates TS, Charlson RJ, Gammon RH (1987) Evidence for the climatic role of marine biogenic sulphur. *Nature (Lond)* **329**: 319–321.
7. Bates TS, Clarke AD, Kapustin VN, Johnson JE, Charlson RJ (1989) Oceanic dimethylsulfide and marine aerosol: difficulties associated with assessing their covariance. *Glob Biogeochem Cyc* **3**: 299–304.

8. Bates TS, Lamb BK (1992) Natural sulfur emissions to the atmosphere of the continental United States. *Glob Biogeochem Cyc* **6**: 431–435.

9. Bates TS, Quinn PK (1997) Dimethysulfide (DMS) in the equatorial Pacific Ocean (1982–1996): evidence of a climate feedback? *Geophys Res Lett* **24**: 861–864.

10. Bekki S, Law KS, Pyle JA (1994) Effect of ozone depletion on atmospheric CH_4 and CO concentrations. *Nature (Lond)* **371**: 595–597.

11. Berger A, Loutre MF, Laskar J (1992) Stability of the astronomical frequencies over the Earth's history for paleoclimate studies. *Science* **255**: 560–566.

12. Berkner LV, Marshall LV (1964) The history of growth of oxygen in the atmosphere. In: Brancazio PJ, Cameron AGW (eds) *The Origin and Evolution of Atmospheres and Oceans*. Wiley, New York, pp. 102–126.

13. Berner RA (1989) Biogeochemical cycles of carbon and sulfur and their effect on atmospheric oxygen over phanerozoic time. *Paleogeog Paleoclimatol Paleoecol* **75**: 97–122.

14. Berner RA (1990) Atmospheric carbon dioxide levels over Phanerozoic time. *Science* **249**: 1382–1386.

15. Berner RA, Canfield DE (1989) A new model for atmospheric oxygen over phanerozoic time. *Am J Sci* **289**: 333–361.

16. Beukes NJ, Klein C (1992) Models for iron-formation deposition. In: Schopf JW, Klein C (eds) *The Proterozoic Biosphere: A Multidisciplinary Study*. Cambridge University Press, Cambridge, pp. 147–151.

17. Blaut M (1994) Metabolism of methanogens. *Ant v Leeuw* **66**: 187–208.

18. Blunier T, Chappellaz J, Schwander J, Stauffer B, Raynaud D (1995) Variations in atmospheric methane concentration during the Holocene epoch. *Nature (Lond)* **374**: 46–49.

19. Bolin B, Crutzen PJ, Vitousek PM, Woodmansee RG, Goldberg ED, Cook RB (1983) Interactions of biogeochemical cycles. In: Bolin B, Cook RB (eds) *The Major Biogeochemical Cycles and their Interactions*. SCOPE, Stockholm, pp. 1–39.

20. Bowden RD, Melillo JM, Steudler PA, Aber JA (1991) Effects of nitrogen additions on annual nitrous oxide fluxes from temperate forest soils in the northeastern United States. *J Geophys Res* **96**(D): 9321–9328.

21. Brasseur G, Granier C (1992) Mount Pinatubo aerosols, chlorofluorocarbons, and ozone depletion. *Science* **257**: 1239–1242.

22. Brown GC, Mussett AE (1981) *The Inaccessible Earth*. Allen & Unwin, Winchester, MA.

23. Chang S, DesMarais D, Mack R, Miller SL, Strathearn GE (1983) Prebiotic organic syntheses and the origin of life. In: Schopf JW (ed.) *Earth's Earliest Biosphere: Its Evolution and Origin*. Princeton University Press, Princeton, pp. 53–92.

24. Chanton JP, Dacey JWH (1991) Effects of vegetation on methane flux, reservoirs, and carbon isotopic composition. In: Sharkey TD, Holland EA,

Mooney HA (eds) *Trace Gas Emissions by Plants*. Academic Press, New York, pp. 65-91.

25. Chapman DJ, Schopf JW (1983) Biological and biochemical effects of the development of an aerobic environment. In: Schopf JW (ed.) *Earth's Earliest Biosphere: Its Origins and Evolution*. Princeton University Press, Princeton, pp. 302-320.

26. Chapman SJ, Kanda K-I, Tsuruta H, Minami K (1996) Influence of temperature and oxygen availability on the flux of methane and carbon dioxide from wetlands: a comparison of peat and paddy soils. *Soil Sci Plant Nutr* **42**: 269-277.

27. Chappallez J, Barnola JM, Raynaud D, Korotkevich YS, Lorius C (1990) Ice-core record of atmospheric methane over the past 160,000 years. *Nature (Lond)* **345**: 127-131.

28. Chappellaz JA, Fung IY, Thompson AM (1993) The atmospheric CH_4 increase since the last glacial maximum. *Tellus* **45**(B): 228-241.

29. Charlson RJ, Lovelock JE, Andreae MO, Warren SG (1987) Oceanic phytoplankton, atmospheric sulphur, cloud albedo and climate. *Nature (Lond)* **326**: 655-661.

30. Charlson RJ, Schwartz SE, Hales JM, Cess RD, Coakley JA Jr, Hansen JE, Hofmann DJ (1992) Climate forcing by anthropogenic aerosols. *Science* **255**: 423-430.

31. Cicerone RJ, Oremland RS (1988) Biogeochemical aspects of atmospheric methane. *Glob Biogeochem Cyc* **2**: 299-327.

32. Conrad R (1988) Biogeochemistry and ecophysiology of atmospheric CO and H_2. *Adv Microb Ecol* **10**: 231-283.

33. Conrad R (1996) Soil micro-organisms as controllers of atmospheric trace gases (H_2, CO_2, CH_4, OCS, N_2O, NO). *Microbiol Rev* **60**: 609-640.

34. Corpe WA, Rheem S (1989) Ecology of methylotrophic bacteria on living leaf surfaces. *FEMS Microbiol Ecol* **62**: 243-250.

35. Cox PA (1995) *The Elements on Earth: Inorganic Chemistry in the Environment*. Oxford University Press, Oxford.

36. Crutzen PJ (1976) Upper limits on atmospheric ozone reductions following increased application of fixed nitrogen to the soil. *Geophys Res Lett* **3**: 169-172.

37. Czepiel PM, Crill PM, Harriss RC (1995) Environmental factors influencing the variability of methane oxidation in temperate zone soils. *J Geophys Res* **100**(D): 9359-9364.

38. Dacey JWH, Drake B, Klug MJ (1994) Stimulation of methane emission by carbon dioxide enrichment of marsh vegetation. *Nature (Lond)* **370**: 47-49.

39. Davidson EA, Schimel JP (1995) Microbial processes of production and consumption of nitric oxide, nitrous oxide and methane. In: Matson PA, Harriss RC (eds) *Biogenic Trace Gases: Measuring Emissions from Soil and Water*. Blackwell Science, Oxford, pp. 327-357.

40. Denier van der Gon HAC, Neue H-U (1996) Oxidation of methane in the rhizosphere of rice plants. *Biol Fertil Soils* **22**: 359-366.

41. Des Marais DJ, Strauss H, Summons RE, Hayes JM (1992) Carbon isotope evidence for the stepwise oxidation of the Proterozoic environment. *Nature (Lond)* **359**: 605–609.

42. Díaz S, Grime JP, Harris J, McPherson E (1993) Evidence of a feedback mechanism limiting plant response to elevated carbon dioxide. *Nature (Lond)* **364**: 616–617.

43. Dlugokencky EJ, Dutton EG, Novelli PC, Tans PP, Masarie KA, Lantz KO, Madronich S (1996) Changes in CH_4 and CO growth rates after the eruption of Mt Pinatubo and their link with changes in tropical tropospheric UV flux. *Geophys Res Lett* **23**: 2761–2764.

44. Dobbie KE, Smith KA (1996) Comparison of CH_4 oxidation rates in woodland, arable and set aside soils. *Soil Biol Biochem* **28**: 1357–1365.

45. Drake BG (1992) The impact of rising CO_2 on ecosystem production. *Wat Air Soil Pollu* **64**: 25–44.

46. Duce RA, Liss PS, Merrill JT, Atlas EL, Buat-Menard P, Hicks BB *et al*. (1991) The atmospheric input of trace species to the world ocean. *Glob Biogeochem Cyc* **5**: 193–259.

47. Duce RA, Tindale NW (1991) Atmospheric transport of iron and its deposition in the ocean. *Limnol Oceanogr* **36**: 1715–1726.   .

48. Ehrlich HL (1990) *Geomicrobiology*, 2nd edn. Marcel Dekker, New York.

49. Fehsenfeld F, Calvert J, Fall R, Goldan P, Guenther AB, Hewitt CN, Lamb B, Liu S, Trainer M, Westberg H, Zimmerman P (1992) Emissions of volatile organic compounds from vegetation and the implications for atmospheric chemistry. *Glob Biogeochem Cyc* **6**: 389–430.

50. Finlayson-Pitts BJ, Ezell MJ, Jayaweera TM, Berko HN, Lai CC (1992) Kinetics of the reactions of OH with methylchloroform and methane: implications for global tropospheric OH and the methane budget. *Geophys Res Lett* **19**: 1371–1374.

51. Finster K, King GM, Bak F (1990) Formation of methylmercaptan and dimethylsulfide from methoxylated aromatic compounds in anoxic marine and freshwater sediments. *FEMS Microbiol Ecol* **74**: 295–302.

52. Fischer AG (1984) The two phanerozoic supercycles. In: Schleifer KH, Stackenbrandt E, Berggren WA, van Couvering JA (eds) *Catastrophes and Earth History: The New Uniformitarianism*. Princeton University Press, Princeton, pp. 129–150.

53. Fisher DA, Hales CH, Filkin DL, Ko MKW, Sze ND, Connell PS, Wuebbles DJ, Isaksen ISA, Stordal F (1990) Model calculations of the relative effects of CFCs and their replacements on stratospheric ozone. *Nature (Lond)* **344**: 508–510.

54. Friedrich B (1985) Evolution of chemolithotrophy. In: Schleifer KHS, Stackebrandt E (ed.) *Evolution of Prokaryotes*. Academic Press, London, pp. 205–234.

55. Fung I, John J, Lerner J, Matthews E, Prather M, Steele LP, Fraser PJ (1991) Three-dimensional model synthesis of the global methane cycle. *J Geophys Res* **96**(D): 13033–13065.

56. Galloway JN, Schleisinger WH, Levy H II, Michaels A, Schnoor JL (1995) Nitrogen fixation: anthropogenic enhancement-environmental response. *Glob Biogeochem Cyc* **9**: 235-252.

57. Gilbert B, Frenzel P (1995) Methanotrophic bacteria in the rhizosphere of rice microcosms and their effect on porewater methane concentration and methane emission. *Biol Fertil Soil* **20**: 93-100.

58. Gillette DA, Stensland GJ, Williams AL, Barnard W, Gatz D, Sinclair PC, Johnson TC (1992) Emissions of alkaline elements calcium, magnesium, potassium and sodium from open sources in the contiguous United States. *Glob Biogeochem Cyc* **6**: 437-457.

59. Gilliland RL (1989) Solar evolution. *Paleogeog Paleoclimatol Paleoecol (Glob Planet Chng Sec)* **75**: 35-55.

60. Guenther A, Hewitt CN, Erickson D, Fall R, Geron C, Graedel T *et al.* (1995) A global model of natural volatile organic compound emissions. *J Geophys Res* **100(D)**: 8873-8892.

61. Hanson RS, Hanson TE (1996) Methanotrophic bacteria. *Microbiol Rev* **60**: 439-471.

62. Hayes JM (1983) Geochemical evidence bearing on the origin of aerobiosis, a speculative hypothesis. In: Schopf JW (ed.) *Earth's Earliest Biosphere: Its Origins and Evolution.* Princeton University Press, Princeton, pp. 291-301.

63. Hines ME, Pelletier RE, Crill PM (1993) Emissions of sulfur gases from marine and freshwater wetlands of the Florida Everglades: rates and extrapolation using remote sensing. *J Geophys Res* **98(D)**: 8991-8999.

64. Holland HD (1978) *The Chemistry of the Atmosphere and Oceans.* Wiley, New York.

65. Hong S, Candelone J-P, Patterson CC, Boutron CF (1996) History of ancient copper smelting pollution during Roman and Medieval times recorded in Greenland ice. *Science* **272**: 246-249.

66. Hudson RJM, Gherini SA, Goldstein RA (1994) Modeling the global carbon cycle: nitrogen fertilization of the terrestrial biosphere and the "missing" CO_2 sink. *Glob Biogeochem Cyc* **8**: 307-333.

67. Hutchin PR, Press MC, Lee JA, Ashenden TW (1995) Elevated concentrations of CO_2 may double methane emissions from mires. *Glob Change Biol* **1**: 125-128.

68. Hütsch BW, Russell P, Mengel K (1996) CH_4 oxidation in two temperate arable soils as affected by nitrate and ammonium application. *Biol Fertil Soils* **23**: 86-92.

69. Kasting JF (1989) Long-term stability of the Earth's climate. *Paleogeog Paleoclimatol Paleoecol* **75**: 83-95.

70. Kasting JF (1992) Proterozoic climates: the effect of changing atmospheric carbon dioxide concentrations. In: Schopf JW, Klein C (eds) *The Proterozoic Biosphere: A Multidisciplinary Study.* Cambridge University Press, Cambridge, pp. 165-168.

71. Kasting JF, Ackerman TP (1986) Climatic consequences of very high carbon dioxide levels in the Earth's early atmosphere. *Science* **234**: 1383–1385.

72. Kasting JF, Holland HD, Kump LR (1992) Atmospheric evolution: the rise of oxygen. In: Schopf JW, Klein C (eds) *The Proterozoic Biosphere: A Multidisciplinary Study*. Cambridge University Press, Cambridge, pp. 159–163.

73. Keeling CD, Chin JFS, Whorf TP (1996) Increased activity of northern vegetation inferred from atmospheric CO_2 measurements. *Nature (Lond)* **382**: 146–149.

74. Kelly DP, Baker SC (1990) The organosulphur cycle: aerobic and anaerobic processes leading to turnover of C1-sulphur compounds. *FEMS Microbiol Rev* **87**: 241–246.

75. Kelly DP, Smith NA (1990) Organic sulphur compounds in the environment: biogeochemistry, microbiology and ecological aspects. *Adv Microb Ecol* **11**: 345–385.

76. Khalil MAK, Rasmussen RA (1989) The potential of soils as a sink of chlorofluorocarbons and other man-made chlorocarbons. *Geophys Res Lett* **16**: 679–682.

77. Kiene RP (1996) Microbiological controls on dimethylsulfide emissions from wetlands and the ocean. In: Murrell JC, Kelly DP (eds) *Microbiology of Atmospheric Trace Gases: Sources, Sinks and Global Change Processes*. Springer, Berlin, pp. 207–225.

78. Kiene RP (1996) Production of methanethiol from dimethylsulfoniopropionate in marine surface waters. *Mar Chem* **54**: 69–83.

79. King GM (1990) Dynamics and controls of methane oxidation in a Danish wetland sediment. *FEMS Microbiol Ecol* **74**: 309–323.

80. King GM (1990) Regulation by light of methane emission from a Danish wetland. *Nature (Lond)* **345**: 513–515.

81. King GM (1992) Ecological aspects of methane oxidation, a key determinant of global methane dynamics. *Adv Microb Ecol* **12**: 431–468.

82. King GM (1994) Associations of methanotrophic bacteria with and methane consumption by the roots of aquatic macrophytes. *Appl Environ Microbiol* **60**: 3220–3227.

83. King GM (1994) Associations of methanotrophs with the roots and rhizomes of aquatic plants. *Appl Environ Microbiol* **60**: 3220–3227.

84. King GM (1996) *In situ* analyses of methane oxidation associated with the roots and rhizomes of a bur reed, *Sparganium eurycarpum*, in a Maine wetland. *Appl Environ Microbiol* **62**: 4548–4555.

85. King GM (1997) Responses of atmospheric methane consumption by soils to global climate change. *Glob Change Biol* **3**: 101–112.

86. King GM, Schnell S (1994) Enhanced ammonium inhibition of methane consumption in forest soils by increasing atmospheric methane. *Nature (Lond)* **370**: 282–284.

87. King GM, Skovgaard H, Roslev P (1990) Methane oxidation in sediments and peats of a subtropical wetland, the Florida Everglades. *Appl Environ Microbiol* **56**: 2902–2911.

88. King GM, Skovgaard H, Roslev P (1990) Methane oxidation in sediments and peats of a subtropical wetland, the Florida Everglades. *Appl Environ Microbiol* **56**: 2902-2911.

89. Komhyr WD, Grass RD, Evans RD, Lenoard RK, Quincy DM, Hofmann DJ, Koenig GL (1994) Unprecedented 1993 ozone decrease over the United States from Dobson spectrophotometer observations. *Geophys Res Lett* **21**: 201-204.

90. Kroeze C, Reijnders L (1992) Halocarbons and global warming. *Sci Tot Environ* **111**: 1-24.

91. Kutzbach J, Bonan G, Foley J, Harrison SP (1996) Vegetation and soil feedbacks on the response of the African monsoon to orbital forcing in the early to middle Holocene. *Nature (Lond)* **384**: 623-626.

92. Langford AO, Fehsenfeld FC (1992) Natural vegetation as a source or sink for atmospheric ammonia: a case study. *Science* **255**: 581-583.

93. Langford AO, Fehsenfeld FC, Zachariassen J, Schimel DS (1992) Gaseous ammonia fluxes and background concentrations in terrestrial ecosystems of the United States. *Glob Biogeochem Cyc* **6**: 459-483.

94. Langner J, Rodhe H (1991) A global 3-dimensional model of the tropospheric sulfur cycle. *J Atm Chem* **13**: 225-263.

95. Law KS, Nisbet EG (1996) Sensitivity of the CH_4 growth rate to changes in CH_4 emissions from natural gas and coal. *J Geophys Res* **101**(D): 14387-14397.

96. Lee DS, Atkins DHF (1994) Atmospheric ammonia emissions from agricultural waste combustion. *Geophys Res Lett* **21**: 281-284.

97. Levy HI, Kasibhatla PS, Moxim WJ, Klonecki AA, Hirsch AI, Oltmans SJ, Chameides WL (1997) The global impact of human activity on tropospheric ozone. *Geophys Res Lett* **24**: 791-794.

98. Liu SC, Cicerone RJ, Donahu TM, Chameides WL (1976) Limitation of fertiliser induced ozone reduction by the long lifetime of the reservoir of fixed nitrogen. *Geophys Res Lett* **3**: 157-160.

98a. Logan GA, Hayes JM, Hieshima GB, Summons RE (1995) Terminal proterozoic reorganization of biogeochemical cycles. *Nature (Lond)* **376**: 53-56.

99. Lovley DR, Klug MJ (1986) Model for the distribution of sulfate reduction and methanogenesis in freshwater sediments. *Geochim Cosmochim Acta* **50**: 11-18.

100. Lovley DR, Woodward JC (1992) Consumption of freons CFC-11 and CFC-12 by anaerobic sediments and soils. *Environ Sci Technol* **26**: 925-929.

101. Mancinelli R (1995) The regulation of methane oxidation in soil. *Annu Rev Microbiol* **49**: 581-605.

102. Martin JH, Coale KH, Johnson KS, Fitzwater SE, Gordon RM, Tanner TJ *et al.* (1994) Testing the iron hypothesis in ecosystems of the equatorial Pacific Ocean. *Nature (Lond)* **371**: 123-129.

103. Martin JH, Gordon RM, Fitzwater SE (1990) Iron in Antarctic water. *Nature (Lond)* **345**: 156-158.

104. Matson PA, Billow C, Hall S, Zachariassen J (1996) Fertilization practices and soil variations control nitrogen oxide emissions from tropical sugar cane. *J Geophys Res* **101**(D): 18533-18545.

105. Matthews E (1994) Nitrogenous fertilisers: global distribution of consumption and associated emissions of nitrous oxide and ammonia. *Glob Biogeochem Cyc* **8**: 411–439.

106. Moore RM, Tokarczyk R (1992) Chloroiodomethane in N. Atlantic waters: a potentially significant source of atmospheric iodine. *Geophys Res Lett* **19**: 1779–1782.

107. Moore TR, Roulet N (1993) Methane flux: water table relations in northern wetlands. *Geophys Res Lett* **20**: 587–590.

108. Moosavi SC, Crill PM, Pullman ER, Funk DW, Peterson KM (1996) Controls on CH_4 flux from an Alaskan boreal wetland. *Glob Biogeochem Cyc* **10**: 287–296.

109. Mosier AR, Parton WJ, Valentine DW, Ojima DS, Schimel DS, Delgado JA (1996) CH_4 and N_2O fluxes in the Colorado shortgrass steep. I, Impact of landscape and nitrogen addition. *Glob Biogeochem Cyc* **10**: 387–400.

110. Munhoven G, Fraçois LM (1996) Glacial–interglacial variability of atmospheric CO_2 due to changing continental silicate rock weathering: a model study. *J Geophys Res* **101**(D): 21423–21437.

111. Murray RW, Leinen M, Murray DW, Mix AC, Knowlton CW (1995) Terrigenous Fe input and biogenic sedimentation in the glacial and interglacial equatorial Pacific Ocean. *Glob Biogeochem Cyc* **9**: 667–684.

112. Oremland RS (1988) The biogeochemistry of methanogenic bacteria. In: Zehnder AJB (ed.) *Biology of Anaerobic Micro-organisms*. Wiley Interscience, New York, pp. 707–770.

113. Oremland RS (1996) Microbial degradation of atmospheric halocarbons. In: Murrell JC, Kelly DP (eds) *Microbiology* of *Atmospheric Trace Gases: Sources, Sinks and Global Change Processes*. Springer, Berlin, pp. 85–101.

114. Overpeck J, Rind D, Lacis A, Healy R (1996) Possible role of dust-induced regional warming in abrupt climate change during the last glacial period. *Nature (Lond)* **384**: 447–449.

115. Owens NJP, Galloway JN, Duce RA (1992) Episodic atmospheric nitrogen deposition to oligotrophic oceans. *Nature (Lond)* **357**: 397–399.

116. Paerl HW, Fogel ML (1994) Isotopic characterization of atmospheric nitrogen inputs as sources of enhanced primary production in coastal Atlantic Ocean. *Mar Biol* **119**: 635–645.

117. Reddy KR, Patrick WH Jr, Lindau CW (1989) Nitrification–denitrification at the plant root-sediment interface in wetlands. *Limnol Oceanogr* **34**: 1004–1013.

118. Reeburgh WS, Whalen SC, Alperin MJ (1993) The role of methylotrophy in the global methane budget. In: Murrell JC, Kelly DP (eds) *Microbial Growth on C1 Compounds*. Intercept, Andover, pp. 1–14.

119. Rennenberg H (1984) The fate of excess sulfur in higher plants. *Ann Rev Plant Physiol* **35**: 121–153.

120. Retallack GJ (1997) Early forest soils and their role in Devonian global change. *Science* **276**: 583–585.

121. Robertson GP, Tiedje JM (1987) Nitrous oxide sources in aerobic soils: nitrification, denitrification and other biological processes. *Soil Biol Biochem* **19**: 187–193.

122. Rogers JE, Whitman WB (1991) Introduction. In: Rogers JE, Whitman WB (eds) *Microbial Production and Consumption of Greenhouse Gases: Methane, Nitrogen Oxides, and Halomethanes.* ASM Press, Washington, DC, pp. 1–6.

123. Rowland FS (1989) Chlorofluorocarbons and the depletion of stratospheric ozone. *Am Sci* **77**: 36–45.

124. Santer BD, Taylor KE, Wigley TML, Johns TC, Jones PD, Karoly DJ, Mitchell JFB, Oort AH, Penner JE, Ramaswamy V, Schwarzkopf MD, Stouffer RJ, Tett S (1996) A search for human influences on the thermal structure of the atmosphere. *Nature (Lond)* **382**: 39–46.

125. Schiller CL, Hastie DR (1996) Nitrous oxide and methane fluxes from perturbed and unperturbed boreal forest sites in northern Ontario. *J Geophys Res* **101**: 22767–22774.

126. Schindler DS, Bayley SE (1993) The biosphere as an increasing sink for atmosphere carbon: estimates from increased nitrogen deposition. *Glob Biogeochem Cyc* **7**: 717–733.

127. Schindler DW, Curtis PJ, Parker BR, Stainton MP (1996) Consequences of climate warming and lake acidification for UV-B penetration in North American boreal lakes. *Nature (Lond)* **379**: 705–708.

128. Schlesinger WH (1991) *Biogeochemistry: An Analysis of Global Change.* Academic Press, New York.

129. Schnell S, King GM (1996) Response of methanotrophic activity in soils and cultures to water stress. *Appl Environ Microbiol* **62**: 3203–3209.

130. Schwartzman DW, Volk T (1989) Biotic enhancement of weathering and the habitability of Earth. *Nature (Lond)* **340**: 457–459.

131. Sebacher DI, Harriss RC, Bartlett KB (1985) Methane emissions to the atmosphere through aquatic plants. *J Envtl Qual* **14**: 40–46.

132. Sigren LK, Byrd GT, Fisher FM, Sass RL (1997) Comparison of soil acetate concentrations and methane production, transport, and emission in two rice cultivars. *Glob Biogeochem Cyc* **11**: 1–14.

133. Söderlund R, Rosswall T (1982) The nitrogen cycles. In: Hutziner O (ed.) *The Handbook of Environmental Chemistry,* Vol. 1/Part B. Springer, Berlin, pp. 60–81.

134. Steudler PA, Bowden RD, Mellilo JM, Aber JD (1989) Influence of nitrogen fertilization on methane uptake in temperate forest soils. *Nature (Lond)* **341**: 314–316.

135. Steudler PA, Melillo JM, Feigl BJ, Neill C, Piccolo MC, Cerri CC (1996) Consequence of forest-to-pasture conversion on CH_4 fluxes in the Brazilian Amazon Basin. *J Geophys Res* **101**(D): 18547–18554.

136. Stouffer RJ, Tett S (1996) A search for human influences on the thermal structure of the atmosphere. *Nature (Lond)* **383**: 39–46.

137. Tate RL (1995) *Soil Microbiology.* Wiley, New York.

138. Tromp TK, Ko MKW, Rodriguez JM, Sze ND (1995) Potential accumulation of a CFC-replacement degradation product in seasonal wetlands. *Nature (Lond)* **376**: 327–330.

139. Vinebrooke RD, Leavitt PR (1996) Effects of ultraviolet radiation on periphyton in an alpine lake. *Limnol Oceanogr* **41**: 1035-1040.

140. Vitousek PM, Howarth RW (1991) Nitrogen limitation on land and in the sea: how can it occur? *Biogeochemistry* **13**: 87-115.

141. Waddington JM, Roulet NT, Swanson RV (1996) Water table control of CH_4 emission enhancement by vascular plants in boreal peatlands. *J Geophys Res* **101**(D): 22775-22785.

142. Walker JCG (1993) Biogeochemical cycles of carbon on a hierarchy of time scales. In: Oremland RS (ed.) *Biogeochemistry of Global Change: Radiatively Active Trace Gases.* Chapman & Hall, New York, pp. 3-28.

143. Watson AJ (1997) Volcanic iron, CO_2, ocean productivity and climate. *Nature (Lond)* **385**: 587-588.

144. Wever R (1991) Formation of halogenated gases by natural sources. In: Rogers JE, Whitman WB (eds) *Microbial Production and Consumption of Greenhouse Gases: Methane, Nitrogen Oxides, and Halomethanes.* ASM Press, Washington, DC, pp. 277-286.

145. Whalen SC, Reeburgh WS (1996) Moisture and temperature sensitivity of CH_4 oxidation in boreal soils. *Soil Biol Biochem* **28**: 1271-1281.

146. Whiting GJ, Chanton JP (1992) Plant-dependent CH_4 emission in a subarctic Canadian fen. *Glob Biogeochem Cyc* **6**: 225-231.

147. Whiting GJ, Chanton JP (1993) Primary production control of methane emission from wetlands. *Nature (Lond)* **364**: 794-795.

148. Williams EJ, Hutchinson GL, Fehsenfeld FC (1992) NO_x and N_2O emissions from soil. *Glob Biogeochem Cyc* **6**: 351-388.

149. Worsley TR, Nance RD (1989) Carbon redox and climate control through earth history: a speculative reconstruction. *Paleogeogr Paleoclimatol Paleoecol (Glob Planet Chng Sec)* **75**: 259-282.

150. Zhuang GS, Yi Z, Duce RA, Brown PR (1992) Link between iron and sulphur cycles suggested by detection of Fe(II) in remote marine aerosols. *Nature (Lond)* **355**: 537-539.

10

Origins and evolution of biogeochemical cycles

Any analysis of the origin and evolution of biogeochemical cycles must address the origins of life, since it is life in its many manifestations that dominates the cycles of most elements. There are, of course, related issues to consider as well. For example, what modes of geochemical cycling existed prior to the advent of life? How have living systems qualitatively and quantitatively changed elemental cycles? What geochemical phenomena or principles affecting elemental cycles have been adapted by or incorporated into organismal metabolism?

In the subsequent sections, biogeochemical cycles are examined from the following perspective. The thermodynamics of energy flow systems establish a framework within which solar radiation sustains reactions among specific chemical components of Earth (dictated by the particular circumstances of its evolution as a planet) resulting in improbable (i.e. non-equilibrium) distributions of matter. Earth's geological characteristics, especially a convective mantle and dynamic crust, impose additional constraints that contribute to non-equilibrium states and material cycles. Within these constraints, life originated from a sequence of high probability events, in contrast to the vanishingly low probability of evolution from an equilibrium distribution of matter. Life has perpetuated itself by coupling solar radiation to biogeochemical cycling and through the evolution of novel biogeochemical pathways that do not occur readily, if at all, in realistic abiological systems.

10.1 Biogeochemical cycles and thermodynamics

One can begin an analysis of the origins of biogeochemical cycles by considering the energetic context within which such cycles first evolved and currently exist. The formalism of irreversible (or non-equilibrium) thermodynamics offers the most useful approach for analysing energy flow systems, of which biogeochemical cycles are but one representative. Classical (or equilibrium) thermodynamics has provided a framework for predicting the kinds of reactions that can occur during the transformations of any given element: reactions with $\Delta G < 0$ are feasible and those with $\Delta G > 0$ are not (see Chapter 1 and Appendix 1). Thus, the oxidation of methane to CO_2 by sulphate is feasible ($\Delta G^{0\prime} = -16.3\,kJ\,mol^{-1}$) and indeed forms the basis for a biogeochemically important anaerobic process; in contrast, ammonium oxidation to nitrate by sulphate is impossible ($\Delta G^{0\prime} = +447.7\,kJ\,mol^{-1}$), and thus far without precedent.

However, in spite of its predictive power, classical thermodynamics offers little insight for understanding the origin of the cyclic behaviours that typify elemental biogeochemistry. In large part this is due to the fact that classical thermodynamics focuses on systems characterised by reversible processes at or very near equilibrium. Equilibrium requires systems that are isolated or in contact with a single reservoir at a fixed temperature, neither of which is characteristic of "real" systems on Earth.

Although various theoretical arguments have facilitated successful extension of classical thermodynamics to biological[4,83,121] and biogeochemical (non-equilibrium)[3,97] processes, physicists and biologists have both at times questioned the application of the second law of thermodynamics (see Appendix 1) to biological systems. In apparent contradiction of the second law, biological evolution results in increasing order (decreasing entropy) through time; likewise, biological evolution has been accompanied by increasing complexity of biogeochemical cycles. The discrepancy between the observed increase in biological complexity and the predicted decrease has been resolved by recognising that Earth is an intermediate in an energy flow system in which the sun and universe represent essentially constant energy sources and sinks, respectively.[84,109] In such cases, both system properties and flows of energy through the intermediate system tend to become constant, and the system exists in a steady state.

While seemingly trivial and obvious, the fact that Earth is an intermediate in an energy flow system is significant for several reasons. First, the second law of thermodynamics applies *sensu strictu* to systems at or very near equilibrium. In addition, the second law requires only that the total entropy of the universe, S_u, increase. Because $dS_u = dS_i + dS_s$ (where $dS_i =$ the entropy change of Earth and $dS_s =$ the source–sink entropy change), $dS_u > 0$ and $dS_s > -dS_i$. This latter inequality means that Earth's entropy (including that of its biological systems) can *decrease* to an extent determined by the *increase* in source–sink entropy. Since only about 0.2% of the total solar energy intercepted by Earth is used in photosynthesis,[43–45] it is evident that $dS_s \gg -dS_i$.

Energy flow systems are best understood in terms of *non-equilibrium* thermodynamic theory, which has provided fundamental insights into the origins of life and biogeochemical cycling. Morowitz[84,85] provides an excellent summary of the points presented briefly here. One basic feature of energy flow systems at steady state is that the system cannot exist in *physical* equilibrium. This is illustrated simply by considering the distribution of

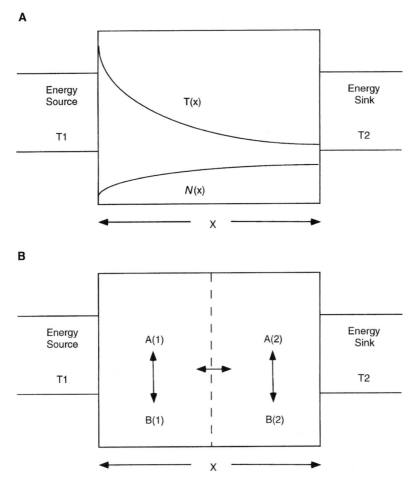

Figure 10.1   A. Distribution of temperature (T) and number density (N) of atoms of a gas in a hypothetical system of linear dimension X intermediate between to isothermal reservoirs with $T_1 > T_2$. B. Conceptual representation of an ideal system as in (A) but with an adiabatic porous barrier placed at the mid-point of the system; barrier pore size diameter is small relative to the mean free path length of two species, A and B, that undergo a temperature-dependent isomerisation reaction[84]

gas molecules of a species A in a closed system intermediate between two infinite thermal reservoirs such that $T_1 > T_2$ (Fig. 10.1A). In such a system, it can be shown readily that the number density of molecules, N_A, within the system varies inversely with the temperature gradient, $N_A \propto 1/T$. Relative to the equilibrium distribution ($N_A[T_1] = N_A[T_2]$), the non-equilibrium, steady-state distribution is associated with an increase in order (decrease in entropy) that results from and is maintained solely by energy flow.

A second important point is that *chemical* equilibrium cannot be obtained in an energy flow system. This can be demonstrated by a simple modification of the preceding illustration (Fig. 10.1B). If species A isomerises reversibly to form species B, then the concentrations of A and B cannot exist in equilibrium at any point in the system. Equilibrium at one point in the system implies equilibrium throughout the system, and the equilibrium constant, $K = B/A$, is temperature independent. Temperature independence requires a reaction enthalpy of zero, which does not generally hold. Thus, disequilibrium characterises the steady state and energy flow condition.

One notable consequence of disequilibrium is that the Gibb's free energy for a reaction is proportional to the extent of departure from equilibrium (see Appendix 1). Since $\Delta G = 0$ at equilibrium, disequilibrium implies a $\Delta G < 0$ for a given reaction (or its back reaction; e.g. $B \rightarrow A$). The maintenance of disequilibrium in an energy flow system at steady state thus provides one of the important preconditions for the evolution of life: a matrix in which chemical energy is available for exploitation.

Morowitz[84] has also demonstrated that systems intermediate between an energy source and sink are characterised by material cycles. In the illustration above with a reversible isomerisation (see Fig. 10.1B), it can be demonstrated that a cycle exists, $A \leftrightarrow B$, with equal but vectorially opposite net flows within the system. That such flows exist is intuitively evident from the concentration gradients that develop within the system, and the fact that even in a steady state (e.g. $d[A]/dt = 0$) there must be diffusive movement of the two molecular species that corresponds to Fick's laws (see Chapter 1). Cycles can also be readily demonstrated for somewhat more complex and biogeochemically relevant cases: the absorption of radiant energy by a system coupled to a thermal sink (Fig. 10.2A); the flow of mass with a high chemical potential into a system and the efflux of mass with a low chemical potential from the system. In the former case, a reaction system characterised by $A \leftrightarrow B \leftrightarrow C \leftrightarrow A$ and $A + h\nu \rightarrow B$ will develop a cycle when irradiated with a flow of heat to the sink. In the latter case, consider a system in contact through a semi-permeable membrane with an effectively infinite isothermal reservoir containing species D and E to which the membrane is permeable (Fig. 10.2B). If the chemical potential of $D > E$ and $A + D \leftrightarrow B + E$, a material cycle, $A \rightarrow B \rightarrow C \rightarrow A$ will develop along with heat flow to the reservoir. In each of these cases energy flow and the development of material flow networks are accompanied by entropy minimisation within the system.

The specific characteristics of material cycles and energy flows in any given system are, of course, critically dependent on the nature of the energy influx, system chemical composition and temperature. For electromagnetic radiation, wavelengths between approximately 300 and 800 nm are especially significant since they promote with considerable specificity electronic reconfiguration in many molecules. Relative to other forms of energy, solar

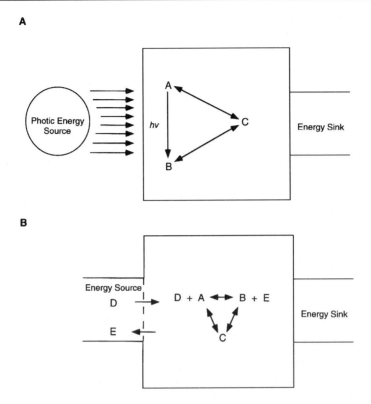

Figure 10.2   A. Conceptual representation of a chemical system connected to an isothermal energy sink and subject to irradiation from a constant source. The reaction $A \rightarrow B$ is photochemically driven; the remaining reactions represent reversible isomerisations. B. Conceptual representation of a chemical system intermediate between two isothermal reservoirs as in Fig. 10.1A. A semi-permeable membrane allows species D and E to flow into or out of the chemical system. Species D exists at a higher chemical potential than E in the source reservoir[84]

energy is likely to enhance chemical reactions in material cycling. The availability of photic energy between wavelengths of 300 and 800 nm may prove a major constraint for the origin and proliferation of life, perhaps relegating to a lifeless state planets or moons distant from a source of these wavelengths, even if liquid water is readily available as is suspected for Jupiter's Europa.[79] Thus, the "habitable zone" concept[63,125] might be modified to include light regimes in addition to criteria that determine the presence of liquid water.

System chemical composition is also a major determinant of the available reactants and potential flow networks. Aside from simply contributing to living mass, some elemental species may function effectively as reaction catalysts, while others act to destabilise flows. In this context, bulk planetary composition is critical since this determines the availability of key elements other than C, H, N, O and S. System temperature is equally significant in that the stability of molecular structures is obviously temperature dependent. Sufficiently elevated temperatures may destabilise mass flows through deleterious effects on molecular structure, thereby defining an upper limit for life.[2,115,123]

Extrapolating from the preceding analysis, one can conclude that the *expected* characteristics of the Earth as an energy flow system include elemental cycling *and* a decrease in entropy. Accordingly, elemental cycling is an *inherent* property of the Earth and its subsystems, and not an *emergent* property dependent on any particular composition or organisation of the biosphere. As stated by Morowitz,[84]

> molecular organisation and material cycles need not be viewed as uniquely biological characteristics; they are general features of all energy flow systems. Rather than being properties of biological systems, they are properties of the environmental matrix in which biological systems can arise and flourish.

Morowitz has also noted that the cyclical flow of matter "is part of the organised behaviour of systems undergoing energy flux", and that "the existence of cycles implies that feedback must be operative in the system. Therefore, the general notions of control theory and the general properties of servo networks must be characteristic of biological systems at the most fundamental level of operation".

The Gaia hypothesis, which states that the biosphere regulates planetary temperature through interactions between biogeochemical cycling and atmospheric composition, also stresses the importance of control theory. However, "gaian" climate control is viewed as an emergent property evident only at the level of the biosphere.[74] While it is undoubtedly true that global climate control involves the biosphere the cybernetic behaviour of the biosphere ("the Gaia hypothesis") can be viewed as a correlate of the principles of non-equilibrium thermodynamics, and the cybernetic behaviour of the biosphere as a consequence of energy flow. It is also evident that living systems are neither the *raison d'être* nor driving force for biogeochemical cycling, but rather that living systems and biogeochemical cycling owe their existence to non-equilibrium chemical conditions established by energy flow.

10.2   Prebiotic Earth and mineral cycles

Although the theories of non-equilibrium thermodynamics[95,96] lead to certain expectations about the evolution of mineral cycles in energy flow systems, it is obvious that the specific characteristics of Earth's mineral cycles are very much dependent on Earth's origins and geological evolution. An active mantle and crustal system insured that early Earth was characterised by exchanges of elements between the atmosphere, hydrosphere and lithosphere, and that many elements were cycled through two or three reservoirs, even if there was little change in specific elemental form.[16] The significance of mantle and crustal cycling can be appreciated by comparisons of Earth with Mars and Venus. Rapid cooling of Mars due to its small size has substantially reduced tectonic activity,[16] while on Venus, periodic and apparently catastrophic crust reformation may promote elemental cycling but preclude life altogether.[92,114]

The geochemical cycling of CO_2 among the atmosphere, lithosphere and hydrosphere as discussed previously (see Chapter 9) provides an apt example of abiological (but temperature-dependent) cycling on Earth in which changes in state occur (e.g. CO_2 as a gas to calcium carbonate as a solid) without accompanying changes in redox status. Living systems certainly affect (accelerate) rates of CO_2 cycling through carbonates and silicates,[98,110] but this cycling also clearly occurs independently of life.

Other elements, such as the halides, various metals and metalloids (e.g. iron, manganese, sodium, magnesium, zinc, mercury), phosphorus and nitrogen are also cycled through lithosphere, atmosphere and hydrosphere as a result of volcanism, hydrothermal venting and tectonic movements of the crust (Fig. 10.3). The age distribution of some elemental deposits or reservoirs within the lithosphere can be described using population-age class models with a mortality/natality component that reflects the extent of cycling.[116,117] Though microbial activities are clearly key controls of the distribution of many elements, for some, cycling is largely an abiological phenomenon to which organisms adapt as necessary. Analogies have also been drawn between the distribution of certain elements and ecological succession, emphasizing again the fact that at least some of the characteristics of biological systems are not uniquely biological, but rather are derived from more universal principles.[117] Important characteristics of the abiological cycles of elements include:

1. acid-based (CO_2) weathering of uplifted crustal minerals;
2. fluvial transport of elements released by weathering;
3. selective changes in the composition of weathered minerals and transported elements due to variations in elemental chemistry;
4. adsorption, exchange, precipitation and aggregation (e.g. clay formation) reactions that differentially affect retention of elements in the hydrosphere;
5. pressure-driven and thermally driven alteration of the distribution of elements among various mineral phases in the crust and mantle, with subsequent changes over time in the availability of elements for exchange with the hydrosphere and atmosphere.

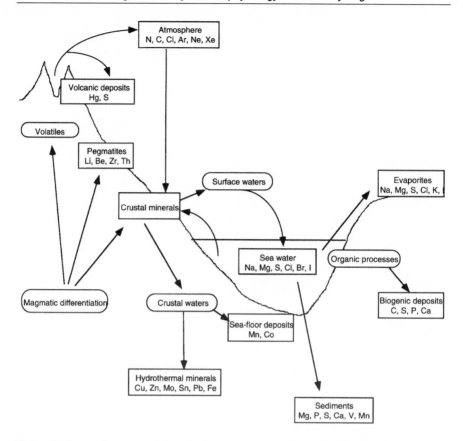

Figure 10.3   Summary of elemental sources and sinks for selected elements, illustrating lithosphere, atmosphere and hydrosphere reservoirs and basic geochemical processes that mobilise or immobilise elements[25a]

Abiological redox transformations are also possible. Models of Earth's formation favour accretion from carbonaceous chondritic meteorites rich in silicates and metal oxides.[16,31] During the thermally driven differentiation of the Earth into core–mantle–crust, numerous reactions likely produced oxidised species of iron, sulphur and carbon, among others, that contributed to redox chemistry. Volcanic and hydrothermal emissions of SO_2 delivered oxidants to the atmosphere and oceans. Photodissociation of water vapour in early Earth's atmosphere undoubtedly provided a small but significant source of molecular oxygen. In addition, and perhaps even more importantly, UV-driven ferrous iron oxidation[11,12] could have been coupled to the reduction of a variety of reactants, including CO_2 (Fig. 10.4). This process

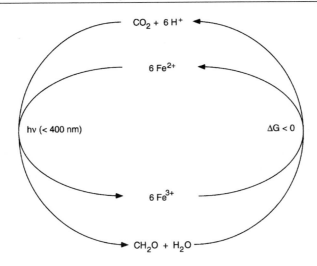

Figure 10.4   A light-driven iron cycle; ferrous iron is photo-oxidised to produce ferric iron and hydrogen; hydrogen is used as a reductant to form simple organics from carbon dioxide. Simple organics are oxidised by ferric iron completing the cycle; note that ΔG for the oxidation reaction is < 0, thereby providing energy for other processes, such as organic polymerisation[29]

provides a plausible abiological explanation for early banded iron formations (see below), and a basis for the evolution of a ferrous iron-based bacterial photosynthesis predating sulphide-based systems (see below).[49,124]

Prebiotic Earth was thus characterised by complex patterns of elemental cycling. The availability of diverse oxidants and reductants, including various forms of organic carbon, may also have supported a range of prebiotic elemental transformations similar to those known currently, including CO_2 fixation (via abiological coupling to pyrite synthesis;[119,120] see also below). As with prebiotic Earth, contemporary elemental cycles remain dependent on, and subject to, disturbance by some strictly geochemical processes (e.g. volcanism, hydrothermal venting). However, it is evident that rates of these processes have decreased over time as Earth's internal heat production has declined (Fig. 10.5A). The relative stabilisation of continental crust accompanying the decline in heat production (Fig. 10.5B) has undoubtedly had a profound impact on the evolution of living systems and their associated elemental cycling. For example, the colonisation of continents by living systems has drastically altered the rates of what were once strictly geochemical processes (e.g. weathering), and contributed to the rise in atmospheric oxygen (see below), which in turn has affected the development of both microbes and metazoans.

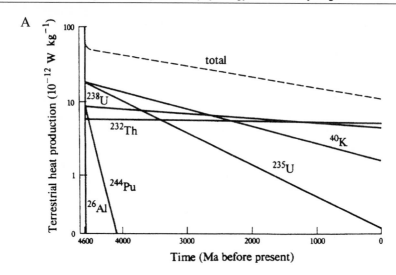

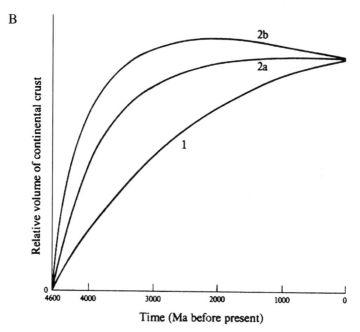

Figure 10.5   A. Terrestrial heat production from radionucleide decay; note extremely rapid loss of heat production from decay of ^{26}Al; this isotope was likely responsible for much of the early heat production after Earth coalesced. B. Three conceptual models for the change in volume of continental crusts over time; model 1 suggests very low early crustal volume and continuous increases while models 2a and 2b suggest a rapid accumulation to near-present values, with a stable volume late in model 2a and a slightly decreasing volume from the Proterozoic to the present in model 2b[16]

10.3   Theoretical perspectives on the origins of life

Theoretical and experimental analyses of the origins of life have occupied a considerable fraction of human thought and effort, and remain central elements in the biological sciences. The magnitude of the challenge posed by life's origin can be illustrated simply by noting the modest level of progress relative to the effort expended during the last century. The pace of progress has quickened during the last decade or so, certainly. A variety of theoretical accounts now exists, some with experimental support (Fig. 10.6). However, limitations of the fossil record, uncertainties in the interpretation of isotopic

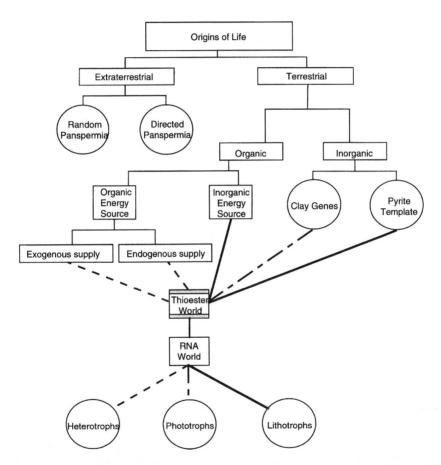

Figure 10.6   Relationships among various theoretical or conceptual models for the origins of life.[77] Dashed and solid lines at the bottom of the figure connect mechanisms with putative metabolic type of first organisms. The thioester world is proposed as an intermediate leading to an RNA world

evidence, and the unavoidable speculation about Earth's history prior to about 3.8 Ga leave ample room for alternative and conflicting propositions. Consensus may yet develop about a *probable* course of events, including a likely microbial phylogeny, but it will be difficult indeed to fill the gap that exists between the point in time for which there is compelling evidence for complex early life, and the preceding developmental period for which evidence is scant or non-existent.[108]

A variety of microfossil, geochemical and isotopic evidence from the Warrawoona rocks of Australia has established the existence of potentially complex microbial communities (in the form of mats) from about 3.5 Gy.[36,53,104,106,107] Recent isotopic evidence from metamorphic formations on Greenland suggests that life could have existed as early as 3.8 Gy.[81,103] Though no data are available for older periods, estimates of the probability of ocean-vapourizing asteroid impacts constrain the window for life's origin to about 4.4–3.8 Gy.[112] Thus, the period during which life first originated, differentiated and proliferated was probably very brief, perhaps no more than 600 million years and possibly as short as a few tens of million years. Such a brief period would have clearly sufficed for global proliferation of progenotes or their genote descendants, since even with a relatively long (by contemporary standards) doubling time of 10–100 days, little more than 3–30 years would be required for a single cell completely to colonise the entire volume of the oceans to the current density of about 1×10^6 cells cm^{-3}. Likewise, differentiation and evolution of a wide diversity of the early microbiota would not have required an especially long period. Recent analyses[32] indicate that substantial differentiation among the bacterial, archaeal and eukaryotic lineages could have occurred during a period of about 200 million years. The geological record also supports substantial and rapid phylogenetic diversification, since isotopic and microfossil evidence from the Warrawoona and other formations[53,104] is consistent with the presence of several major microbial functional groups (e.g. methanogens, photosynthetic bacteria). Although the exact timing of differentiation among the major domains remains controversial, there appears little reason to doubt a rapid pace of evolution prior to 3.8–3.5 Gy.

Nonetheless, the pace of life's evolution after its origin is but one issue. Adherents to the concept of panspermia regard any period before 3.5 Ga as too short for the origin of life itself.[56,57] Panspermia, first proposed by Arrhenius about 100 years ago,[89,91] is based on the assertion that life originated elsewhere, and was transported to Earth. The feasibility of transport has received tentative support from observations of putative microfossils in a Martian meteorite.[50,78] The lack of time for evolution on Earth is supported by estimates of the extremely low probability (4×10^{-83}) of randomly assembling from a pool of 20 different amino acids an enzyme with a specific sequence 100 residues in length. Assuming one abiological assembly reaction s^{-1} cm^{-3} in an ocean with a contemporary volume, up to

6×10^{51} years would be required for acquistion of a single enzyme. Since bacteria contain on the order of 10^3 enzymes or proteins and the age of the Universe is only about 10^{10} years, random assembly on Earth is judged impossible.

However, if not on Earth, when and where did life originate? Assuming that Earth is one-half to one-third the age of the Universe, then there is no place sufficiently old to meet the constraints of the above calculations. Some take this as support for divine creation; some invoke a many n-dimensional Universe that provides for ages and experience not directly discernible by human observation.[56] Obviously, these latter arguments depart from conventional scientific approaches.

Others have recognised the inherent limitations of the time estimates above, and proposed that the need for an extraordinarily long "incubation" period is obviated by invoking plausibly small pools of amino acids (e.g. 10 or so), short catalytic peptides (10–20 residues) with some tolerance for sequence variability,[28,29] and non-random searches of only portions of the total sequence space (due to the existence of chaotic attractors).[64] Under these conditions, much slower reaction rates (e.g. 10^5-fold slower) in only a fraction of the ocean are more than sufficient to account for rapid (e.g. $< 10^6$ years) evolution of complex, possibly living, organic systems. Eigen *et al.*[38] have taken a more empirical approach to the problem, and developed an estimate of the age of the genetic code based on a comparative analysis of tRNA sequences. Their results suggest that the genetic code does not predate the age of the Earth; this would tend to obviate an origin of life on Earth outside of the solar system.

Of course, the time required for the origin of life on Earth and its diversification begs the question of the mechanisms. We summarise several widely discussed proposals below (see also Fig. 10.6), but emphasise that any successful theory must ultimately account for a number of basic phenomena:

1. Replicative information flow is unidirectional from DNA to RNA to proteins, and proteins are essential as catalysts; early information flow may have been based on polynucleotide (ribozyme) catalysts.

2. The universality and conservation of chirality in biopolymers (i.e. only L-amino acids are incorporated into proteins; D-sugars predominate in catabolism and anabolism).[27,113]

3. The universality of ATP as the currency for energy conservation and utilisation; the centrality of phosphorylated intermediates in metabolism.

4. The universality of typically scarce metals as required components of enzymes, co-factors and electron transport systems.

5. The phylogenetic relationships of extant Archaea, Bacteria and Eucarya as deduced by molecular approaches (although these

relationships are currently uncertain, and phylogenetic analysis may not be able to illuminate very early organismal history).[40,71,82]

6. The (early?) evolution of two distinct and exclusive membranes (e.g. fatty acid lipids in Bacteria and Eucarya, isoprenoid lipids in Archaea), membrane transport systems and membrane-bound systems for energy transduction.

7. The central role of thermophily and sulphur metabolism as characteristics of taxa at the base of archeal, bacterial and eucaryal domains;[1,88] but see Forterre[40] for an alternative view.

8. The inconsistency of retrograde metabolic evolution with the hypothesised transition from RNA to DNA, and with the biosynthetic pathways of certain amino acids and deoxyribonucleotides;[72,119]

9. The availability of chemical and electromagnetic energy sources consistent with plausible prebiotic environmental conditions.

10.3.1 The Oparin–Haldane theory

Perhaps the most widely known theory of life's origin derives from Darwin's "warm pond" concept, as modified subsequently (and independently) by Oparin and Haldane.[48,90] (Bernal[5] includes a translation of Oparin's original paper and a copy of Haldane;[48] see also references in de Duve[29] and Lazcano *et al.*[72] and Fig. 10.7.) The Oparin–Haldane theory proposes that life originated in a prebiotic "soup" consisting of diverse organics synthesised by various chemical mechanisms,[21] including input from comets and chondritic meteors, both of which are organic rich.[23,76,87] Although cometary sources may have resulted in millimolar to molar organic concentrations in the early oceans,[23,24] the importance of comets is not without controversy.[91] Meteoric organic sources deserve attention nonetheless, since recent analyses have documented chiral abiologically-formed amino acids in the Murchison meteorite; such amino acids could have played a role in the origin of chirality prior to the advent of life.

The formation of *coacervates* in the soup is another key aspect of early evolution according to proponents of the Oparin–Haldane theory (see Lazcano *et al.*[72]). Coacervates consist of proteinaceous (or lipidic) membranes that enclose an organic-rich medium in which crude metabolic pathways develop. Competition and selection among coacervates provides the basis for progressive increases in complexity, including the acquisition of RNA- and then DNA-based genetic systems for control of replication and metabolism. At some indeterminate point, lipid biosynthesis led to the replacement of proteinaceous membranes.

Proponents of the Oparin–Haldane theory suggest that the primeval soup contained sufficiently abundant and diverse substrates that metabolism was entirely heterotrophic.[58] Substrate depletion and Darwinian competition drive the diversification of metabolism and the

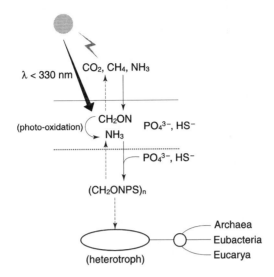

Figure 10.7   Simplified conceptual representation of the Oparin–Haldane model for origins of life. UV, lightning discharges or other energy sources result in organic matter synthesis from the constituents of a reducing atmosphere. Some photooxidation occurs in the upper layers of a primordial ocean with phosphate and sulphur incorporation and polymerisation taking place in lower layers. Evolution leads to a heterotroph initially. Phylogenetic development is unclear

evolution of metabolic pathways, ultimately leading to phototrophs or chemolithotrophs once the original organics are sufficiently dilute or the rate of organic utilisation exceeds the rate of supply.[41,42,55]

The plausibility of a prevital "soup" has been championed by Miller[80] and others, who have demonstrated the abiotic synthesis of various organics from gases presumably abundant in Earth's early atmosphere (e.g. CH_4, H_2, NH_3). Numerous variations of this work using diverse gas mixtures and energy sources have produced a wide array of simple to complex organics, including amino acids, sugars and nucleic acids, many of which play important roles in intermediary metabolism (e.g. alanine). In addition, both Oparin's and Fox's work on coacervates have demonstrated the plausibility of coacervate formation, and the ability of coacervates to grow and divide by purely abiological means.[72]

In spite of such modifications, the Oparin–Haldane theory remains problematic. First, models of Earth's early atmosphere raise doubts about the abundance of CH_4, H_2 and NH_3.[21,31] The more likely neutral composition, dominated by CO_2 and N_2, greatly limits the prospects for organic synthesis *á la* Miller. While it is true that organic synthesis occurs with a neutral atmosphere under a variety of conditions, the products (e.g. simple acids and sugars) have not been viewed as central to metabolic evolution.

Second, while UV radiation can promote certain organic syntheses, in general, photochemical oxidation at the ocean surface likely would have seriously constrained the accumulation of any prebiotic organics formed in the atmosphere or water column. Polymer hydrolysis is also kinetically and thermodynamically favoured in aqueous media. This greatly constrains the types and amounts of precursors for coacervate development and cellularisation.

Third, while many organics have been synthesised abiologically, the syntheses often involve contrived conditions probably remote from those that existed on early Earth; in addition product yields are often complex tars, with only small proportions of the most important biometabolites. Moreover, certain metabolites, including the critical sugar ribose, have not yet been obtained in significant yields. Thus, there appears no reasonable route for synthesis of RNA or DNA from soup precursors.[111] The prospects for evolution of a nutritionally fastidious progenote in a fully sufficient medium are therefore remote, as is the likelihood of retrograde evolution of metabolic pathways.

It is also noteworthy that Oparin–Haldane organisms appear to exist strictly as a consequence of abiological organic syntheses, with no significant contribution to biogeochemical cycles of mass and energy until late in their evolutionary history (perhaps subsequent to depletion of the primordial soup). Thus, the phenomenon of biogeochemical cycling in this instance happens subsequent to the origins of life, not contemporaneously. The Oparin–Haldane theory has been modified in response to these and other criticisms. For example, the locus of chemical synthesis has been extended to include evaporating lagoons and drying muds in addition to the oceanic water column. Lagoons and other shallow systems are attractive as sites of increased reactant concentrations, but the problem of UV oxidation remains. Synthesis in association with muds may eliminate UV sensitivity, but also implicitly invokes important concepts in other theories (see below). Morowitz *et al.*[86] have suggested that coacervates (vesicles in their usage) developed an autotrophic metabolism based on the early incorporation of phototransducers into membranes. This proposal provides an alternative to the original heterotrophic models. It also stresses the importance of early cellularisation; that is, the development of membrane-based segregation of evolving chemosynthetic systems from the bulk environment.

With some modifications, the Oparin–Haldane theory is compatible with current notions of an "RNA world"[60,61,72] that is thought to have existed prior to the advent of proteins as enzymes, and DNA as the genetic medium. Briefly, the RNA world is conceptualised as a replicating biochemical system in which RNA with catalytic activity (ribozymes) functioned in catabolism and biosynthesis in lieu of proteins; RNA also functioned in information storage and replication. The RNA world may be been preceded by a catalytic–genetic nucleic acid system with a polypeptide backbone substituting for modern ribose-based phosphate ester linkages.[10] An early peptide nucleic acid (PNA) world solves problems arising from the implausibility of prebiotic syntheses of (deoxy)ribose. The plausibility of an RNA world is supported by the now well-established diverse catalytic activities of RNA;[39,93] recent work has also led to suggestions that if certain derivatives of ribonucleic acids were present in the RNA world, the range of catalysis may have been much greater than at present.[100] It has also been shown recently that PNA catalyses the polymerisation of complementary RNA,[10] which suggests that a PNA to RNA transition could have occurred with no loss of genetic information.

Nonetheless, neither the original Oparin–Haldane theory nor its more recent variants are especially satisfying accounts of the origins of life. Major deficiencies at present include the inability of the theory to account for the currently accepted phylogenetic and biochemical differentiation of Archaea, Bacteria and Eucarya; inability to account for the presumed thermophilic characteristic of the earliest common ancestor of these groups; and inability to account for the origins of chirality.[19,119] Thus, even if other limitations are addressed, the theory is inconsistent with the living record whose origins it attempts to explain.

10.3.2 Clays and life

Cairns-Smith[19] and others[30,94] have rejected many of the basic tenets of the Oparin–Haldane theory. In response to polymer instability in solution, and strong doubts about the plausibility of organic synthesis in general and nucleic acids specifically, it has been proposed that biochemical evolution was facilitated by clays produced from igneous rock weathering (Fig. 10.8).

Cairns-Smith's multi-stage model is based on clay minerals or "crystal genes" with replicable defects in their crystal structure. Certain crystal genes increase in abundance through a form of inorganic evolution based on selection for those that are synthesised most effectively. Subsequently, assemblages of clays that promote mutual synthesis and propagation result in the formation of "complex vital muds". A subset of these clay minerals acquires a system for phototransduction. Light energy drives the synthesis of simple organics, ultimately leading to multi-step complex syntheses. Chirality is introduced at this point, arising primarily from the inherent chirality of clay crystals themselves. Continued evolution of clays and their associated organic syntheses lead to production of structural polymers, including polynucleotides that constitute a secondary genetic system. Further polynucleotide evolution results in partial control of a subset of the organic synthesis systems, for example peptide synthesis. The evolution of protein enzymes and cellularisation with lipid membranes promote the development of novel mechanisms for organic synthesis and genetic control. At this stage, dependence on clay minerals is lost, and evolution proceeds towards an early ancestral form (progenote).

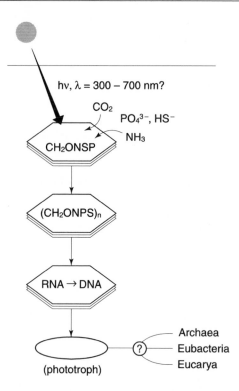

Figure 10.8   Simplified conceptual representation of the Cairns-Smith clay-based model for origins of life. Simple organic synthesis occurs on clay surfaces by a phototransducing system. The clays serve as a "genetic" template that is coupled to organic polymerisation. Synthesis of ribose and then ribonucleic acids leads to an RNA world, a transition to DNA as the genetic material and perhaps the origins of a phototroph. Phylogenetic development is unclear

Cairns-Smith's proposal has a number of strengths. First, a variety of clays catalyse the formation of complex organics, including polypeptides, sugars and fatty acids under conditions conceivable for early Earth; energy sources include light (UV and higher wavelengths) and high temperature. At least partial, stereospecific amino-acid binding and polymerisation have also been demonstrated.[30] Thus, there is no obvious impediment for clay-based organic synthesis, and the depolymerisation that occurs in aqueous solutions is avoided, at least to the extent that products remain bound to clay surfaces. Although clay surface charges are negative, it is evident that they bind many organics, even those that are negatively charged. Further, organic synthesis on clays could well account for the introduction of chirality due to the well known and constant chirality found in many clay minerals.[19,30]

Second, rapid attenuation of UV by clays in the sediments of shallow lagoons or drying muds minimises photochemical organic matter oxidation that constrains product accumulation in the water column. Early acquisition of a photosynthetic system coupled to organic synthesis further reduces any dependencies on soup-supplied organics while introducing autotrophy near the beginning of the sequence of evolutionary events. Autotrophy would likely promote the gradual, step-wise acquisition of metabolic pathways, in contrast to the less likely retrograde acquisition of metabolism embodied in the Oparin–Haldane theory.

Third, replication, transmission of heritable defect structures and structural evolution have been documented for various clay minerals, thus establishing in principle the ability of clays to function as inorganic genes with an associated potential for selective organic synthesis. In addition, the subsequent evolution of a clay-bound RNA world, followed by protein (enzyme) evolution and transition to a DNA-based genetic system, appears plausible, and is supported in part by the fact that ribose may have been initially available from synthesis catalysed by clays.

Fourth, the ability of clays to concentrate cations and phosphate facilitates the introduction of metal catalysts into the evolution of enzymatic activity, and phosphorus into biosynthesis and energy metabolism.[30] Phosphate may have also played an early role in clay stabilisation and the selection of replicating crystal genes. In contrast, no such roles for phosphorus are obvious in the dynamics of coacervates, and the limited concentrations of many metals in aqueous solution may pose problems for evolution of metal catalysis in a primordial soup.

Though not explicitly proposed by Cairns-Smith, elemental cycles may have developed contemporaneously with clay-based life, rather than after the fact. This is because autotrophically produced organics would have been mineralised subsequent to detachment from clay surfaces and transport into bulk solution, thus creating a cycle of synthesis and degradation. Organic degradation in solution could also have selected for strong binding to clay surfaces, and for efficient nutrient or substrate utilisation.

Cairns-Smith's theory has a number of weaknesses as well. In contrast to the presence of a global ocean early in Earth's history, the abundance of sites suitable for clay evolution is entirely uncertain. The proposed requirement for photic energy may prove especially limiting. Of equal importance, the temporal stability of sites suitable for clay evolution is unknown. Temporal stability has not been addressed in the development of any of the major theories, yet it is clear that the geochemical matrix of the early oceans and sediments was very dynamic, which raises several basic questions: how sensitive might clay life have been to disturbances such as burial and resuspension or fluctuations in substrate supply? Would rapid burial in shallow systems preclude the evolution of a photosystem?

There is also little information available on the kinds of clays that might have existed prior to 3.5 Ga. This is significant because the properties of clays vary substantially among the numerous types that have been characterised. Since not all clays may be suitable as crystal genes, some additional attention needs to be given to the early genesis of clays themselves.

Cairns-Smith proposes that cellularisation occurs late in the evolutionary scheme for clays. However, Morowitz *et al.*[86] have argued that segregation of evolving, prebiotic systems from their bulk environment is a necessary, critical early step in evolution. Whether or not the clay theory can be modified along the lines of the pyrite model discussed below is uncertain.

A more serious problem arises from the inability of the current clay-life model to explain microbial phylogeny and the presumed traits of early life. For example, the origins of thermophily, sulphur metabolism, and the differentiation of Bacteria and Archaea are not immediately apparent. Moving the locus of clay life to hydrothermal vents or similar systems is not likely to solve the problem without creating serious complications (e.g. paucity of clays within the vent structures themselves, absence of light, etc.). Clays overlying certain structures could be heated, as in the Guaymas Basin (e.g. Jørgensen[59] *et al.*), but a non-photic energy source would have to be invoked. As a result, the proposal for clay-based life improves on the Oparin–Haldane theory, but falls short of providing a robust path to a living system, and from that point to a progenote consistent with molecular phylogenetic predictions.

10.3.3 Pyrite and the origins of life

Wächtershäuser[119] has provided the most biochemically detailed theory to date, while rejecting the Oparin–Haldane theory entirely. In spite of its detailed biochemistry, Wächtershäuser's theory is deceptively simple (Fig. 10.9). Pyrite synthesis provides energy for organic synthesis, and the resulting pyrite surface provides the initial "habitat" for growth. Organic synthesis involved surface chemistries, polymerisation and perhaps phosphate from the outset. Autocatalytic cycles based on metabolites that exist now as coenzymes (e.g. thiamine pyrophosphate) developed rapidly and were accompanied by both horizontal and vertical extensions (i.e. spatial growth) of the synthetic systems. Competition among synthetic systems optimised efficiency and kinetics. Cellularisation (or rather semi-cellularisation) ensued early with the advent of isoprenoid- rather than fatty acid-based membranes. Transport of non-ionic substrates through such membranes was not initially limiting. Complete cellularisation followed, along with incorporation of pyrite crystals within the early organisms. Metabolism then shifted from pyrite surfaces exclusively to what represented an early cytoplasm. This shift was accompanied by the

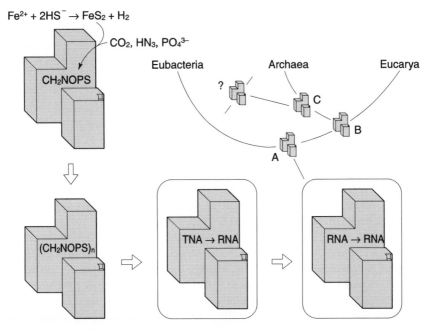

$$Fe^{2+} + 2HS^- \rightarrow FeS_2 + H_2$$

CO_2, HN_3, PO_4^{3-}

Figure 10.9   Simplified conceptual representation of Wächtershäuser's pyrite-based model for origins of life. Pyrite synthesis provides the driving force for organic synthesis; pyrite also provides a surface for stabilising and protecting synthetic reactions, leading ultimately to complex polymerisations and early cellularisation (incorporating pyrite grains). Development of a "tribonucleic" acid precedes RNA- which then leads to DNA-based genetic systems

evolution of soluble coenzymes, and facilitated by mechanisms for substrate transport across membranes by proteins. Information storage and genetic control of metabolism originated with the evolution of "tribonucleic acids", triose-based polymers of purines at first, followed later by the syntheses of ribose, ribonucleic acids, pyrimidines, the precursors of tRNA, rRNA and, mRNA, and, of course, DNA.

Aside from numerous specific predictions about temporal patterns for the acquisition of metabolites and the organisation of biochemical pathways, Wächtershäuser's theory includes an analysis of the differentiation of Bacteria, Archaea and Eucarya. Although the analysis was developed subsequent to and influenced by Woese's three domain proposal,[126,127] it provides a resolution of the genetic similarities among domains, and the incompatibility between archaeal isoprenoid lipids and the fatty acid lipids of bacteria and eukaryotes. The central, indeed necessary, role of elevated temperature in organic synthesis coupled to pyrite formation also provides a logical explanation for the thermophilic, sulphur-metabolizing character of the putative progenote at the root of the phylogenetic tree.

Unlike the Oparin–Haldane and Cairns-Smith theories, there is at present little direct evidence to support complex organic synthesis in association with pyrite formation. The basic mechanism for generating reducing power (as molecular hydrogen) has been verified,[27,33,99] and amide bond synthesis has been demonstrated,[66] both at temperatures (100 °C) within the range of hydrothermal vents. In addition, the formation of an activated thioester and acetic acid from CO, methanethiol and various combinations of FeS and NiS, and catalytic metals has been demonstrated. Claims for sulphide- (but not pyrite-) based amino acid synthesis and polymerisation from simple precurors have also been presented.[67] None the less, Wächtershäuser's detailed and compelling biochemistries are arguably more distant than those of other theories. It remains to be seen whether pyrite-coupled organic formation will prove sufficient to account for complex, heterogeneous autocatalytic systems.

In spite of the need for further verification of the modes and range of organic synthesis, the pyrite theory has a variety of strengths. In addition to its ability to explain phylogenetic differentiation, the proposed binding of organics to pyrite surfaces occurs via phosphate, resulting in much stronger surface associations than are possible for clays; early selection for phosphate esters through strong surface bonding also provides a basis for the central role of phosphorus in intermediary metabolism.

The incorporation of metal catalysts into chemical pathways is facilitated by the availability of high metal concentrations in the hydrothermal environments most conducive to evolution of pyrite-based life. The Fe–S clusters occurring in ferridoxins, which are considered very ancient electron transport proteins, the presence of nickel in urease and the zeolite-like structure of chaperonins (also ancient proteins) evoke associations with metal sulphide-rich vents,[88] and recent observations indicate that nickel and other metal sulphides may prove important as catalysts for pyrite-linked organic synthesis. The connections among pyrite synthesis, metals, phosphorus and other simple metabolites further suggest that the evolution of pyrite-based life was intimately associated with a number of elemental cycles, and that some of the most basic principles of biogeochemical cycling were established prior to the semicellular state.

Because of the chemoautotrophic nature of early pyrite-based life, metabolic pathways were likely acquired in a stepwise progression leading from simple to complex.[119] This progression is most evident and compelling in Wächtershäuser's proposals for the evolution of coenzymes. It differs markedly from the less plausible retrograde evolution proposed by others (see above). Wächtershäuser[119] also describes a potentially viable mechanism for the synthesis of ribose from surface-bound triose phosphate (recall that ribose synthesis significantly constrains the Oparin–Haldane theory). Pyrite-based life is thus compatible with an RNA world, and may not require introduction of PNA as an RNA precursor.

Among the weaknesses of Wächtershäuser's theory is the apparent requirement for hydrothermal or other high temperature systems. While this requirement is based on chemical considerations and is consistent with phylogenetic predictions, individual vents are relatively short-lived systems characterised by spatially sharp thermal gradients.[25,118] Temporal instability and the potentially narrow zone suitable for Wächtershäuser's proposed chemistries could severely limit the loci and time available for evolution of pyrite-based life. Assumption of a hot early ocean (e.g. $> 60\,°C$) provides a possible solution to the problem, since the origin of life would then be constrained more by the availability of sulphide and pyrite than by vents *per se*. High oceanic temperatures and an abundance of sulphide are both compatible with the geophysical context of the early Archean era.

Perhaps a more important issue concerns the source of the phosphate that appears necessary for initial surface bonding of organics to pyrite. While a strength of Wächtershäuser's proposal is the central role of phosphate very early in evolution, contemporary vent fluids are strongly depleted in phosphate due to its reaction with basalts at high temperatures in the subsurface charging zones that feed hydrothermal vents.[118] If phosphate depletion proves typical of ancient vents, then it may be difficult to explain how life could have started there.

Morowitz *et al.*[86] have raised a somewhat different problem. They invoke the principle of continuity to suggest that neither clay nor pyrite templates for genes or organic binding can be accommodated within cellular organisms. Morowitz *et al.* acknowledge that clays and pyrite might have contributed to prevital organic synthesis, but they propose that these contributions preceded cellularisation, and that metabolic evolution occurred only after cellularisation. In this scheme, there is little role for either clays or pyrite, and certainly no role that remains through the acquisition of either energy metabolism or a genetic system.

Of course, invocation of the principle of continuity represents a philosophical position rather than a theoretical necessity. Continuity may provide an appropriate constraint at some point subsequent to cellularisation without invalidating a somewhat different mode of organic evolution prior to that. Substantial changes in environmental conditions could promote saltatory changes in modes of microbial evolution, just as infrequent but catastrophic environmental changes (e.g. meteor impacts; oxygenation of the atmosphere) have altered patterns of evolution in the macrobiota. In addition, it must be noted that Wächtershäuser's proposed pathways for metabolic evolution are largely consistent with the principle of continuity.

Overall then, the case for a pyrite-based origin for life appears somewhat more compelling than that for origin in a primordial soup, even though the mechanisms for pyrite-coupled complex organic synthesis and cellularisation remain to be documented, vents may prove too ephemeral and the source of phosphate is unclear. On the other hand, the range of organic syntheses associated with pyrite appears amenable to experimental analysis, thereby providing a means to test basic elements of the theory.

10.3.4   The "thioester world"

Another rather different sulphur-based proposal has been offered in response to deficiencies recognised in both the Oparin–Haldane theory and the conceptual framework of the RNA world. de Duve[29] has described the "thioester world" as a predecessor to an RNA world (Fig. 10.10). The

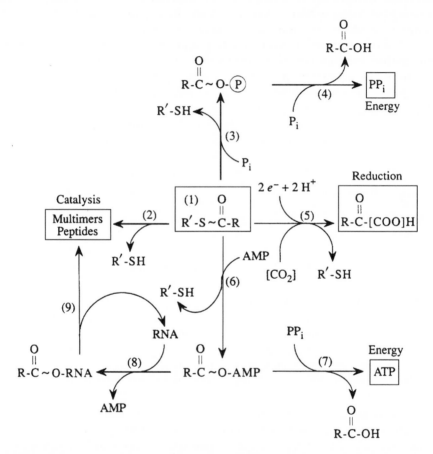

Figure 10.10   Synopsis of chemical transformations in the thioester world illustrating: (1) a pool of thioesters; (2) polymerisation of protoenzymes; (3) generation of high-energy phosphate esters; (4) generation of pyrophosphate, a primordial energy carrier; (5) thioester-based organic synthesis reactions; (6) formation of high-energy adenylate derivatives; (7) production of ATP; (8) generation of acyl-RNA complexes (e.g. amino-charged tRNA); (9) peptide formation[29]

rationale is that, as noted above, there exists no obvious prebiotic mechanism to account for the production of RNA from a primordial soup. The thioester world provides a transition consistent with the principle of continuity because thioesters (e.g. those involving coenzyme A) play a fundamental and ubiquitous role in extant biochemistry.

In the thioester world, catalytic polypeptides (short sequences of amino acids) arise spontaneously from thioester derivatives of amino acids. The amino acids are produced by any of a variety of prebiotic syntheses in the atmosphere and oceans, or are derived from meteoric input. One such process might be based on the production of hydrogen from UV oxidation of ferrous iron at the surface of the primordial ocean (Fig. 10.11). The ferric iron resulting from the reaction precipitates from solution, possibly accounting for the earliest banded iron formations (BIFs). Ferric iron can also participate in a reaction resulting in the formation of thioesters (Fig. 10.12). Other routes for thioester formation include reactions on surfaces and in hot, acidic solutions.[65] Phosphorolytic attack of thioesters by inorganic phosphate

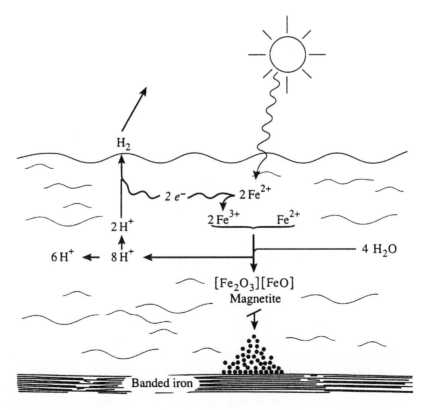

Figure 10.11   Light-driven hydrogen production and iron oxidation resulting in early banded iron formations[29]

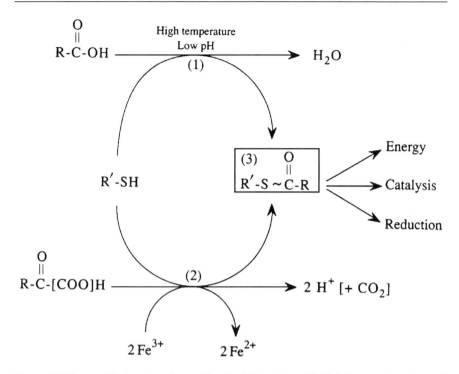

Figure 10.12   Mechanisms for synthesis of thioesters. (1) High temperature, low pH spontaneous synthesis from thiols and organic acids; (2) oxidative synthesis from thiols and α-keto organic acids based on ferric iron reduction[29]

provides a mechanism for introducing high-energy phosphate esters into metabolism, perhaps leading to the synthesis of pyrophosphate as the first unit of energy currency, followed later by the synthesis of AMP and ATP.

The production of various catalytic multimers, thioesters and phosphorylated organics forms the basis for a "protometabolism" in the thioester world.[28,29] Protometabolic pathways rapidly form networks, likely including cyclic mass flow, that are stabilised by interactions among metabolites, and perhaps are spontaneously organised into complex energy-dissipating structures.[37,51,64,70,96] These protometabolic pathways could have occurred in association with clays or pyrite, or perhaps in association with iron oxide flocs produced by UV photooxidation.

It is tempting to speculate that globally ubiquitous iron oxides were incorporated into vesicles or coacervates at an early evolutionary stage, but subsequent to enrichment of the oceans with simple organic acids and amines. The "intracellular" thioester world was maintained and extended by diffusion of sulphide across primitive membranes. An intracellular iron redox cycle, perhaps driven by light and sulphide, could have supported chemosynthesis. While initially dependent on wavelengths ≤ 400 nm, acquisition of a porphyrin phototransducer could have led to an iron-based photosynthetic system analogous to that documented by Widdel *et al*.[124] for a bacterial anaerobic, iron-oxidizing phototroph. Alternatively (or in addition), iron respiration and photometabolism could have occurred on the external surface of vesicles, contributing to a vectorial flow of both protons and thioesters into vesicles, thereby giving rise to an early protonmotive transport system.

In addition to serving as a plausible precursor for the RNA world, the thioester world is attractive in a number of respects. It includes a possible example of very early, global-scale geochemical (iron, carbon, sulphur) cycling driven by what were to become core biological processes. The thioester world may also include an early example of interactions between biochemical processes and the atmosphere, since thioester-linked hydrogen production could have contributed reducing equivalents that were critical for any atmospheric organic syntheses. Tantalising linkages can also be made between the dynamics of various thioesters and polypeptide synthesis, the metabolism of high-energy phosphates, and iron-based photosynthesis and metabolism.

Nonetheless, the thioester world leaves unanswered some basic questions or problems. For example, it is not clear how the thioester world can be reconciled with current understanding of microbial phylogeny. It is also not clear what range of catalytic activites is possible for small multimers with only a limited range of amino acids available. Are such catalysts more or less suitable than ribozymes for establishing protometabolic networks? Cellularisation appears problematic. From what sources were lipids derived, and what was their composition? How was a flow of mass and energy sustained across membranes? Also unclear are the origins of chirality in sugars and amino acids. These and other questions pose considerable conceptual and experimental challenges. However, their resolution could ultimately unify elements of the several theories for the origin of life into a coherent, comprehensive framework amenable to unequivocal testing.

10.4   Evolution of biogeochemical cycles

Irrespective of its origin, the advent of life heralded major changes in elemental cycles. Aside from the colonisation of terrestrial surfaces, perhaps the most important changes occurred as a result of the evolution of oxygenic photosynthesis and respiration. Indeed, the history of atmospheric oxygen and its dynamics provides a useful context for considering the evolution of biogeochemical cycles generally, and for many elements specifically.

The paramount importance of oxygen deserves specific comment. The chemistry of oxygen suits it uniquely for its biogeochemical role.

Oxygen is the most abundant element at Earth's surface, and ranks behind only hydrogen and helium in the universe.[22,105] Oxygen occurs in several combined forms dominated by iron and silicon oxides and, to a lesser extent, sulphate deposits. The oceans dominate aqueous pools of combined oxygen (total abundance about 1.4×10^{12} Tg oxygen), while the atmosphere contains about 1.2×10^9 Tg of molecular oxygen. Biospheric cycling of oxygen between the atmosphere and hydrosphere occurs relatively rapidly, with turnover times of about 5000 and 5 000 000 years, respectively.

Oxygen ranks second after fluorine in its electronegativity (3.5 vs. 4). It is exceeded in its reduction potential ($E^0 = 1.229$ V) by only fluorine, chlorine, a few elemental species, and a small number of other highly oxidised species (e.g. perchlorate, permanganate). However, unlike many oxidants, molecular oxygen is relatively stable, reacting either slowly or not at all with most organics and many reduced inorganics under typical, biologically relevant conditions.

The stability of molecular oxygen is profoundly important. It allows biological systems control over reaction rates and directions, and thus control over the coupling of oxygen to energy conservation. Although oxygen utilisation has exacted a cost that can be measured in sophisticated membrane-bound electron transport systems and enzymes for the detoxification of peroxide and superoxide, it is apparent that kinetic instability and relatively low abundance have constrained the evolution of more powerful oxidant systems.

The development of an oxygen-rich atmosphere on Earth can therefore be viewed as an inevitable outcome of evolution in an aqueous medium containing biochemical systems capable of catalysing the photolysis of water. One can speculate further that an oxygen-rich atmosphere may be an expected outcome for any terrestrial planet with an abundance of water, illumination between about 300 and 800 nm, and characteristics otherwise suitable for organic life. Extrapolation from extant animals and from the fossil record and paleoatmospheric reconstructions also indicates that some form of multicellular or metazoan life might be expected once oxygen abundance exceeds a critical threshold concentration.

Changes in the modes and rates of evolution in biogeochemical cycling are thus intimately linked to the history of atmospheric oxygen. Changes in mode can be illustrated by considering the nitrogen cycle. Although closure of the nitrogen cycle currently depends on denitrification and two obligately aerobic steps (ammonia and nitrite oxidation; see Chapter 1), a closed cycle can be constructed in principle from nitrogen fixation, assimilation, mineralisation and a series of anaerobic transformations based on bicarbonate, sulphate, ferric iron and nitrate reduction (Fig. 10.13). Of these reactions only the first three are considered phylogenetically ancient and known from extant organisms. The remaining redox transformations are thermodynamically feasible, but thus far unreported, perhaps due to kinetic

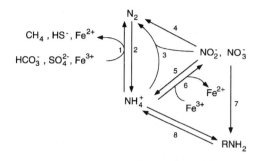

Figure 10.13   Hypothetical anaerobic nitrogen cycle based on the following thermodynamically permissible reactions: (1) ammonium oxidation to dinitrogen by carbon dioxide, sulphate or ferric iron (no evidence at present; possibly kinetically limited); (2) dinitrogen fixation by various organic or inorganic reductants (known); (3) ammonium oxidation by nitrite or nitrate producing dinitrogen (known); (4) denitrification (known); (5) nitrite or nitrate respiration (known); (6) ferric iron oxidation of ammonium to nitrite or nitrate (no evidence at present); (7) nitrate assimilation (known); (8) ammonium assimilation and dissimilation (known)

limitations. In addition, even if such reactions did occur, they would be constrained by electron-acceptor recycling. It is unclear whether volcanic SO_2 emissions or photochemical and possibly photosynthetic oxidation of ferrous iron and sulphide would have been sufficient to maintain closure. Prior to the advent of oxygen then, nitrogen could have been lost from the atmosphere and accumulated in the oceans. The rise of oxygen ensured closure for the nitrogen cycle by promoting the evolution of ammonia and nitrite oxidisers, and perhaps denitrifiers as well, which some argue were derived subsequent to the evolution of aerobic respiration.[58]

In an analogous manner, the availability of nitrogenous, manganic and ferric oxides could have facilitated closed biogeochemical cycles for other elements, especially sulphur, for which both anaerobic photosynthetic and chemolithotrophic transformations are known.[75] The extent to which anaerobic cycling could have occurred depends, of course, on the extent to which electron acceptors were routinely available. As noted above, this may have been a major constraint, at least until the evolution of sulphide- and iron-oxidizing photosynthesis. Although the exact timing of these processes is uncertain, the geological record suggests that oxygenic photosynthesis likely evolved soon after its anoxic precursors, and that both were widespread by 3.5–3.0 Gy, if not earlier.[42,45,46]

The advent in time of oxygenic photosynthesis and the history of oxygen concentrations in the atmosphere have been inferred from several lines of evidence (see Fig. 9.4). The earliest dates for photosynthesis and oxygen production are based on microfossils 3.5 Gy old or younger that are consistent with the structures of extant cyanobacteria. After correcting for the effects of abiological processes, the isotopic (^{13}C) depletion of organic and inorganic carbon coincident with these microfossils is similar to that from modern photosynthetic systems, which implies comparable biochemical processes.

The massive BIFs deposited from about 3.8 to 2 Gy are attributed to episodic oxidation of oceanic ferrous iron resulting in precipitates of haematite and the mixed oxide magnitite.[8,68,122] While photochemical and anaerobic photosynthetic oxidation[12,49,124] could have contributed to the formation of some early BIFs (perhaps 3.8–3.5 Gy), the later formations are almost certainly due to oxygen production. During most of the period of BIF formation, the atmosphere was presumably anoxic or essentially so. In spite of that, it has been proposed that the oceans or shallow lagoonal systems could have contained relatively oxygen-rich "oases" sustained by locally high photosynthesis.[62] Such oases could have promoted the evolution of aerobic metabolism long before substantial oxygenation of the atmosphere and prior to the proliferation of aerobes. The "red beds", which are extensive iron oxide deposits beginning at about 2 Gy, provide evidence for continental weathering[54,122] and therefore oxygen accumulation in the atmosphere to levels of perhaps 0.01 PAL (1% of present atmospheric level). The roughly coincidental disappearance of BIFs and deposition of red beds suggests nearly complete removal of reduced species from seawater by this point, resulting in a slightly oxidising atmosphere and neutral to oxidising oceans.[62]

Microfossil and geological evidence from presumed freshwater Archean lakes in the Tumbiana formation strongly supports oxygenic photosynthesis by 2.7 Gy, with an earlier date for the evolution of photosynthesis implied by the ephemeral nature of the lakes.[18]

Sulphur isotopic evidence (^{34}S) from various Archean and contemporary sulphide deposits is consistent with oxygenation of the atmosphere between 2.5 and 2 Gy, and accumulation of oceanic sulphate concentrations to near modern levels.[47] A combination of sulphur isotope data and an estimate of the timing for the divergence of the non-photosynthetic sulphide-oxidising bacteria supports an increase in atmospheric oxygen to values > 0.05–0.18 PAL by 1–0.6 Gy.[20]

The timing and magnitude of this increase is further supported by shifts in the isotopic composition of organic matter and carbonate, and increases in organic matter burial rates that are derived in part from the isotopic data. It is particularly important to note that changes in organic matter burial play a major role in determining the excess of oxygen production over oxygen utilisation for mineralisation. Organic matter burial in turn is controlled by a number of biological (e.g. rate of organic matter production, rate of heterotrophic consumption, types of heterotrophs available) and geological factors (e.g. continental weathering and erosion, flux of sediment from continents, tectonic and orogenic activity).

A rise in oxygen concentration above 0.1 PAL near the end of the Proterozoic agrees with the presumed requirements of larger metazoans. The explosive metazoan radiation at the beginning of the Cambrian[128] has been attributed to changes in oxygen concentration. This radiation is thought by some to have been preceded by a restructuring of marine biogeochemical cycles.[73] Specifically, photosynthesis and aerobic respiration in the late Proterozoic are viewed as surficial phenomena in a stratified ocean with anoxic deep waters where organic matter is mineralised by sulphate reduction. The evolution of algal predators capable of forming rapidly sinking faecal pellets allowed a shift in the locus of organic mineralisation from the water column to sediments and oxygenation of subsurface waters. This in turn promoted expansion of the marine fauna, stimulating especially benthic macro-organisms, and altered both the dynamics and significance of sulphate reduction relative to aerobic respiration.

The available data collectively support an earlier evolution of oxygenic photosynthesis, perhaps by about 3.5 Gy. The oceans were progressively oxidised over a period of about 1.5 Gy, with complete oxidation by 2 Gy.[63] The atmosphere was then progressively oxidised, although concentrations sufficient to sustain unicellular eukaryotes, that is $\sim 1\%$ of the present atmospheric level (~ 0.01 PAL), and metazoans (~ 0.1 PAL) may not have appeared until about 1.5–1 Ga and 0.6 Ga, respectively.[63,99] Contemporary levels of oxygen appear to have been established by about 144 My ago during the Cretaceous period according to Berner.[6]

While the rate of oxygen accumulation through the Proterozoic may have been regulated by geological factors that control organic matter burial,[62] oxygen dynamics through the Phanerozoic (0.6 Gy to present) are more complex, including both biological and geochemical processes.[6,7,69] In addition to continental weathering and oxidation of reduced organic matter and sulphur, the evolution of terrestrial plants with high carbon:phosphorus ratios and lignin has affected rates and patterns of organic matter burial and oxygen accumulation. The evolution of cellulolytic and lignolytic fungi; the incidence of fire at elevated oxygen concentrations (> 1 PAL); decreases in phosphorus sinks resulting from changes in the chemistry of hydrothermal vent fluids as the Earth cooled; and periodic changes in the assembly and disaggregation of continents all contribute to oxygen regulation and account to varying degrees for stability over the last 100–200 My.[101] Of particular interest are speculative correlations among biological innovation and evolution, episodes of glaciation and increases in the C:P ratios of buried carbon; these parameters have been proposed as elements of a global temperature regulation system that supplements the CO_2–silicate control system (see Chapter 9).

While it has been possible to reconstruct the histories of carbon, oxygen and sulphur and certain groups of microbes, the histories of other elements and groups remain more ambiguous. For instance, patterns in sulphur isotopic data throughout the Archean (3.8–2.5 Gy) can be attributed to either biological sulphate reduction in the presence of low sulphate concentrations or supply of sulphide primarily from abiological sources such as hydrothermal vents.[47] More definitive evidence for the widespread occurrence of sulphate-reducing bacteria is not available until about 2.3–2.2 Gy, at which point the sulphate had accumulated to high concentrations as a result of oxidation of the oceans and atmosphere.[102] It is thus unclear whether sulphate reduction originated early in response to the availability of oxidised sulphur species produced by anaerobic photosynthesis, or later in response to changes in planetary redox status. Similar questions can be asked for other functional groups of bacteria. Did anaerobic photosynthetic bacteria oxidise reduced nitrogen species (e.g. ammonia) and thereby facilitate early evolution of denitrification, or did this latter process depend on the prior evolution of aerobic respirers, the availability of nitrate from nitrification and an oxygenic atmosphere?[13–15,34,35,58] At what point did methane-oxidising bacteria originate? Hayes[52] infers an earlier origin from isotopic evidence. However, methanotrophs are not well known from marine systems with the exception of mytilid symbionts (see Chapter 7), which may suggest evolution of methanotrophy in terrestrial systems subsequent to oxidation of the atmosphere (e.g. post 2 Gy).

Few of these questions or discrepancies can be answered with data from the fossil record. In most cases, there are no products or substrates preserved from which one can derive isotopic data. As a consequence, phylogenetic inference may be the only tool available. Analysis of the 16S rRNA "molecular clock" in conjunction with fossil records of symbiont-bearing molluscs and sulphur isotopic data have helped establish the timing for the origins of certain sulphide-oxidising bacteria.[20] Whether 16S rRNA analysis alone will prove sufficiently reliable for other groups remains to be seen since the accurate calibration of molecular clocks is a subject of some debate, and calibrations for one group may not be extrapolated for others.[17,40] In some instances, certain key enzymes may prove as valuable as 16S rRNA. For example, Coyne and Tiedje[26] suggest that comparisons of dissimilatory nitrite reductases may help resolve the origins of denitrification.

References

1. Achenbach-Richter L, Gupta R, Stetter KO, Woese CR (1987) Were the original eubacteria thermophiles? *Syst Appl Microbiol* **9**: 34–39.
2. Ahern TJ, Klibanov KM (1985) The mechanism of irreversible enzyme inactivation at 100 °C. *Science* **228**: 1280–1284.
3. Aoki I (1989) Holological study of lakes from an entropy viewpoint — Lake Mendota. *Ecol Mod* **45**: 81–93.

4. Bermudez J, Wagensberg J (1986) On the entropy production in microbiological stationary states. *J Theor Biol* **122**: 347–358.

5. Bernal JD (1967) *The Origin of Life*. Weidenfeld & Nicolson, London.

6. Berner RA (1989) Biogeochemical cycles of carbon and sulfur and their effect on atmospheric oxygen over phanerozoic time. *Paleogeog Paleoclimatol Paleoecol* **75**: 97–122.

7. Berner RA, Canfield DE (1989) A new model for atmospheric oxygen over phanerozoic time. *Am J Sci* **289**: 333–361.

8. Beukes NJ, Klein C (1992) Models for iron-formation deposition. In: Schopf JW, Klein C (eds) *The Proterozoic Biosphere: A Multidisciplinary Study*. Cambridge University Press, Cambridge, pp. 147–151.

9. Blöchl E, Keller M, Wächtershäuser G, Stetter KO (1992) Reaction depending on iron sulfide and linking geochemistry and biochemistry. *Proc Natl Acad Sci USA* **89**: 8117–8120.

10. Böhler C, Nielsen PE, Orgel LE (1995) Template switching between PNA and RNA oligonucleotides. *Nature (Lond)* **376**: 578–581.

11. Borowska ZK, Mauserall DC (1987) Efficient near ultraviolet light induced formation of hydrogen by ferrous hydroxide. *Orig Life* **17**: 251–259.

12. Braterman PS, Cairns-Smith AG, Sloper RW (1983) Photooxidation of hydrated Fe^{2+}-significance for banded iron formations. *Nature (Lond)* **303**: 163–164.

13. Broda E (1975) The history of inorganic nitrogen in the biosphere. *J Mol Evol* **7**: 87–100.

14. Broda E (1977) The position of nitrate respiration in evolution. *Orig Life* **8**: 173–174.

15. Broda E (1988) Two kinds of lithotrophs missing in nature. *Zeit Allg Mikrobiol* **17**: 1977.

16. Brown GC, Mussett AE (1981) *The Inaccessible Earth*. Allen & Unwin, Winchester, MA.

17. Brown JR, Doolittle WF (1995) Root of the universal tree of life based on ancient aminoacy-tRNA synthetase gene duplications. *Proc Natl Acad Sci USA* **92**: 2441–2445.

18. Buick R (1992) The antiquity of oxygenic photosynthesis — evidence from stromatolites in sulphate-deficient archaean lakes. *Science* **255**: 74–77.

19. Cairns-Smith AG (1982) *Genetic Takeover and the Mineral Origins of Life*. Cambridge University Press, Cambridge.

20. Canfield DE, Teske A (1996) Late proterozoic rise in atmospheric oxygen concentration inferred from phylogenetic and sulphur-isotope studies. *Nature (Lond)* **382**: 127–132.

21. Chang S, DesMarais D, Mack R, Miller SL, Strathearn GE (1983) Prebiotic organic synthesis and the origin of life. In: Schopf JW (ed.) *Earth's Earliest Biosphere: Its Evolution and Origin*. Princeton University Press, Princeton, pp. 53–92.

22. Chapman DJ, Schopf JW (1983) Biological and biochemical effects of the development of an aerobic environment. In: Schopf JW (ed.) *Earth's Earliest Biosphere: Its Origins and Evolution*. Princeton University Press, Princeton, pp. 302–320.

23. Chyba C, Sagan C (1992) Endogenous production, exogenous delivery and impact-shock synthesis of organic molecules: an inventory for the origins of life. *Nature (Lond)* **355**: 125–132.

24. Chyba C, Sagan C (1997) Comets as a source of prebiotic organic molecules for the early Earth. In: Thomas PJ, Chyba CF, McCay CP (eds) *Comets and the Origin and Evolution of Life*. Springer, New York, pp. 147–173.

25. Corliss JB (1990) Hot springs and the origins of life. *Nature (Lond)* **347**: 624.

25a. Cox PA (1995) *The Elements on Earth: Inorganic Chemistry in the Environment*. Oxford University Press, Oxford.

26. Coyne MS, Tiedje JA (1990) Distribution and diversity of dissimilatory NO_2^- reductases in denitrifying bacteria. In: Revsbech NP, Sørensen J (eds) *Denitrification in Soil and Sediment*. Plenum, New York, pp. 21–35.

27. Cronin JR, Pizzarello S (1997) Enantiomeric excesses in meteoric amino acids. *Science* **275**: 951–955.

28. de Duve C (1987) Selection by differential molecular survival: a possible mechanism of early chemical evolution. *Proc Natl Acad Sci USA* **84**: 8253–8256.

29. de Duve C (1991) *Blueprint for a Cell*. Neil Patterson, Burlington, NC.

30. Degens ET (1989) *Perspectives on Biogeochemistry*. Springer, Berlin.

31. Delsemme A (1997) The origin of the atmosphere and of the oceans. In: Thomas PJ, Chyba CF, McCay CP (eds) *Comets and the Origin and Evolution of Life*. Springer, New York, pp. 29–67.

32. Doolittle RF, Feng D-F, Tsang S, Cho G, Little E (1996) Determining divergence times of the major kingdoms of living organism with a protein clock. *Science* **271**: 470–477.

33. Drobner E, Huber H, Wächtershäuser G, Rose D, Stetter KO (1990) Pyrite formation linked with hydrogen evolution under anaerobic conditions. *Nature (Lond)* **346**: 742–744.

34. Egami F (1973) A comment to the concept on the role of nitrate fermentation and nitrate respiration in an evolutionary pathway of energy metabolism. *Zeit Allg Mikrobiol* **13**: 177–181.

35. Egami F (1977) Anaerobic respiration and photoautotrophy in the evolution of prokaryotes. *Orig Life* **8**: 169–171.

36. Ehrlich HL (1990) *Geomicrobiology*, 2nd edn. Marcel Dekker, New York.

37. Eigen M (1971) Selforganization of matter and the evolution of biological macromolecules. *Naturwissensch* **58**: 465–523.

38. Eigen M, Lindemann BF, Tietze M, Winkler-Cwatitsch R, Dress A, von Haeseler A (1989) How old is the genetic code? Statistical geometry of tRNA provides an answer. *Science* **244**: 673–679.

39. Ekland EH, Bartel DP (1996) RNA-catalyzed RNA polymerization using nucleoside triphosphates. *Nature (Lond)* **382**: 373–376.

40. Forterre P (1995) Thermoreduction, a hypothesis for the origin of prokaryotes. *CR Acad Sci Paris* **318**: 415–422.

41. Friedrich B (1985) Evolution of chemolithotrophy. In: Schleifer KH, Stackenbrandt E (eds) *Evolution of Prokaryotes*. Academic Press, London, pp. 205–234.

42. Fuchs G, Stupperich E (1985) Evolution of autotrophic CO_2 fixation. In: Schleifer KH, Stackenbrandt E (eds) *Evolution of Prokaryotes*. Academic Press, London, pp. 235–251.

43. Gates DM (1980) *Biophysical Ecology*. Springer, New York.

44. Gest H (1980) The evolution of biological energy transduction systems. *FEMS Microbiol Lett* **7**: 73–77.

45. Gest H, Schopf JW (1983) Biochemical evolution of anaerobic energy conversion: the transition from fermentation to anoxygenic photosynthesis. In: Schopf JW (ed.) *Earth's Earliest Biosphere: Its Origins and Evolution*. Princeton University Press, Princeton, pp. 135–148.

46. Schopf JW (ed.) *Earth's Earliest Biosphere: Its Origins and Evolution*. Princeton University Press, Princeton.

47. Habicht KS, Canfield DE (1996) Sulphur isotope fractionation in modern microbial mats and the evolution of the sulphur cycle. *Nature (Lond)* **382**: 342–343.

48. Haldane JBS (1929) *The Origin of Life*. Rationalist Annual.

49. Hartman H (1984) The evolution of photosynthesis and microbial mats: a speculation on the Banded Iron Formations. In: Crawford B (ed.) *Microbial Mats: Stromatolites*. Alan R. Liss, New York, pp. 449–453

50. Harvey RP, McSween HY Jr (1996) A possible high-temperature origin for the carbonates in the martian meteorite ALH84001. *Nature (Lond)* **382**: 49–51.

51. Hassell MP, Comins HN, May RM (1994) Species coexistence and self-organizing spatial dynamics. *Nature (Lond)* **370**: 290–292.

52. Hayes JM (1983) Geochemical evidence bearing on the origin of aerobiosis, a speculative hypothesis. In: Schopf JW (ed.) *Earth's Earliest Biosphere: Its Origins and Evolution*. Princeton University Press, Princeton, pp. 291–301.

53. Hayes JM, Kaplan IR, Wedeking KW (1983) Precambrian organic geochemistry, preservation of the record. In: Schopf JW (ed.) *Earth's Earliest Biosphere: Its Origins and Evolution*. Princeton University Press, Princeton, pp. 93–134.

54. Holland HD (1992) Chemistry and evolution of the proterozoic ocean. In: Schopf JW, Klein C (eds) *The Proterozoic Biosphere: A Multidisciplinary Study*. Cambridge University Press, Cambridge, pp. 169–172.

55. Horvath RS (1974) Evolution of anaerobic energy-yielding metabolic pathways of the prokaryotes. *J Theor Biol* **47**: 361–371.

56. Hoyle F (1982) The universe: past and present reflections. *Ann Rev Astron Astrophys* **20**: 1–35.

57. Hoyle F, Wickramasinghe NC (1978) *Lifecloud*. Harper & Row, New York.

58. Jones CW (1985) Evolution of autotrophic CO_2 fixation. In: Schleifer KH, Stackenbrandt E (eds) *Evolution of Prokaryotes*. Academic Press, London, pp. 175–204.

59. Jørgensen BB, Zawacki LZ, Jannasch HW (1990) Thermophilic bacterial sulfate reduction in deep-sea sediments at the Guaymas Basin hydrothermal vent site (Gulf of California). *Deep-Sea Res* **37**: 695–710.

60. Joyce GF (1989) RNA evolution and the origins of life. *Nature (Lond)* **338**: 217–224.

61. Joyce GF, Schwartz AW, Miller SL, Orgel LE (1987) The case for an ancestral genetic system involving simple analogues of the nucleotides. *Proc Natl Acad Sci USA* **84**: 4398–4402.

62. Kasting JF, Holland HD, Kump LR (1992) Atmospheric evolution: the rise of oxygen. In: Schopf JW, Klein C (eds) *The Proterozoic Biosphere: A Multidisciplinary Study*. Cambridge University Press, Cambridge, pp. 159–163.

63. Kasting JF, DP Whitmire, RT Reynolds (1993) Habitable zones around main sequence stars. *Icarus* **101**: 1–21.

64. Kauffman SA (1993) *Origins of Order: Self-organization and Evolution of Complex Systems*. Oxford University Press, Oxford.

65. Keefe AD, Newton GL, Miller SL (1995) A possible prebiotic synthesis of pantetheine, a precursor to coenzyme A. *Nature (Lond)* **373**: 683–685.

66. Keller M, Blöchl E, Wächtershäuser G, Stetter KO (1994) Formation of amide bonds without a condensation agent and implications for origin of life. *Nature (Lond)* **368**: 836–838.

67. Kimoto T, Fujinaga T (1990) Non-biotic synthesis of organic polymers on H_2S-rich sea-floor: a possible reaction in the origin of life. *Mar Chem* **30**: 179–192.

68. Klein C, Beukes NJ (1992) Time distribution, stratigraphy, and sedimentologic setting, and geochemistry of precambrian iron-formations. In: Schopf JW, Klein C (eds) *The Proterozoic Biosphere: A Multidisciplinary Study*. Cambridge University Press, Cambridge, pp. 139–146.

69. Kump LR (1989) Chemical stability of the atmosphere and ocean. *Paleogeog Paleoclimatol Paleoecol* **75**: 123–136.

70. Küppers B (1975) The general principles of selection and evolution at the molecular level. *Progr Biophys Molec Biol* **30**: 1–22.

71. Lake JA (1988) Origin of the eukaryotic nucleus determined by rate-invariant analysis of rRNA sequences. *Nature (Lond)* **331**: 184–186.

72. Lazcano A, Fox GE, Oró JF (1992) Life before DNA: the origin and evolution of early archaean cells. In: Mortlock RP (ed.) *The Evolution of Metabolic Function*. CRC Press, Boca Raton, pp. 237–296.

73. Logan GA, Hayes JM, Hieshima GB, Summons RE (1995) Terminal proterozoic reorganization of biogeochemical cycles. *Nature (Lond)* **376**: 53–56.

74. Lovelock JE (1979) *Gaia: A New Look at Life on Earth*. Oxford University Press, Oxford.

75. Madigan MT, Martinko JM, Parker J (1997) *Biology of Micro-organisms*, 8th edn. Prentice Hall, Upper Saddle River, NJ.

76. McKay CP, Borucki WJ (1997) Organic synthesis in experimental impact shocks. *Science* **276**: 390–392.

77. McKay CP (1997) Life in comets. In: Thomas PJ, Chyba CF, McCay CP (eds) *Comets and the Origin and Evolution of Life*. Springer, New York, pp. 273-282.

78. McKay DS, Gibson EK Jr, Thomas-Kerpta KL, Vali H, Romanek CS, Clemett SJ, Chillier XDF, Maechling CR, Zare RN (1996) Search for past life on Mars: possible relic biogenic activity in Martian meteorite ALH84001. *Science* **273**: 924-930.

79. McKinnon WB (1997) Sighting the seas of Europa. *Nature (Lond)* **386**: 765-766.

80. Miller SL, Orgel LE (1974) *The Origins of Life on Earth*. Prentice-Hall, Englewood Cliffs, NJ.

81. Mojzsis SJ, Arrhenius G, McKeegan KD, Harrison TM, Nutman AP, Friend CRL (1996) Evidence for life on Earth before 3,800 million years ago. *Nature (Lond)* **384**: 55-59.

82. Mooers AØ, Redfield RJ (1996) Digging up the roots of life. *Nature (Lond)* **379**: 587-588.

83. Morowitz HJ (1960) Some consequences of the application of the Second Law of thermodynamics to cellular systems. *Biochim Biophys Acta* **40**: 340-345.

84. Morowitz HJ (1968) *Energy Flow in Biology*. Academic Press, New York.

85. Morowitz HJ (1992) *Beginnings of Cellular Life: Metabolism Recapitulates Biogenesis*. Yale University Press, New Haven.

86. Morowitz HJ, Deamer DW, Smith T (1991) Biogenesis as an evolutionary process. *J Mol Evol* **33**: 207-208.

87. Mukhin LM, Gerasimov ML, Safonova EN (1990) Origin of precursors of organic molecules during evaporation of meteorites and mafic terrestrial rocks. *Nature (Lond)* **340**: 46-48.

88. Nisbet EG, Fowler CMR (1996) Some like it hot. *Nature (Lond)* **382**: 404-405.

89. Ohmoto H, Felder RP (1987) Bacterial activity in the warmer, sulphate-bearing, Archaen oceans. *Nature (Lond)* **328**: 244-246.

90. Oparin AI (1957) *The Origin of Life on Earth*, 3rd edn. Academic Press, New York.

91. Oró J, Lazcano A (1997) Comets and the origin and evolution of life. In: Thomas PJ, Chyba CF, McCay CP (eds) *Comets and the Origin and Evolution of Life*. Springer, New York, pp. 3-27.

92. Parmentier EM, Hess PC (1992) Chemical differentiation of a convecting planetary interior: consequences for a one plate planet such as Venus. *Geophys Res Lett* **19**: 2015-2018.

93. Piccirilli JA, Vyle JS, Caruthers MH, Cech TR (1993) Metal ion catalysis in the *Tetrahymena* ribozyme reaction. *Nature (Lond)* **361**: 85-88.

94. Ponnamperuma C, Shimoyama A, Friebele E (1982) Clay and the origin of life. *Orig Life* **12**: 9-40.

95. Prigogine I (1967) *Thermodynamics of Irreversible Processes*. Wiley, New York.

96. Prigogine I, Nicolis G (1971) Biological order, structure and instabilities. *Quart Rev Biophys* **4**: 107-148.

97. Reeburgh WS (1983) Rates of biogeochemical processes in anoxic sediments. *Annu Rev Earth Planet Sci* **11**: 269-298.

98. Retallack GJ (1997) Early forest soils and their role in Devonian global change. *Science* **276**: 583–585.

99. Robbins EI, Porter KG, Haberyan KA (1985) Pellet microfossils: possible evidence for metazoan life in Early Proterozoic time. *Proc Natl Acad Sci* **82**: 5809–5813.

100. Robertson MP, Miller SL (1995) Prebiotic synthesis of 5-substituted uracils: a bridge between the RNA world and the DNA-protein world. *Science* **268**: 702–705.

101. Robinson JM (1989) Phanerozoic O_2 variation, fire, and terrestrial ecology. *Paleogeog Paleoclimatol Paleoecol* **75**: 223–240.

102. Schidlowski M (1979) Antiquity and evolutionary status of bacterial sulfate reduction: sulfur isotope evidence. *Orig Life* **9**: 299–311.

103. Schidlowski M (1988) A 3,800-million-year isotopic record of life from carbon in sedimentary rocks. *Nature (Lond)* **333**: 313–318.

104. Schidlowski M, Hayes JM, Kaplan IR (1983) Isotopic inferences of ancient biochemistries: carbon, sulfur, hydrogen and nitrogen. In: Schopf JW (ed.) *Earth's Earliest Biosphere: Its Origins and Evolution*. Princeton University Press, Princeton, pp. 149–187.

105. Schlesinger WH (1991) *Biogeochemistry: An Analysis of Global Change*. Academic Press, New York.

106. Schopf JW (1992) Evolution of the proterozoic biosphere: benchmarks, tempo, and mode. In: Schopf JW, Klein C (eds) *The Proterozoic Biosphere: A Multidisciplinary Approach*. Cambridge University Press, Cambridge, pp. 585–600.

107. Schopf JW (1992) Paleobiology of the archean. In: Schopf JW, Klein C (eds) *The Proterozoic Biosphere: A Multidisciplinary Study*. Cambridge University Press, Cambridge, pp. 25–39.

108. Schopf JW, Hayes JM, Walter MR (1983) Evolution of Earth's earliest ecosystems: recent progress and unsolved problems. In: Schopf JW (ed.) *Earth's Earliest Biosphere: Its Origins and Evolution*. Princeton University Press, Princeton, pp. 361–383.

109. Schrödinger E (1944) *What is Life?* Cambridge University Press, Cambridge.

110. Schwartzman DW, Volk T (1989) Biotic enhancement of weathering and the habitability of Earth. *Nature (Lond)* **340**: 457–459.

111. Shapiro R (1984) The improbability of prebiotic nucleic acid synthesis. *Orig Life* **14**: 565–570.

112. Sleep NH, Zahnle KJ, Kasting JF, Morowitz HJ (1989) Annihilation of ecosystems by large asteroid impacts on the early Earth. *Nature (Lond)* **342**: 139–142.

113. Spach G (1984) Chiral versus chemical evolutions and the appearance of life. *Orig Life* **14**: 433–437.

114. Steinbach V, Yuen DA (1992) The effects of multiple phase transitions on Venusian mantle convection. *Geophys Res Lett* **19**: 2243–2246.

115. Stetter KO, Fiala G, Huber R, Huber G, Segerer A (1986) Life above the boiling point of water? *Experientia* **42**: 1187–1191.

116. Vezier J (1988) Solid Earth as a recycling system: temporal dimensions of global tectonics. In: Lerman A, Meybeck M (eds) *Physical and Chemical Weathering in Geochemical Cycles.* Kluwer, Dordrecht, pp. 357-372.

117. Vezier J, Laznicka P, Jansen SL (1989) Mineralization through geologic time: recycling perspective. *Am J Sci* **289**: 484-524.

118. Von Damm KL (1990) Seafloor hydrothermal activity: black smoker chemistry and chimneys. *Annu Rev Earth Planet Sci* **18**: 173-204.

119. Wächtershäuser G (1988) Before enzymes and templates: theory of surface metabolism. *Microbiol Rev* **52**: 452-484.

120. Wächtershäuser G (1988) Pyrite formation, the first energy source for life: a hypothesis. *Syst Appl Microbiol* **10**: 207-210.

121. Wake LV, Christopher RK, Rickard PAD, Andersen JE, Ralph JB (1977) A thermodynamic assessment of possible substrates for sulphate-reducing bacteria. *Aust J Biol Sci* **1977**: 155-172.

122. Walker JCG, Klein C, Schidlowski M, Schopf JW, Stevenson DJ, Walter MR (1983) Environmental evolution of the archaean-early proterozoic Earth. In: Schopf JW (ed.) *Earth's Earliest Biosphere: Its Origins and Evolution.* Princeton University Press, Princeton, pp. 260-290.

123. White RH (1984) Hydrolytic stability of biomolecules at high temperatures and its implications for life at 250 °C. *Nature (Lond)* **310**: 430-432.

124. Widdel F, Schnell S, Heising S, Ehrenreich A, Assmus B, Schink B (1993) Ferrous iron oxidation by anoxygenic phototrophic bacteria. *Nature (Lond)* **362**: 834-836.

125. Williams DM, Kasting JF, Wade RA (1997) Habitable moons around extrasolar giant planets. *Nature (Lond)* **385**: 234-236.

126. Woese C (1987) Bacterial evolution. *Microbiol Rev* **51**: 221-271.

127. Woese C, Kandler O, Wheelis ML (1990) Towards a natural system of organisms: proposal for the domains Archaea, Bacteria and Eucarya. *Proc Natl Acad Sci* **87**: 4576-4579.

128. Wray GA, Levinton JS, Shapiro LH (1996) Molecular evidence for deep Precambrian divergences among metazoan phyla. *Science* **274**: 568-573.

Appendix 1

Thermodynamics and calculation of energy yields of metabolic processes

Understanding which bacterial processes predominate under a given set of circumstances requires an understanding of the energetics of dissimilatory metabolism and growth. There are two aspects involved: (i) considerations based on chemical thermodynamics; and (ii) kinetic constraints on chemical reactions and the coupling of those reactions to growth. Thermodynamics provides a framework for predicting whether or not a specific reaction can occur spontaneously, and the magnitude of energy exchanges (production or consumption) between reactants and their environment; thermodynamics also predicts whether a reaction may be usefully coupled to biological work and the extent of disorder (entropy change) associated with a reaction. However, thermodynamics does not directly predict the kinetic characteristics or mechanisms of reactions.

In contrast, the analysis of kinetics deals exclusively with mechanism and rate. Lacking a suitable mechanism, a thermodynamically permissible reaction may proceed only slowly, if at all. Reaction rates also depend on the proportion of reactants that are sufficiently energetic to participate in a given reaction. This proportion is a function of temperature and "activation energy" as specified by the Boltzmann distribution law. Activation energy, E_a, refers to a minimum or threshold molecular energy state that is required in order for colliding molecules to undergo a specified reaction. Reaction rates (or rate constants) are related to temperature and activation energy according to: $k = A e^{-E_a/RT}$ where A and E_a are characteristic reaction constants, R is the gas law constant and T is temperature in K. In general, the higher the activation energy, the slower the reaction.

Enzymes, which function as catalysts, act by lowering activation energies, thereby enhancing reaction rates. Enzymes provide reaction mechanisms that facilitate processes that might not occur otherwise. For example, hydrogen and oxygen can coexist in a gas or liquid without reacting, even though reduction of oxygen by hydrogen to form water should occur spontaneously according to thermodynamic calculations. Addition of bacterial hydrogenases to a stable hydrogen and oxygen mixture promotes rapid activity. There are numerous other examples in which microbial enzymes greatly accelerate chemical reactions, the energy from which supports growth and maintenance. It has become almost axiomatic that, if a proposed reaction is thermodynamically permissible, there is an enzyme or microbe to exploit it.

The repertoire of microbial catalysts or enzymes is not unlimited, however. Although oxidation of N_2 by O_2 to produce nitrate should occur spontaneously and provide energy for growth, the reaction is extremely slow due to a high activation energy and the absence of suitable catalytic enzymes. Likewise, reduction of N_2 by H_2 to form ammonia is favourable thermodynamically, but does not proceed at all under standard abiological conditions, again as a result of a high activation energy. A complex microbial enzyme, nitrogenase, catalyses ammonia production from N_2 and H_2, but this is an energetically expensive rather than energy-producing process. Even so, the microbial process is considerably more efficient than the commercial chemical process used for industrial ammonia production. Such limitations notwithstanding, microbes are remarkable chemists that fully exploit their environment, extracting the chemical energy from numerous reactions for use in an extraordinary array of biological processes.

The principles of equilibrium thermodynamics can be applied readily to understand the energetics of metabolic processes and the biogeochemical behaviour of micro-organisms. The origins of life, metabolic diversity, microbial competition, and mass and energy cycles can and should be understood in the context of thermodynamics. The concepts of internal energy, entropy and free energy are especially important. These concepts and basic thermodynamic principles are considered here in outline form. For more thorough and formal treatments, readers are urged to consult other sources.[1-3]

One of the most useful thermodynamic parameters for predictive purposes in biological as well as physical–chemical systems is the Gibbs free energy, G. Gibbs free energy is an extensive state function, meaning that the value of G depends on the state and mass of the system under consideration. Gibbs free energy is a function of several parameters: internal energy (E or U); enthalpy (H); entropy (S); temperature (T).

Internal energy (U) for a system at rest (kinetic energy $= 0$) and in the absence of external energy fields (e.g. gravity; potential energy $= 0$) is defined as the sum of translational, rotational, vibrational, electronic, relativistic and interactional energies of the molecules within a specified system. Changes in the internal energy of a system, ΔU, are related to ·heat exchanges between the system and its surroundings and work done by or on the system according to the first law of thermodynamics: $\Delta U = q + w$, where q = heat flow and w = work (usually described in terms of pressure–volume changes); U has units of joules.

Enthalpy, another state function, has units of joules and is defined as: $H = U + PV$, where P and V represent pressure and volume. The expression for enthalpy does not constitute a new thermodynamic law, but rather is derived from the first law for systems in which work occurs as a result of pressure–volume changes. Consider a closed system that is heated at constant pressure. For such a system,

$$\Delta U = U_2 - U_1 = q_p + P(V_2 - V_1),$$

where q_p = heat flow at constant pressure, and

$$q_p = U_2 - U_1 - P(V_2 - V_1) = (U_2 - PV_2) - (U_1 - PV_1) = H_2 - H_1 = \Delta H.$$

More generally, for any change of state, the change in enthalpy is expressed as

$$\Delta H = U_2 + P_2 V_2 - (U_1 + P_1 V_1) = \Delta U + \Delta(PV).$$

For most biological systems, the term $\Delta(PV)$ is small since volume changes of water are small with modest heating at low or modest pressures; thus, ΔH is approximately equal to ΔU.

The enthalpy changes observed during the combustion of biological materials, or which occur in association with many biological processes, have been estimated from various types of calorimetry. For example, bomb calorimetry measures *heats of combustion* in a constant volume system, and has been applied successfully and extensively in ecological studies to develop energy flow budgets for both simple and complex ecosystems. These budgets have facilitated analyses of both energy and mass cycling, and the roles of various functional groups of organisms in each. Calorimetric analyses have also proven useful in understanding the energetics of microbial growth and estimating growth efficiencies. Calorimetry has been especially valuable for estimating *heats of formation*. Standard heats of formation, typically represented as ΔH_f^0, can in turn be used to estimate the standard enthalpies of various reactions from Hess' Law of Constant Heat Summation:

$$\Delta H^0 = \Sigma \Delta H_f^0 \text{ (products)} - \Sigma \Delta H_f^0 \text{ (reactants)}. \tag{A.1.1}$$

As an example, consider a calculation of the ΔH^0 for methane oxidation by aerobic respiration:

$$CH_4 + 2O_2 \longrightarrow CO_2 + 2H_2O$$

Based on values of ΔH^0 for CO_2, H_2O, CH_4 and O_2 we have that $\Delta H^0 = (-391.51 + [2 \times -285.83]) - (-74.60 + [2 \times 0.00]) = -890.57 \text{ kJ mol}^{-1}$. Note that $\Delta H_f^0 = 0.00$ for oxygen by definition, since it is taken in its standard state. While enthalpy calculations are useful in a number of contexts, they perhaps have their greatest utility for microbiologists and biogeo-chemists in calculating Gibbs free energy changes (ΔG).

Gibbs free energy (with units of joules) is a measure of the maximum amount of energy available during a transition from one state to another in an isothermal system at constant pressure (the Hemholtz energy, A, is similar but requires constant volume). Such conditions typify most living systems, and are especially applicable for micro-organisms which, due to their small size, are nearly always in thermal equilibrium with their environment. However, Gibbs free energy cannot be discussed properly without first considering the second law of thermodynamics and entropy.

The first law of thermodynamics requires conservation of energy during transitions of state for a given system. Thus, in accordance with the first law, the kinetic energy (about 0.7 J) of an egg dropped from a height of 1 m on to a floor is transferred to the surroundings of the egg by several mechanisms upon impact; a portion of the kinetic energy is also expended in rupturing the calcium apatite structure of the shell and redistributing the egg contents. Heating the egg in its new state might result in an internal energy equivalent to that of the original state, but clearly, even "all the king's horses and all the king's men" cannot restore the egg. More aptly stated, the conservation of energy during changes of state provides no consistent predictive capability for the direction and spontaneity of change. Many salts dissolve spontaneously in enderthermic reactions that involve heat transfer from the solvent, with no change in internal energy (i.e. $\Delta U = 0$); others dissolve spontaneously in exerthermic reactions that involve heat transfer to the solvent, again with no change in internal energy. Some other parameter then must provide an indication of the directionality and spontaneity of system change.

The second law of thermodynamics has been stated in various forms, each of which addresses the spontaneity or permissibility of change. From Clausius we have: "it is impossible for a system to undergo a cyclic process whose sole effects are the flow of heat into the system from a cold reservoir and the flow of an equal amount of heat out of the system to a hot reservoir", an obvious expression of common experience. The second law is typically presented more formally in terms of the state variable entropy, S, with units of $J K^{-1}$. Entropy, or, more precisely, entropy change (ΔS), provides an indication of the capacity for changes in a system as well as the direction. Spontaneous, irreversible change (typical of biological systems) is accompanied by an increase in entropy, $\Delta S > 0$, while at equilibrium $\Delta S = 0$, and S is maximised relative to some initial state. Accordingly, heat flows spontaneously from hot to cold reservoirs and diffusion occurs from high to low concentrations.

It is useful to conceptualise entropy changes by considering changes of state in some system that interacts with its surrounding (e.g. through PV work); the system by definition is in thermal and mechanical but not material equilibrium, while the surroundings are in thermal, mechanical and material equilibrium. If the system and surrounding are isolated together, then the sum of the system and its surroundings can be denoted the "universe", and since entropy is an extensive state function,

$$S_u = S_s + S_{su} \quad \text{and} \quad dS_u = dS_s + dS_{su}$$

where the subscripts refer to universe, system and surroundings, respectively. For an infinitesimal, reversible heat exchange at constant pressure in the system due to some hypothetical endothermic reaction, $dQ_s = -dQ_{su}$ since the surroundings must supply the heat. dS is defined as dQ/T (or the heat exchange at a given temperature), hence from the first law and definition for entropy,

$$dS_{su} = dQ_{su}/T = -dQ_s/T = -(dU + P\, dV_s)/T.$$

At equilibrium $dS_u = 0$, therefore

$$dS_{su} - (dU + P\, dV_s)/T = 0 = (dU + P\, dV_s) - T\, dS_{su} = dH - T\, dS_{su}.$$

For an irreversible process within the system, the total entropy must be positive, that is

$$dS_u = dS_s + dS_{su} > 0$$

For the surroundings, which are in thermodynamic equilibrium, $dS_{su} = dQ_{su}/T$. For an irreversible process $dS_s \neq dQ_s/T$, but since $dQ_s + dQ_{su} = 0$, $dS_s > -dS_{su} = -dQ_{su}/T = dQ_s/T$. It follows that, in a closed system in thermal and mechanical equilibrium, $dS_s > dQ_s/T$ for an irreversible process. Again from the first law (and PV work at constant pressure),

$$T\, dS > U + P\, dV \quad \text{or} \quad U + P\, dV - T\, dS < 0 \quad \text{or} \quad dH - T\, dS < 0.$$

The final expression, which consists of three state functions, provides the definition for Gibbs free energy,

$$G = H - TS \quad \text{or} \quad \Delta G = \Delta H - T\Delta S$$

It follows from the above that at equilibrium $\Delta G = 0$, and that for spontaneous processes $\Delta G < 0$; further, processes characterised by $\Delta G > 0$ violate the second law and do not occur spontaneously. These relationships are especially important in biological and biogeochemical processes because they provide a basis for predicting the direction of change as well as the quantitative extent of change. In addition, free energy changes can be directly related to specific chemical and biochemical reactions, including electrochemical transformations, as outlined below.

Although isolated systems performing pressure–volume work are a remote abstraction of living organisms, the fact that $\Delta G = 0$ for processes at equilibrium provides a means to relate thermodynamic properties to the biochemical reactions that characterise living systems. Consider the following reaction,

$$aA + bB \longrightarrow cC + dD$$

where a, b, c and d refer to stoichiometric reaction coefficients. As with Hess' law (A.1.1), it can be shown for such a reaction that

$$\Delta G = \Sigma G_f \text{ products} - \Sigma G_f \text{ reactants.} \tag{A.1.2}$$

It can be further shown that, in general, $G = G_f^0 + nRT \ln[X]$ where $n =$ number of moles and $[X] =$ concentration.

For the reaction above then, $\Delta G = G_C + G_D - (G_A + G_B)$ which takes the following form:

$$\Delta G = \Delta G^0 + RT \ln\{([C]^c [D]^d)/([A]^a [B]^b)\}$$

where
$$([C]^c [D]^d)/([A]^a [B]^b) = Q \tag{A.1.3}$$

The term "Q" provides a measure of the displacement of a reaction from equilibrium since it is based on actual reactant and product concentrations. In the event that reactants and products are present at equilibrium concentrations $K = Q$, and since $\Delta G = \Delta G^0 + RT \ln Q$, $\Delta G^0 = -RT \ln K$.

Since H^+ is often a reactant or product in biogeochemical reactions, ΔG^0 is ususally modified by the addition of the G_f^0 for H^+ at a concentration of 10^{-7} M, or the physiologically common pH of 7. In this case, the resulting term is referred to as $\Delta G^{0'}$. The ΔG^0 for many reactions of biogeochemical interest can be calculated from (A.1.2) using compilations of G_f^0 that are available for numerous organic and inorganic species (Table A.1).

The thermodynamic relationships outlined above can be extended to electrochemical reactions. These reactions are characterised by changes in the valence state of reactants and products. An electron donor (or reductant) is oxidised by an electron acceptor (oxidant) according to the general reaction scheme:

$$A^{(x)} + B^{(y)} \longrightarrow A^{(x-n)} + B^{(y+n)} \tag{A.1.4}$$

where x and y are valence states and $n =$ number of electrons transferred. Reactions of this type (also know as *redox* reactions) are profoundly important in biological and biogeochemical systems. Dissimilatory metabolism is entirely dependent on redox reactions, and certain assimilatory processes likewise involve redox transformations (e.g. desaturation of fatty acids). Further, an array of abiological redox reactions plays critical roles in the mobilization and immobilization of numerous elements. Ferric iron reduction by hydrogen sulphide and the formation of pyrite are key redox processes in iron deposition in marine sediments.

Table A.1   The free energy of formation from elements ΔG_f^0. For the most stable form of the elements $\Delta G_f^0 = 0$. Data from Thauer *et al.*[4]

Substance	State*	ΔG_f^0 (kJ mol^{-1})
H$^+$	aq	40.01 (pH 7)
H$_2$O	l	−237.57
CO$_2$	aq	−394.90
CH$_4$	g	−50.82
Methanol	aq	−175.56
Ethanol	aq	−181.84
Formaldehyde	aq	−130.72
Formate	aq	−351.54
Acetate	aq	−369.93
Propionate	aq	−361.08
Butyrate	aq	−353.21
Fumarate	aq	−604.21
Succinate	aq	−691.35
Lactate	aq	−519.56
α-D-glucose	aq	−917.61
L-alanine	aq	−371.53
Glycine	aq	−370.78
NH$_4^+$	aq	−79.61
NO	g	+86.73
NO$_2^-$	aq	−37.29
NO$_3^-$	aq	−111.45
N$_2$O	g	+104.33
HS$^-$	aq	−12.06
SO$_3^{2-}$	aq	−486.04
SO$_4^{2-}$	aq	−745.82
Fe^{2+}	aq	−85.05
Fe^{3+}	aq	−10.47

*l, liquid; aq, aqueous solution; g, gas phase.

Chemical weathering by oxidation of the mineral cinnabar mobilises mercury.

The energetics of redox reactions can be analysed by separating them into "half-reactions" of the type,

$$(1)\ A^{(x)} \longrightarrow a^{(x-n)} + n\,e^-$$

$$(2)\ B^{(y)} + n\,e^- \longrightarrow B^{(y+n)}$$

the sum of which is obviously equation (A.1.4). Since each half-reaction consumes or produces electrons, it is possible to establish in principle (and often in practice) electrochemical cells consisting of separate reductant and oxidant connected by a conducting bridge that permits a flow of current at an electrical potential or voltage, designated E, that is characteristic of the

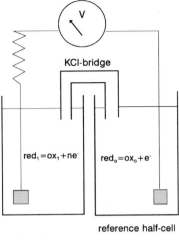

reference half-cell

total reaction: $ox_1 + nred_o = red_1 + nox_o$

Figure A.1   Principle of measuring redox potentials. If the reference electrode (to the right) represents a standard hydrogen electrode it is kept at pH $= 0$ and bubbled with H_2 at atmospheric pressure

two half-reactions (Fig. A.1). Since voltage represents a force acting through a distance, or energy, E for an electrochemical reaction can be related to ΔG as,

$$\Delta G = -nEF, \qquad (A.1.5)$$

where F is the Faraday proportionality constant in JV^{-1} From this relationship, it can be deduced that,

$$E = (RT/nF)\ln K - (RT/nF)\ln Q \qquad (A.1.6)$$

and that when products and reactants are in their standard states ($Q = 1$),

$$E_0 = (RT/nF)\ln K$$

As a matter of convention, E_0 for any given half-reaction is measured relative to the potential for the hydrogen half-reaction: $H_2 \rightleftharpoons 2H^+ + 2e^-$ with hydrogen at 1 atm, $H^+ = 1$ M and temperature $= 25\,^\circ$C; the potential for this reaction is by definition 0.000 V. In practice, potentials are measured relative to a more convenient reference reaction, typically based on mercury reduction: $Hg_2Cl_{2(s)} + 2e^- \longrightarrow 2Hg_{(l)} + 2Cl^-$. This reaction is incorporated into the calomel reference electrode. Half-cell potentials measured relative to a calomel reference are then corrected to that of a hydrogen reference.

The E_0 (and thus ΔG^0) for reactions of biological interest can be calculated from compilations of half-reaction potentials (Table A.2) by a simple

Table A.2 Standard potentials (E_0') at pH 7 for some redox pairs. Data compiled from Stumm and Morgan[3] and Thauer *et al.*[4]

Redox pairs	E_0' (V)
Some cellular e^- transfer systems	
Cytochrome a ox/red	+0.38
Cytochrome c_1 ox/red	+0.23
Ubiquinone ox/red	+0.11
Cytochrome b ox/red	+0.03
APS/AMP + HSO_3^-	−0.06
FAD/FADH	−0.22
Cytochrome c_3 ox/red	−0.29
NAD/NADH	−0.32
Flavodoxin ox/red	−0.37
Ferredoxin ox/red	−0.39
Some important organic redox pairs	
Fumarate/succinate	+0.03
Glycine/acetate + NH_4^+	−0.01
Dihydroxyacetone phosphate/ glycerolphosphate	−0.19
Pyruvate/lactate	−0.19
CO_2/acetate	−0.29
CO_2/pyruvate	−0.31
CO_2/formate	−0.43
Some important inorganic redox pairs	
O_2/H_2O	+0.82
Fe^{3+}/Fe^{2+}	+0.77
NO_3^-/N_2	+0.75
MnO_2/$MnCO_3$	+0.52
NO_3^-/NO_2^-	+0.43
NO_3^-/NH_4^+	+0.38
SO_4^{2-}/HS^-	−0.22
CO_2/CH_4	−0.24
S^0/HS^-	−0.27
H_2O/H_2	−0.41

addition process. Note that potentials compiled for biological reactions typically are at pH = 7 and 25 °C.

References

1. Battley EH (1987) *Energetics of Microbial Growth*. Wiley, New York.
2. Levine IN (1988) *Physical Chemistry*, 3rd edn. McGraw-Hill, New York.
3. Stumm W, Morgan JJ (1996) *Aquatic Chemistry*. Wiley, New York.
4. Thauer RK, Jungerman K, Decker K (1977) Energy conservation in chemotrophic anaerobic bacteria. *Bact Rev* **41**: 100–180.

Index

Abiological element cycling 250–254
Abundance, relative, soil water 88–89
Acetate formation, energetics 23
Acetoclastic methanogens 16
Acetogens (and acetate formation) 11–12
H_2/CO_2 26–27
Actinorhizal N_2-fixing symbionts 178
Activation energy 284
Advection 29, 39
Aerobic conditions
communities, bioenergetics 25
respiration 12–14
sewage treatment works, microbial mats 157
see also Oxygenic photosynthesis
Agar, hydrolysis 44
Aggregates in water columns 76
formation 73–74
Albedo 210
Algae
benthic 128
pelagic, as source of detritus 118
Ammonium and ammonia 128–130, 217–219
assimilation 20
atmosphere and 217–219
availability and uptake
in aquatic sediments 130–131, 137–138
in soil 96
denitrification 57
methanotrophs inhibited by 233
mobility 59
oxidation, see Nitrification; Oxidation
production in aquatic sediments 128–130
as ruminant microbe N source 173

Amoebae, rhizomastigid, methanogens in 181
Amylose and amylopectin, hydrolysis 44
Anabaena spp. as symbionts 178
Anabolic metabolism, see Assimilatory metabolism
Anaerobic conditions
communities, bioenergetics in 25–27, 28
mineralisation 27–28
N_2 cycling 272–273
protists, hydrogen scavengers in 180–181
respiration 14–16
see also Anoxygenic photosynthesis
Angiosperms, N_2-fixing nodules 178
Animals, see Fauna
Anoxic layers in stratified water columns 158–159, 161–162
Anoxygenic photosynthesis 18
in cyanobacterial mats 155–156
and evolution of atmosphere 273
Anthropogenic factors
halomethane production 224–225
soil biogeochemistry changes 109, 111
Aquatic systems, see Plankton; Sediments; Water
Aquifex spp. 197
Archaea in extreme environments 197–198
Archean, early, atmosphere during 210–211, 274
Assimilatory/anabolic metabolism 19–21
dissimilatory and, coupling 6–7
in water columns 65

Atmosphere (in biogeochemical
      cycling) 203–244
   as elemental reservoir 203–206
   evolution 208–213, 271–276
   gases, climate change and 213–233,
      234
   structure 206–207
ATP synthesis 9–10, 24–25
Autotrophs 7–8
   in aquatic sediments 127–128
   assimilatory metabolism 20–21
   as symbionts 179–183
   see also specific types
   Azolla spp., cyanobacterial symbionts
      in 178

Bacteria, basic relevant properties 4–6
Banded iron formation (BIF)
   biogeochemical cycle evolution and
      274
   thioester world and 269
Basalt as subsurface environment 194
Beggiatoa mats 144–147
   mineral cycling 147–148
Benthic algae 128
Bioenergetics of microbial metabolism
      21–29, 284–292
Biofilms and microbial mats 142
Biological oxygen demand and
      microbial biomass degradation
      47
Biomass, microbial, degradation 45–46
Biosphere
   interactions with
      troposphere/stratosphere 207
   methane dynamics 216
   as reservoir for Earth's elements 203,
      205
Biota
   atmospheric gases affected by 210
   microbial, in water columns
      characteristics 64–67
      formation/sources 73–74

suspended particles and their 75
   see also Life
Bioturbation 36
   macrofaunal, in aquatic sediments
      136
Bivalves, symbionts in 182–183
Black band disease 156
Boltzmann distribution law 284
Brines, submarine 190–192

Cairns-Smith's theory of clay-based
      origin of life 261–264, 267
Calorimetric estimations of enthalpy
      286
Carbohydrates
   fermentation, see Fermentations
   polymeric, see Polysaccharides
Carbon 54–57
   budget, N. Sea microbial loop 72
   organic
      dissolved (DOC), in aquatic
         sediments 122–131
      polymers, see Polymers
      see also Organic compounds
   redox reactions 58
   substrate as source of energy and of 7
Carbon, flow/cycling 54–57
   aquatic sediments 122–128
   microbial mats 147–148
      of cyanobacteria 151
   terrestrial
      global change and 109
      lignin and 106
   water-columns 62, 69–70, 76–78
   stratified 160
Carbon-1 compounds, assimilation by
      autotrophs 8, 20
   see also specific C-1 compounds
Carbon dioxide
   assimilation
      methanotrophs 16, 20
      phototrophs 16–18
   atmospheric (and hydrospheric and
      lithospheric)

consumption/depletion 205
  controls for 208-209
  and evolution of atmosphere
    208-210, 251
  global climate change and 214-215
  soil biogeochemistry 110-111
  wetland plant production and 226
  see also Hydrogen/carbon
    dioxide-methanogens
Carbon disulphide 220
Carbon monoxide, atmospheric 211,
  223-224
Carbon:nitrogen ratio
  microbial biomass degradation and
    45-46
  terrestrial plants, 87 104-105
  water columns 76-78
Carboniferous period, lignin and 105
Carbonyl sulphide 220
Catabolism, *see* Energy metabolism
Catalysts
  enzymes as 284-285
  metal, and origins of life 266
Cation concentration ability of clays 263
Cellobiase 52
Cellobiohydrase 52
Cellularisation in origin of life
  in Cairns-Smith's theory 264
  in Wächtersháuser's theory 264
Cellulose, hydrolysis 43, 52-53
  enzymes involved 51-52
  by symbionts 175
  in ruminants 171
Cellulosome 52-53
CFCs 224
Chemical equilibria/disequilibria in
    energy flow systems 247-248
Chemodenitrification 99-100
Chemolithotrophs/chemoautotrophs 8,
  144-147
  aerobic respiration 13-14
  in aquatic sediments 128
  (hyper)thermophilic 196-197
  in microbial mats

in cyanobacterial mats 155-156
  lithotroph-based mats 144-147
  origin of life and 266
  symbiotic 181-183
Chemo-organotrophs,
    (hyper)thermophilic 196-197
Chirality 257
Chitin and chitosan, hydrolysis 43
Chlorofluorocarbons 224
Chlorophyll 17
Chroococcian genus *Synechoccus* 64
Clays and life 261-264, 267
Climate change, global (and greenhouse
    effect) 205, 213-233
  prehistoric 208
  soil carbon and associated elements
    and 109
  trace gases and 213-233
Clostridia spp.
  fermentations 11
  thermocellum, cellulosome 52-53
Cloud condensation nuclei and sulphur
    gases 220
Coacervates 258
Coagulation theory, aggregate
    formation 74
Colloidal particles in water columns 74
  degradation 72
Colony forming seawater isolates 66-67
Colourless sulphur bacteria
  in cyanobacterial mats 145
  mats based on 144-147
Columbia River Basalt Group as
    subsurface environment 194
Communities (microbial), structure
  bioenergetics and 25-29
  transport mechanisms and 29-40
Constant heat summation, Hess' law
  286
Copper-enriched particles, global
    deposition 206
Corals, stony, black band disease 156
Counts, bacterial, in water columns
  64-65

Cryosphere as reservoir for Earth's
elements 203
Culture of planktonic bacteria 66–67
Cyanella 167, 179
Cyanobacteria 148–156, 178–179
mats of 148–156
mineral cycling 150–156
permanent 143
planktonic analogue of, in stratified
water columns 162
structure/constituents 145, 149–150
oxygenic photosynthesis 17–18
primary production 63–64
in stratified water columns 162
symbiotic N_2-fixing 178–179
Cycads, cyanobacterial symbionts
178–179
Cycles, biogeochemical/mineral 43–61,
122–139, 150–156, 203–283
in aquatic sediments 122–139
atmosphere, *see* Atmosphere
comparison 53–59
evolution 271–276
in microbial mats 147
of cyanobacteria 150–156
origins 245–271
in soils, *see* Soils
thermodynamics and 245–250
in water columns 76–78
see also specific minerals/elements

Degradation of organic polymers, *see*
Polymers
Denitrification 57
ammonium 57
Aquifex spp. 197
nitrate (and in general) 14, 55, 57
in soil 93–96, 99
in stratified water columns 161
Desulfurococcaceae 197
Detritus
aquatic sediments, *see* Sediments
mineralisation 45–46, 49, 53

Diatoms
in cyanobacterial mats 149
cyanobacterial symbionts in 179
Diffusion 30–33
communities controlled by 32–39
limited 32–33
in stratified water columns 161
Dihydrogen, *see* Hydrogen
Dimethyl sulphide, atmospheric H_2S
and 219–223
Dimictic lakes 158
Dinitrogen (N_2)
atmospheric, and global climate
change 217–219
mobility/availability 59
reduction 20–21, 55
Dinitrogenase 20–21
Dinitrogenase reductase 20–21
Dissimilatory metabolism, *see* Energy
metabolism
Dissolved organic matter (DOM)
in aquatic sediments 120, 122–131
carbon (DOC) 122
in water columns 67–71
composition 67–68
degradation 69–72
origin 69
Diversity, microbial
extreme environments 189–190,
194–195
soil 101–102
Drying, soil biogeochemistry and 110

Electrochemical reactions, *see* Redox
reactions
Electron transport phosphorylation
9–10
Endocellulase (endo-1,4-β-glucanase) 52
Endosymbiosis, *see* Symbiosis
Energetics of microbial metabolism
21–29, 284–292
Energy
activation 284

conservation 287
flow, biogeochemical cycles and
    245-250
free, *see* Free energy
internal, Gibb's free energy as
    function of 285
oxygen coupled to conservation of
    272
Energy budget 7
Energy metabolism
    (dissimilatory/catabolic
    metabolism) 6-19
assimilatory metabolism and,
    coupling 6-7
Energy source, substrate as 7
Enthalpy 286
Entropy changes 287-288
Enzymes
    as catalysts 284-285
    hydrolytic 50-53
        in aquatic sediments 118, 120
        in water columns 71-72
Escherichia coli, solute uptake 34
Ethanol fermentation to acetate,
    energetics 23
Eubacteria, anoxygenic photosynthesis
    18
Eukaryotes
    as symbionts 167
    symbiosis of, *see* Symbiosis
Eutrophication, nitrogen, soil 110
Evolution
    atmosphere 208-213, 271-276
    biogeochemical cycles 271-276
    life 258
Evolutionary responses
    of primary producers' to terrestrial
        environment 87
    rumination as 174
Excretion of metabolites 35-36
Exoenzymes 50-51
    exocellulase (exo-1,4-β-glucanase) 52
    in water columns 71

Extraterrestrial origin of life 256-257,
    258
Extreme environments 5, 188-202

Faecal pellets, zooplankton, as source of
    detritus 118
Fatty acid fermentation to acetate,
    energetics 23
Fauna
    aquatic, aquatic sediments and
        118-120, 136
    soil, response to water 102
    see also Herbivores; Invertebrates;
        Mammals; Metazoans;
        Zooplankton
Fermentations 10-12
    energetics 22-23
    in communities of fermenting
        bacteria 25-27
    symbiotic 168-175
        postgastric 169, 174
        pregastric 169, 170-174
Fick's laws
    first law 30-31
    second law 30
Flocculation 75-76
Flux
    of elements in Earth's reservoirs
        203-205
    of organic particles in water columns
        73
    soil water 88
Fossil fuel, mineralisation 49
Frankia spp. 178
Free energy
    Gibb's, *see* Gibb's free energy
    standard 22-23
Freons 224
Freshwater systems
    sediments, *see* Sediments
    sulphur cycling in 220-223
    symbiotic methanogenesis 181
    see also Lakes; Wetlands

Fungi
  cyanobacterial symbionts in 179
  as rivals 6
  soil organic matter cycling and 101,
    108

Gaia hypothesis 250
Gallionella spp. 157
Gases
  atmospheric, and climate change
    213–233, 234
  soil, transport 233
  see also specific gases
Gene(s), clay 261–264, 267
Generation times 5
Gibb's free energy 285, 287–288
  chemical disequilibria and 248
Glucose fermentation 10–12
β-Glucosidase 52
Green non-sulphur bacteria, anoxygenic
    photosynthesis 19
Green sulphur bacteria
  anoxygenic photosynthesis 18–19
  in cyanobacterial mats 150
Greenhouse effect, *see* Climate change
Growth
  in culture of planktonic bacteria
    66–67
  soil microbes, water potential 92
Growth yields
  energetics 24
  planktonic bacteria 66

Haemoglobin, root nodules 177
Haldane's theory of origin of life
    258–261
Halobacteria, anoxygenic
    photosynthesis 19
Halocline 158
Halomethanes 224–225
Halophilic microbes (in general)
    189–193
HCFCs 224

Heat, Earth's internal production 253,
    254
  see also Hess' law
Heliobacteria, anoxygenic
    photosynthesis 19
Herbivores
  detoxification of plant secondary
    metabolites 168
  ruminating, symbiotic fermentations
    169–175
  soil biogeochemistry and 110
Hess' law of constant heat summation
    286
Heterotrophs 7
  communities, bioenergetics 25
  eukaryotic, symbionts in 178–179
  in mineral cycles 43
    in carbon cycles in aquatic
      sediments 122–128
HFCs (and their decomposition)
    224–225
High temperature, *see* Hyperthermal
    environments
Homoacetogens, *see* Hydrogen/carbon
    dioxide-acetogens
Hot conditions, *see* Hyperthermal
    environments
Humic material, degradation 48–50
  in water columns 67–68
Hydrocarbons, degradation 48–50
  soil 109
Hydrochlorofluorocarbons 224
Hydrofluorocarbons (and their
    decomposition) 224–225
Hydrogen (H_2)
  production 35
    energetics 23
  scavenging, in anaerobic protists
    180–181
Hydrogen sulphide, atmospheric
    219–223
Hydrogen/carbon dioxide
    (H_2/CO_2)-acetogens
    (homoacetogens) 11

bioenergetics in communities
containing 26-27
Hydrogen/carbon dioxide
(H_2/CO_2)-methanogens 16
bioenergetics in communities
containing 25-26
as symbionts
in protists 180-181
in ruminants 171
Hydrogenosome 180-181
Hydrolysis, organic polymers 6, 43-53
in aquatic sediments 118, 120
dissolved organic carbon produced
via 122-123
dissolved organic nitrogen
produced via 128-130
in ruminants by symbionts 170-171
in water columns 71-72
Hydrosphere
CO_2, *see* Carbon dioxide
interactions with
troposphere/stratosphere 207
as reservoir for Earth's elements 203
Hydrothermal environments, *see*
Hyperthermal/hydrothermal
environments
Hyperhaline systems 189-193, 195-196
see also Halobacteria; Halocline
Hyperthermal/hydrothermal
environments 192-193
microbial mats 143
origin of life in 267

Inorganic redox pairs, examples 292
Invertebrates
in extreme environments 192-193
symbionts in/on
chemotrophic 182-183
phototrophic 179-180
Iron
availability 59
biogeochemical cycle evolution and
274
cycling 58

in aquatic sediments 139
oxidation, *see* Oxidation
reduction 14-15, 58
thioester world and 269-271
Iron bacteria, mats 157
Iron pyrites/sulphide in origin of life
264-268

Kinetic constraints on microbial
metabolism 21

Labile pool of dissolved organics in
water columns 71
Lactate metabolism 10-11
Lakes
sediments, *see* Sediments
stratified water columns 158
Laminarins, hydrolysis 44
Land use and soil biogeochemistry 111
Leghaemoglobin 177
Legumes, N_2 fixation 177-178
Leptothrix spp. 157
Lichens 179
Life
evolution 258
origins of 264-271
see also Biota
Light and cyanobacterial mat thickness
150
see also entries under Photo-
Lignin 48, 105-109
Limnic systems, *see* Freshwater systems
Lithosphere
CO_2, *see* Carbon dioxide
evolution 208
interactions with
troposphere/stratosphere 207
as reservoir for Earth's elements 203
Lithotrophs, *see* Chemolithotrophs
Litter components, decomposition 49
lignin:nitrogen ratios and 106-107

Mammals, symbiotic digestions
168-175

Manganese
  in aquatic sediments 138–139
  oxidation by / reduction of, *see*
      Reduction
  oxidation of 13–14, 58
  redox reactions involving 57–58
Mannan, hydrolysis 43–44
Mariager Fjord 159–161
Marine systems
  hypersaline 190–192
  sediments, *see* Sediments
  sulphur cycling between atmosphere
      and 220–221
  symbiotic methanogenesis 181
Mat(s), microbial 142–157
  extreme environments 190
  origin / development 142–143
Matrix, soil, water potential due to
      89–90
Membranes, evolution 258
Mercury, global impact 206
Meromictic lakes 158
Metabolism 6–29
  assimilatory / anabolic / energy, *see*
      Assimilatory metabolism
  dissimilatory / energy / catabolic, *see*
      Energy metabolism
  diversity 6
  energetics of 21–29, 284–292
  volume-specific rate 5
  see also Protometabolism *and specific*
      areas of metabolism
Metal catalysts and origins of life 266
Metazoans in extreme environments
      192–193
Meteoric origins of life 256, 258
Methane 215–217, 225–233
  consumption 217, 230–233
    by microbes, *see* Methanotrophs
  global climate change and 215–217
  production 217, 225–231
    by microbes, *see* Methanogenesis
  see also Halomethanes

Methanogenesis 16, 35–36
  climate change and 227–231
  H_2 / CO_2, *see* Hydrogen / carbon
      dioxide-methanogens
  hypersaline 195
  lake sediments 120–121
  soil 101–102
  stratified water columns 160–161
  by symbionts
    in protists 180–181
    in ruminants 171
  wetland 227–229
Methanotrophs 230–233
  climate change and 230–233
  CO_2 assimilation 16, 20
  evolution 276
  extreme environments and 191–192
  symbiotic 181–183
Methylated sulphur species 220
Microbial loop, planktonic 62–63
  N. Sea, carbon budget 72
Microphytes, *see* Plankton
Mineral cycles, *see* Cycles
Mineralisation 43–61
  anaerobic 27, 28
  water columns 64–67, 70–71
    stratified 161
Mixed acid fermentation 11
Mollusc, bivalves, symbionts in 182–183
Monod equation 33–34
Monomictic lakes 158
Mosses, cyanobacterial symbionts in
      178
Mutualism 166

Nematodes, symbiotic sulphur bacteria
      183
Nitrate
  aquatic sediments and 134–137
  mobility / availability / distribution 59
    in stratified water columns 161
  production 13, 55, 57
  reduction, *see* Denitrification

Nitric oxides (NO$_x$), soil 96–100
Nitrification 13, 57, 130–134
  aquatic sediments 130–134
    denitrification dependence on
        134–135
  soil 93–94, 100–101
  stratified water columns 161
  see also Chemodenitrification
Nitrite
  distribution in stratified water
       columns 161
  production 13
Nitrogen
  in aquatic sediments, dissolved
       organic and inorganic
       (DON/DIN)) 128–130, 133
  Earth's reservoirs, distribution 204
  microbial biomass degradation and
       45–46
  mineralisation 57
  redox reactions 58
  for ruminant microbes, source 173
  soil, eutrophication 110
  wetland plant production and
       methane emission and 227–229
  see also Carbon:nitrogen ratio;
    Dinitrogen
Nitrogen fixation 20–21, 175–179,
    217–218
  aquatic sediments 138
  climate change and 217–218
  cyanobacterial mats 152
  extreme environments 193
  symbiotic 175–179
  ruminants 173
Nitrogen flow/cycling/cycling 56,
    93–99, 128–138, 217–219
  aquatic sediments 128–138
  cyanobacterial mats 151–152
  global climate change and 217–219
  mobility/availability of molecules 59
  prehistoric, atmospheric O$_2$ and
    272–273

terrestrial 93–99, 104–106
water columns 76–78
  stratified 160
Nitrogenase 20–21
Nitrous oxide (N$_2$O)
  atmospheric, climate change and
    218–219
  soil 96–99
North Sea microbial loop, carbon
    budget 72
Nostoc spp. 178–179
Nucleic acid, *see* Peptide nucleic acid;
    RNA; Tribonucleic acid

Oils, crude, degradation 49
Oligotrophs, obligatory, in water
    columns 66
Oparin's theory of origin of life 258–261
Organic compounds
  mineralisation, *see* Mineralisation
  in origin of life
    in Cairns-Smith's theory 261–264
    in Oparin–Haldane's theory
       258–261
    in Wächtershäuser's theory
       264–268
  polymeric, degradation, *see* Polymers
  redox pairs, examples 292
  reduction 15–16
Organic matter
  aquatic sediments 117–131
  dissolved, *see* Dissolved organic
    matter
  global climate change and 214–217
  particulate, *see* Particles
  terrestrial plants, response to 103–109
  water columns 67–72
    composition 67–68
    origin 69
    turnover 69–72, 121
Origins
  of biogeochemical cycles 245–271
  of life 264–271

Oscillatorian *Trichodesmium* 63–64
Oxidation
  in carbon cycle 55
  iron 13–14, 58
    prebiotic 252–253
  manganese 13–14, 58
  methane
    extreme environments 191–192
    wetlands 231
    see also Methanotrophs
  nitrogen compounds and in nitrogen
      cycle (e.g. NH_4^+) 13, 55, 57
    in soil 93, 96, 99, 101
    see also Nitrification
  sulphur compounds (sulphide etc.)
      and in sulphur cycle 13, 56
    evolution of biogeochemical cycles
        and atmosphere and 274
    in mats 144–147, 155
    in stratified water columns 161
    by symbionts 181–183
Oxidation–reduction, *see* Redox
Oxygen 214–217
  in aerobic respiration 12–14
  in aquatic sediments,
      availability/gradients, 37–39
    nitrification and 130–132
    oxygen consumption and 37–39
  atmospheric 214–217, 271–276
    evolution and prehistoric trends
        210–212, 271–276
    global climate change and 214–217
  microbial biomass degradation and
      biological demand for 47
  in microbial mats
    cycling 147–148
    production and concentration
        gradients 145–147, 152–153
    uptake/consumption 147, 153–154
  in soil, denitrification and 93, 95
  see also Anoxic layers
Oxygenic photosynthesis 17–18
  in cyanobacterial mats 156

evolution of (and evolution of
    atmosphere) 211, 273–275
Ozone
  stratospheric, destruction 219
  tropospheric 223

Panspermia 256–257
Particles/particulate organic matter
  in aquatic sediments (coarse and fine)
      117–121
  of terrestrial plants, responses to 103
  in water columns 73–76
    degradation 72, 121
    suspended, *see* Suspended particles
Peatland, sulphur cycling 221
Pectin, hydrolysis 44
Peptide nucleic acid in origins of life
    261
Peptidoglycan, hydrolysis 44
Phanerozoic
  oxygen dynamics in transition from
      Proterozoic to 211–212
  oxygen dynamics through 275
Phosphate (in origins of life)
  clays ability to concentrate 263
  organic–pyrite bonding and 267
  thioester world and 269–270
Phosphorus
  cycling
    aquatic sediments 138
    water columns 76–78
  microbial biomass degradation and
      45–46
Phosphorylation, ADP 9–10
  see also ATP synthesis
Photo(auto)trophs (and phototrophy) 8,
    16–18
  aquatic sediments and 128
  in microbial mats 149
    competition with lithotrophs 156
    sulphide oxidation 155
  in stratified water columns 161–162
  as symbionts 179–180

symbionts in (in eukaryotic
    phototrophs) 178–179
see also Light
Photolysis
Aquifex spp. 197
in water columns 67–70
Photosynthesis 16–18
anoxygenic, *see* Anoxygenic
    photosynthesis
in cyanobacterial mats 152–154,
    155–156
evolution of (and evolution of
    atmosphere) 211, 273–275
origin and evolution 263
oxygenic, *see* Oxygenic
    photosynthesis
Plankton (implying predominantly
    microphytes) in water columns
    62–84
in stratified columns, as analogue of
    cyanobacterial mats 162
see also Diatoms
Plants/vegetation
aquatic
    macroscopic, nitrification in aquatic
        sediments and 136
    microscopic, *see* Plankton
terrestrial
    organic matter from, responses to
        103–109
    responses to environment 87–88
    roots, *see* Rooted plants; Roots
    species composition, interactions
        with successional change
        and dynamics of microbial
        nitrogen transformations 96
wetland, production, nitrogen and
    methane emission and 226–230
Pogonophorans, chemotrophic
    symbionts 183
Polymers (organic), degradation 6,
    43–53
in aquatic sediments 118, 120

dissolved organic carbon produced
    via 122–123
dissolved organic nitrogen
    produced via 128–130
symbiotic 167–175
Polynucleotide hydrolysis in aquatic
    sediments 128
Polysaccharides, hydrolysis 43–44
enzymes involved 51–52
by symbionts 167–175
Pore size, soil, water potential and
    90–91
Pore space, water-filled 91
$NO/N_2O/N_2$ and 98
Prebiotic/prevital Earth
mineral cycles 250–254
"soup" of organics in
    Oparin–Haldane's theory
    258–261
Primary production
in soil, evolutionary response of
    producers 87
in water columns, prokaryotic 63–64
Primeval soup in Oparin–Haldane's
    theory 258–261
Prochlorophytes 18, 64
Production
primary, *see* Primary production
wetland plant, methane emission and
    226–230
Proteolysis in aquatic sediments 128
Proterozoic, atmospheric oxygenation
    211–212, 275
Protists 180–181
as symbionts 169
symbionts in 179–181
    hydrogen scavenging 180–181
    sulphur compound-oxidising
        182–183
Protometabolism in thioester world 270
Protozoan grazing in water columns 65
Purple non-sulphur bacteria
anoxygenic photosynthesis 18

Purple non-sulphur bacteria (*cont.*)
  hydrogen scavenging in anaerobic
    protists 180–181
Purple sulphur bacteria
  anoxygenic photosynthesis 18
  in microbial mats 156
    in cyanobacteria-based mats
      149–150, 155
Pycnocline 158
Pyrites in origin of life 264–268
Pyruvate metabolism 10–11

Radionuclide decay, Earth's internal
    heat production from 254
Redox potentials (of redox
    couples/pairs) 27–29, 292
  measurement principles 291
Redox reactions
  abiological 252–253
  in aquatic sediments 139
  in dissimilatory metabolism 9, 57–59
  energetics 289–291
Reduction
  in carbon cycle 54
  iron 14–15, 58
  manganese (=oxidation by
      manganese) 14–15
    in aquatic sediments 138
  nitrogen compounds 14, 20–21
    in nitrogen cycle 55
    see also Denitrification
  sulphur compounds (e.g. sulphate)
      and in sulphur cycle 13–14, 56
    evolution of sulphate reducers 276
    H_2S in atmosphere and 219–223
    methanogenesis and 228
    subsurface environments 194
    by symbionts in anaerobic protists
      180–181
Reservoirs for Earth's elements 203–206
Respiration 12–16
  aerobic 12–14
  anaerobic 14–16

Rhizobia 176–177
Rhizomastigid amoebae, methanogens
    in 181
Rhizosphere
  nitrification in aquatic systems and
    136
  nitrogen fixation 176–178
Ribonucleic acid, *see* RNA
Ribosomal RNA, *see* RNA, ribosomal
Rice culture, methane and 226
RNA, ribosomal
  (hyper)thermophiles 196
  planktonic bacteria 66–67
  sulphur-oxidising bacteria,
      establishing their timing of
      origin 276
RNA world (in origins of life) 260–261
  thioester world as precursor to
    268–271
Root(s)
  denitrification and 96–97
  nodules, N_2 fixation 176–178
  see also Rhizosphere
Rooted plants
  on fringes of aquatic systems
    121–122, 127–128
  response to water 102
Ruminants, symbiotic fermentations
    169–175

Salinity, *see* Brines *and entries under*
    Halo-
Sea, *see* Marine systems
Sea urchins, symbiotic fermentation 175
Sediments, aquatic (marine and
    freshwater/lake) 117–141
  comparison of freshwater and marine
    117–122
  deeply buried 195
  detritus 118
    mineralisation 49–50
  diffusion controlled communities
    36–39

Sewage treatment works, aerobic, microbial mats 157
Size of bacteria 4-5
Snow particles, marine 75
Soils 85-116
    diffusion controlled communities 36-37
    methane consumption 231-233
    responses of biogeochemistry to disturbance/change 109-111
Solutes
    excretion 35-36
    uptake 32-35
Soup, prebiotic/primeval/prevital, in Oparin-Haldane's theory 258-261
Sphagnum spp., cyanobacterial symbionts in 178
Sponges, cyanobacterial symbionts in 179
Springs, hot 196, 198
Standard free energy 22-23
Starch, hydrolysis 44
    in ruminants 172
Stratification, aquatic systems 120-121, 158-162
    see also Zonation
Stratosphere
    interactions with other spheres 207
    ozone destruction 219
    structure 207
Stress, water
    ammonia oxidation and denitrification and 99
    physiology 91-92
Stromatolithic mats 143
Submarine hypersaline systems 190-192
Substrates
    as carbon and energy source 7
    for hydrolysis of organic polymers 43-50
Subsurface environments, terrestrial 193-195

Successional change in terrestrial ecosystems 110
    plant species composition and dynamics of microbial nitrogen transformations and their interactions with 96
Succinate fermentation 12
Sulphate, reduction, *see* Reduction
Sulphide
    oxidation, *see* Oxidation
    production in stratified water columns 160-161
    see also specific sulphides
Sulphur
    assimilation 21
    cycling 58
        global climate change and 219-223
        in microbial mats 147-148, 151
        in stratified water columns 160
    iron availability and 59
    isotopic date as evidence of oxygenation of atmosphere 274, 276
    in origin of life 268-271
    oxidation, *see* Oxidation
Sulphur bacteria 144-147
    anoxygenic photosynthesis 18-19
    colourless, *see* Colourless sulphur bacteria
    green, *see* Green sulphur bacteria
    in mats 144-147, 156
        of cyanobacteria 145, 149-150, 155
    purple, *see* Purple sulphur bacteria
    in stratified water columns 161-162
    symbiotic 180-183
        hydrogen scavenging 180, 181
Sulphureta 156
Superlabile pool of dissolved organics in water columns 71
Suspended particles in water columns 73-76
    microbial biota 75
Symbiosis 166-187

Symbiosis 166–187 (*cont.*)
  definition 166–167
  by eukaryotes 167, 169
  by prokaryotes 167–187
Synechoccus spp. 64
Syntrophy 35–36, 166–167

Temperature
  high, *see* Hyperthermal environments
  soil biogeochemistry and changes in
      110
  stratification in aquatic systems
      (thermoclines) 120, 158
Terrestrial systems
  plants, *see* Plants
  soils, *see* Soils
  subsurface environments 193–195
Thermal changes and stratification, *see*
      Temperature
Thermodynamics 21–23, 245–250,
      284–292
  biogeochemical cycles and 245–250
  first law 287
  microbial metabolism and 21–23,
      284–292
  second law 287–288
Thermophilic microbes, *see*
      Hyperthermal environments
Thermoproteales 197
Thioester world 268–271
Transpiration 102
Transport (in the environment)
  gases in soil, methane consumption
      and 233
  mechanisms 29–40
Tribonucleic acid and pyrite-based
      origin of life 265
Trichodesmium spp. 63–64
Trickling filters of aerobic sewage
      treatment works, microbial mats
      157
Trifluoroacetate and trifluoromethane
      224–225

Troposphere
  interactions with other spheres 207
  ozone 223
  structure 206–207
Turbulence 29, 39–40

UV and origins of life 260, 263

Vegetation, *see* Plants
Vents, hydrothermal 192–193, 196, 198
Vestimentiferans, chemotrophic
      symbionts 183
Volcanoes, global/atmospheric effects
      205–206, 208
Volume-specific metabolic rate 5

Wächtershäuser's theory of origin of
      life 264–268
Water (in aquatic systems), columns
      62–84
  stratified 120–121, 158–162
  see also Plankton; Sediments
Water (in soil) 88–102, 110
  content 86–87
      interactions with water potential
          and biogeochemistry
          93–102, 232–233
      methane consumption and 232–233
  as master variable 88–102
  stress, *see* Stress
Water potential (in soils) 89–91
  interactions with water content and
      biogeochemistry 93–102,
      232–233
  intra-/extracellular 92
  methane consumption and 232–233
Waxes, hydrocarbon, degradation 109
Wetlands 225–230
  methane
      emission 225–231
      oxidation 231

nitrogen 227–229
sulphur cycling 220–223

Xylan, hydrolysis 44

Zonation

of bacterial plate in stratified water
  columns 161–162
in cyanobacterial mats 150
see also Stratification
Zooplankton, faecal pellets as source of
  detritus 118